PLANETARY AND PROTO-PLANETARY NEBULAE:
FROM IRAS TO ISO

ASTROPHYSICS AND SPACE SCIENCE LIBRARY

A SERIES OF BOOKS ON THE RECENT DEVELOPMENTS
OF SPACE SCIENCE AND OF GENERAL GEOPHYSICS AND ASTROPHYSICS
PUBLISHED IN CONNECTION WITH THE JOURNAL
SPACE SCIENCE REVIEWS

Editorial Board

R.L.F. BOYD, *University College, London, England*

W. B. BURTON, *Sterrewacht, Leiden, The Netherlands*

L. GOLDBERG, *Kitt Peak National Observatory, Tucson, Ariz., U.S.A.*

C. DE JAGER, *University of Utrecht, The Netherlands*

J. KLECZEK, *Czechoslovak Academy of Sciences, Ondřejov, Czechoslovakia*

Z. KOPAL, *University of Manchester, England*

R. LÜST, *European Space Agency, Paris, France*

L. I. SEDOV, *Academy of Sciences of the U.S.S.R., Moscow, U.S.S.R.*

Z. ŠVESTKA, *Laboratory for Space Research, Utrecht, The Netherlands*

VOLUME 135

PROCEEDINGS

PLANETARY AND PROTO-PLANETARY NEBULAE: FROM IRAS TO ISO

PROCEEDINGS OF THE FRASCATI WORKSHOP 1986,
VULCANO ISLAND, SEPTEMBER 8–12, 1986

Edited by

ANDREA PREITE MARTINEZ

Istituto di Astrofisica Spaziale,
Frascati, Italy

D. REIDEL PUBLISHING COMPANY

A MEMBER OF THE KLUWER ACADEMIC PUBLISHERS GROUP

DORDRECHT / BOSTON / LANCASTER / TOKYO

Library of Congress Cataloging in Publication Data

Planetary and proto-planetary nebulae: From IRAS to ISO.

(Astrophysics and space science library; v. 135)
Includes bibliographies and indexes.
1. Planetary nebulae—Congresses. I. Preite Martínez, Andrea. II. Title: Proto-planetary nebulae: From IRAS to ISO. III. Series.
QB855.5.P52   1987   523.1'135   87-9672
ISBN 90-277-2517-9

Published by D. Reidel Publishing Company,
P.O. Box 17, 3300 AA Dordrecht, Holland.

Sold and distributed in the U.S.A. and Canada
by Kluwer Academic Publishers,
101 Philip Drive, Assinippi Park, Norwell, MA 02061, U.S.A.

In all other countries, sold and distributed
by Kluwer Academic Publishers Group,
P.O. Box 322, 3300 AH Dordrecht, Holland.

All Rights Reserved
© 1987 by D. Reidel Publishing Company, Dordrecht, Holland
No part of the material protected by this copyright notice may be reproduced or utilized in any form or by any means, electronic or mechanical including photocopying, recording or by any information storage and retrieval system, without written permission from the copyright owner

Printed in The Netherlands

TABLE OF CONTENTS

Preface ix

List of Participants xi

SESSION I. Chairman: L.Rodriguez

S.R.Pottasch: Infrared emission from young planetary nebulae — 1

M.Perinotto: Advances in knowledge of planetary nebulae from UV astronomy — 13

B.Stenholm, A.Acker: Status of the spectroscopic survey of planetary nebulae — 25

SESSION II. Chairman: K.Hunger

A.Acker: Announcement - "The Strasbourg-ESO Catalogue of galactic planetary nebulae" — 35

A.Leene, C.Y.Zhang, S.R.Pottasch: IRAS Additional Observations of planetary nebulae — 39

P.F.Roche: Spectroscopy at 8-13μm as a probe of the carbon to oxygen ratio in planetary nebulae — 45

SESSION III. Chairman: A.Renzini

L.F.Rodríguez: Protoplanetary nebulae — 55

W.E.C.J. van der Veen, H.J.Habing, T.Geballe: The evolution from Miras to planetary nebulae as derived from observations of Miras and OH/IR stars — 69

S.R.Pottasch: The position of PN nuclei in the H-R diagram: present status — 79

M.Peimbert: On the nitrogen and helium enrichment of the interstellar medium — 91

M.Grewing, L.Bianchi, M.Gutekunst: Temperatures, luminosities and mass loss rates for PN nuclei — 101

G.Silvestro, M.Robberto: An outflow model for bipolar planetary
    nebulae and the case of NGC 6302

SESSION IV. Chairman: S.R.Pottasch

D.Schoenberner: Mass loss and the transition of an AGB star to a
    central star of a planetary nebula ... 113

F.D'Antona, I.Mazzitelli, F.Sabbadin: Observational constraints to
    the theory of planetary nebulae evolution ... 121

SESSION V. Chairman: M.Perinotto

J.Koeppen: Rambling along a Schoenberner track: nebular evolution
    and NLTE stellar atmospheres ... 131

K.Hunger, U.Heber: Evolution of hot subdwarfs - an empirical
    approach ... 137

D.Schoenberner: Excitation of planetary nebulae and evolution of
    their central star ... 143

L.Bianchi, M.Grewing, C.Falcetta, M.Baessgen: Ionisation and
    dynamical structure of planetary nebulae ... 153

R.Viotti: Formation and structure of nebulae around symbiotic
    objects based on radio to x-rays observations (a review) ... 163

SESSION VI. Chairman: I.J.Danziger

L.Rosino, T.Iijima, S.Ortolani, A.Mammano: The peculiar planetary
    nebula NGC 2346 and its nucleus ... 171

R.Costero, M.Tapia, R.Mendez, J.Echevarría, M.Roth, A.Quintero,
    J.Barral: NGC 2346. Visible and IR observations of several
    mass-loss episodes? ... 183

M.de Muizon, A.Preite Martinez, M.Heydari-Malayeri:
    Infrared and optical spectroscopy of the suspected planetary
    nebula He 2-77 ... 185

A.Preite Martinez, S.R.Pottasch: Contribution of nebular emission
    lines to IRAS photometric survey fluxes ... 197

L.B.d'Hendecourt, A.Léger: Infrared characteristics of Poly-
    cyclic Aromatic Hydrocarbons and the interpretation of IR
    astronomical spectra ... 203

TABLE OF CONTENTS

P.Persi, A.Preite Martinez, M.Ferrari-Toniolo, L.Spinoglio:
    Near-Infrared photometry of IRAS planetary nebulae   221

A.Leene, S.R.Pottasch: The effect of line emission on the IRAS
    data of planetary nebulae   233

SESSION VII. Chairman: L.Rosino

P.Harrington: Modeling the thermal emission from dust in planetary
    nebulae   239

P.Lenzuni, A.Natta, N.Panagia: Evolution of dust in planetary
    nebulae   249

A.Salama, P.De Bernardis, S.Masi, G.Moreno: Observations of the
    spectrum of the interplanetary dust emission   255

SESSION VIII. Chairman: P.Harrington

M.F.Kessler: The Infrared Space Observatory (ISO) project   261

FINAL DISCUSSION. Chairman: M.Peimbert

Discussion introduced by: S.R.Pottasch   271
                           V.Weidemann   272
                           L.F.Rodríguez   274
                           R.S.Clegg   275
                           P.Harrington   276
                           N.Panagia   278
                           M.F.Kessler   279

INDEX OF SUBJECTS   281

INDEX OF ASTRONOMICAL OBJECTS   285

PREFACE

There are two questions that we can ask ourselves in order to describe this workshop. The first question is a double question: why a conference on this subject and why a workshop?

The first idea of organizing this workshop came while reading the scientific objectives of one of the instruments onboard the ISO satellite (a phase A document concerning the IR camera). On going through the scientific motivations for building the instrument I realized with surprise that no mention was made of Planetary Nebulae (PN). At present this is no longer true. There is a chapter indicating the capabilities of the camera in the PN field and what we can reasonably expect from that instrument. But it was at this moment that the first idea of organizing a workshop on the subject of PN came.

Of course there are other, stronger motivations. The first one is that I think this is the right moment after IRAS. I think we all spent the last two or three years working on IRAS data. IRAS represented a corner-stone for those working on Planetary Nebulae: the amount of data that came out of the instruments onboard the satellite was enormous and opened up new ways of looking at planetary nebulae, as well as at other fields. At the same time, IRAS is also responsible for a side effect on the scientific community, in the sense that in the last year or two a great amount of ground-based observations were made in this field connected with IRAS results.

But of course the number of questions that have arisen whilst trying to interpret and understand what the IRAS results mean, are greater than the answers we got. So that is why I think that after 2 to 4 years of work it is necessary to check something, to check if the hypothesis we made in interpreting the IR data are correct, and possibly to discover something more.

As far as ISO is concerned, one needs scientific motivations for building a given instrument, and when the satellite is launched, and if the instrument works, results will be, or can be, fortunately much different from what expected.

The third motivation for planning a conference is that we attend quite a number of conferences but only one conference is devoted to Planetary Nebulae every five years. I think that the amount of new data gathered in recent years suggests that, besides the IAU Symposia on Planetary Nebulae, this topic deserves other conferences in between time. It is true that there are sessions devoted to PN in other conferences on related topics, but I think that the amount of data and the amount of thinking behind the subject of PN deserves more frequent and more specific meetings.

These were the best motivations for organizing the workshop. Of course there is also our involvement in one of the ISO instruments (the ISO IR-camera) in Frascati.

Why then a workshop? The choice was not to follow the standard pattern of a conference. I think that some times conferences are too formal, there is not enough time for discussion, and contributions are compressed. Participants do not feel free to say what they want. The format of a workshop will allow people to express themselves freely, and I hope they will do so.

I do not think we need an answer to the question: why a workshop in Vulcano? We can all give the appropiate answer after inspecting the place.

Andrea Preite Martinez

List of participants.

M. BADIALI, Istituto di Astrofisica Spaziale, Frascati, Italy
L. BIANCHI, Osservatorio Astronomico, Pino Torinese, Italy
M. BUSSO, Osservatorio Astronomico, Pino Torinese, Italy
V. CALOI, Istituto di Astrofisica Spaziale, Frascati, Italy
R. CLEGG, University College, London, UK
?. CRISTALDI, Osservatorio Astronomico, Catania, Italy
F. D'ANTONA, Osservatorio Astronomico, Rome, Italy
J. DANZIGER, E.S.O., Garching bei Munchen, FRG
M. de MUIZON, Sterrewacht, Leiden, The Netherlands
M. FERRARI-TONIOLO, Istituto di Astrofisica Spaziale, Frascati, Italy
R. FUSCO-FEMIANO, Istituto di Astrofisica Spaziale, Frascati, Italy
F. GIOVANNELLI, Istituto di Astrofisica Spaziale, Frascati, Italy
L. GREGGIO, Dipartimento di Astronomia, Bologna, Italy
M. GREWING, Astron. Institut, Tubingen, FRG
R. GRUENWALD, Observatoire de Meudon, Meudon, France
P. HARRINGTON, Astronomy Program, Univ. Maryland, USA
L. d'HENDECOURT, G.P.S., Universite' Paris 7, Paris, France
K. HUNGER, Institut fur Theor. Physik, Kiel, FRG
S. KARAKUŁA, University of Lodz, Lodz, Poland
M.F. KESSLER, ESA ESTEC, The Netherlands
J. KOEPPEN, Inst. f. Theor. Astrophys., Heidelberg, FRG
A. LEENE, Kapteyn Astronomical Inst., Groningen, The Netherlands
P. LENZUNI, Osservatorio di Arcetri, Florence, Italy
F. MATTEUCCI, E.S.O., Garching bei Munchen, FRG
A. MONETI, Osservatorio di Arcetri, Florence, Italy
D. MONK, University College, London, UK
G. NATALI, Istituto di Astrofisica Spaziale, Frascati, Italy
A. NATTA, Centro per l'Astronomia IR-CNR, Florence, Italy
N. PANAGIA, Space Tel. Sci. Institute, Baltimore, USA
M. PEIMBERT, Instituto de Astronomia, UNAM, Mexico
M. PERINOTTO, Osservatorio di Arcetri, Florence, Italy
P. PERSI, Istituto di Astrofisica Spaziale, Frascati, Italy
S.R. POTTASCH, Kapteyn Astronomical Inst., Groningen, The Netherlands
A. PREITE MARTINEZ, Istituto di Astrofisica Spaziale, Frascati, Italy
A. RENZINI, Dipartimento di Astronomia, Bologna, Italy
P. ROCHE, University College, London, UK
L.F. RODRIGUEZ, Instituto de Astronomia, UNAM, Mexico
L. ROSINO, Istituto di Astronomia, Padova, Italy
F. SABBADIN, Osservatorio Astrofisico di Asiago, Asiago, Italy
A. SALAMA, Istituto di Fisica, Univ. di Roma, Roma, Italy
F. SCALTRITI, Osservatorio Astronomico, Pino Torinese, Italy
G. SILVESTRO, Ist. di Fisica Generale, Univ. di Torino, Torino, Italy
B. STENHOLM, Lund Observatory, Lund, Sweden
C. TOTARO, Universita' di Messina, Messina, Italy
R. VIOTTI, Istituto di Astrofisica Spaziale, Frascati, Italy
W.E.C.J. van der VEEN, Sterrewacht, Leiden, The Netherlands
V. WEIDEMANN, Inst. f. Theoretische Physik, Kiel, FRG

INFRARED EMISSION FROM YOUNG PLANETARY NEBULAE

S.R. Pottasch
University of Groningen
the Netherlands

ABSTRACT

   Measurements of planetary nebulae in the far infrared, especially from IRAS, are presented and discussed. The dust temperature and intrinsic luminosity are found to vary as the nebula evolves. The source of energy which heats the dust is discussed and it is shown that heating by nebular Lyman α is usually insufficient, especially in young nebulae. The problem as to whether the far infrared emission only comes from the ionized region of the nebula or whether the neutral material is also important is argued. The dust mass of the nebula is found to evolve, the dust to gas mass ratio having a high value for very young nebulae. It is argued that the dust is being constantly destroyed.
   The dust emission spectra of many young nebulae are presented and they are found to fall in three distinct categories. Finally an estimate of the spatial distribution of young nebulae near the galactic center, as deduced from IRAS measurements, is given.

I. INTRODUCTION

   Infrared emission has been observed from planetary nebulae for almost 20 years (Gillett et al., 1967). It was initially conjectured that the radiation is due primarily to emission by heated dust. This is now thought to be partly true, but line emission by abundant ions also plays an important role, at least in the larger, more evolved nebulae.
   Broad band continuum measurements (between 8 and 20 µm) have been made from the ground for some of the stronger nebulae (Cohen and Barlow, 1974, 1980). A few broad band far infrared airplane measurements followed (Moseley, 1980). In 1983, the IRAS made an almost complete survey of the sky with a sensitivity not previously available. Several kinds of data have been produced by this satellite and they will be discussed in this review.
   The IRAS point source catalogue lists the fluxes of almost 850 of the roughly 1250 known planetary nebulae. This is quite remarkable because there are now more planetary nebulae whose far infrared flux is

known than there are nebulae whose Hβ flux or radio continuum flux density has been measured. The 400 nebulae which are not observed by the IRAS fall into several categories. First there are nebulae larger than 3 arc minutes which are not recorded in the point source catalogue even though they may be strong sources. Secondly, there are the intrinsically large evolved, nebulae which emit much less of their energy in the far infrared. Thirdly, some of the 'nebulae' are misidentified and are in reality some kind of star or even a plate flaw. The IRAS positions are almost always within 15" to 20" of the optical position so that the identification is easy to make. Coupled with the very characteristic energy distribution of planetary nebulae in the far infrared (the peak flux density almost always occurs at 25 μm or 60 μm), the IRAS measurements can be used to confirm the correctness of an identification. If a point source does not have an IRAS measurement or if the IRAS measurement peaks in the 12 μm band, the object may not be a planetary nebula and confirming evidence should be required.

A detailed review of this subject has recently been published (see Proceedings of the Calgary workshop on Late Stages of Stellar Evolution). Therefore only a summary of the present situation will be here.

II. THE TEMPERATURE AND LUMINOSITY OF THE DUST

A. Dust Temperature

It is now clear that the observed far infrared continuum radiation is emitted by dust, if for no other reason than there is not a reasonable alternative. But it is not clear exactly what the composition of this dust is, not what its emissivity is. These are, in fact some of the properties that are to be determined from the far IR measurements.

In order to describe the continuum spectral energy distribution a parameter T_D, which will be called 'dust temperature' may be introduced. It is that temperature for which a blackbody gives the best fit to the observed emission. The points at 25 μm and 60 μm are given the greatest weight in determining this temperature for two reasons. Firstly they are usually the strongest and therefore best determined flux densities, and secondly, they are the bands least likely to be contaminated by line emission (Pottasch, 1986).

There is a clear relation between the dust temperature and size of the nebula. The small, young nebulae have the highest temperatures, about 300K. As the nebula evolves the dust temperature becomes lower. The very largest nebulae have temperatures of the order of 50K. The relationship shows very little scatter. Even OH/IR stars conform to this relation.

This empirical relation can be explained if it is assumed that the dust is heated directly by radiation from the star, assuming that the size of the dust grain and its composition and density does not change with the evolution. Thus this agreement does not necessarily prove that direct stellar heating is important, but we will presently come to other arguments that this is true.

The relationship described above provides a means of determining the distance to nebulae since T_D can be determined independently of the distance, while the nebular radius is directly proportional to it. Unfortunately the scatter in the relation, much of which is intrinsic and is due to the differences of absolute stellar luminosity, limits the accuracy of this method to a factor two.

B. Far infrared luminosity

A large amount of energy is emitted in the infrared. It is roughly comparable to the total energy emitted by the central star. The younger, smaller nebulae have higher total luminosity than the older nebulae. This is partly expected because after the shell burning ceases in the central star, the stellar luminosity will decrease. There is a great deal of scatter in the relationship between infrared luminosity and size of the nebula. Part of this is due to uncertainty in the nebular distance. The rest of the scatter is unexplained and may have to do with evolutionary effects.

The question arises as to what fraction the far infrared luminosity makes in the total luminosity emitted by the star. The stellar luminosity can only be determined when the temperature and the radius of the star can be determined. The radius is determined from visual magnitude and the temperature can be determined either from the Zanstra method or the Energy balance method. The results are qualitatively as follows. For the small nebulae, $R_N \leqslant 4 \times 10^{16}$ cm, it is difficult to determine the stellar luminosity because it is hard to measure the visual magnitude of the central star in the small bright nebula. But it appears that the far infrared is at least 50% of the total stellar luminosity. It is a somewhat lower percentage nebulae with radius up to 10^{17} cm but usually still more than 40% of the total. Only for the nebulae larger than 10^{17} cm does the fraction go below 40% reaching less than 10% for nebulae greater than 10^{18} cm.

When a large fraction of the total energy appears in the far infrared, the validity of the stellar temperature determination is brought into question. The problem is complicated however. Consider for example, the determination of the Zanstra temperature. When stellar radiation is first absorbed by the gas, remitted by the gas and then absorbed by the dust (e.g. Lyman α) the temperature determination is not affected. Similarly there is also no effect if radiation from the star on the longwave side of the Lyman limit is directly absorbed by the dust. Only when stellar photons with energy high enough to ionize hydrogen are directly absorbed by the dust will the Zanstra temperature be affected (it will always be too low because some ionizing photons will not be counted). But it is difficult to quantify this effect because some of the necessary data is not available. The absorptivity of the dust in the far ultraviolet, for example, is poorly known. The only general statement which can be made is that the central stars of young nebulae, which are most affected, are often cool and therefore emit much of their energy longward of the Lyman limit. From these general energy consideration it is expected that the stellar temperatures may sometimes have to be increased by 10 to 15% and in a few extreme cases by 25%.

III SOURCE OF FAR INFRARED RADIATION

A. Radio continuum versus far infrared

An insight into the source of the far infrared radiation can be had by plotting the radio continuum emission, which is produced by free-free transitions in the ionized nebular gas, against the far infrared energy. Such a plot is shown as Fig. 1.

The dashed line in the figure is the expected infrared radiation if all the Lyman α radiation produced in the nebula is absorbed by the dust and reemitted in the infrared. (for the low density case; at densities substantially greater than 10^4 cm^{-3}, 2 quantum emission is suspressed so that about 50% more energy is available as Lyman α emission). From the figure, it can be seen that there are some nebulae whose infrared emission could be explained if all the Lyman α radiation were absorbed by the dust. For the majority of nebulae, however, there is not enough energy available in Lyman α emission to provide the heating. This is especially true of the younger nebulae. For nebulae with radii less than 4×10^{17} cm the infrared excess is quite high, varying between 4 and 16. It is clear that for all these nebulae another source of dust heating beside Lyman α is necessary.

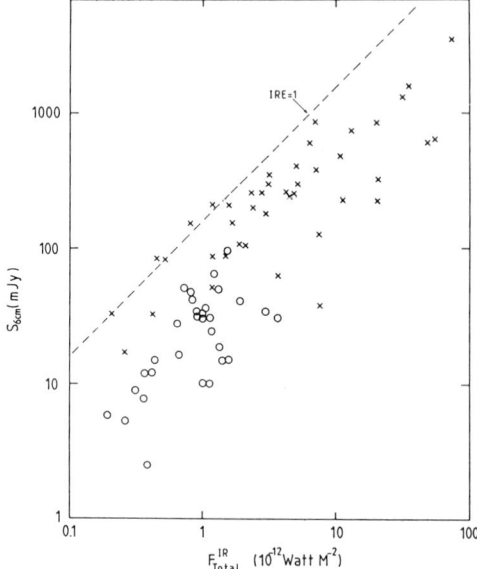

Fig. 1 The 6 cm radio flux density of individual nebulae is plotted against their total infrared flux. The bright nearby nebulae are given as crosses and the galactic center nebulae as open circles. The line shows the infrared emission which would be produced if all the Lyman α radiation generated in the nebula heated the dust.

As discussed in the previous section, direct absorption of starlight by the dust is a likely possibility. Other possibilities are available, the most probable being resonance lines of abundant ions which, like Lyman α, are scattered many times before leaving the nebula and therefore have a greater chance of being absorbed by the dust. The strongest of these resonance lines in the CIV doublet at λ 1549 Å, which has been measured in many nebulae. By comparing the C^{3+} abundance obtained from this doublet with that obtained from a recombination line of this ion it is possible to estimate how much of the doublet line intensity is being absorbed by the dust. It appears to be less than the Lyman α energy, usually considerably less. It seems difficult to avoid the conclusion that the dust in the smaller nebulae obtains a substantial fraction of its energy by direct absorption of starlight.

As mentioned earlier, a point of great importance is whether the starlight absorbed by the dust is on the long or shortwavelength side of Lyman continuum emission. To explore this further, the spatial extent of the far infrared emission must first be discussed.

B. Spatial extent of far infrared emission

There is very little known about the spatial extent of small planetary nebulae in the far infrared. The spatial resolution of the IRAS observations was only sufficient to resolve larger nebulae, with an extent greater than 2 arc minutes. Two young nebulae have now been mapped at several wavelengths in the far infrared between 8.7 μm and 23 μm with a spatial resolution of the order of 1 arc sec. These are NGC 7027 (Aitken and Roche, 1983; Bentley, 1982) and BD + 30 3639 (Bentley et al., 1984). That they are young is not only shown by their small size and high electron density, but by their low ionized mass as well (Pottasch, 1980). This also indicates that they are embedded in a more extensive region of neutral or molecular material. The CO emission from the material surrounding NGC 7027 is well studied (Mufson et al., 1975) and extends an order of magnitude beyond the ionized matter. The two nebulae differ in that the exciting star of NGC 7027 has an extremely high temperature (> 300000 K) while the exciting star of BD + 30 3639 has a relatively low temperature (about 30000 K). They both have a high infrared excess, about 4 and 8 respectively.

In both cases the far infrared radiation comes only from the ionized region. There appears to be no thermal (continuum) emission by dust outside the ionized region at wavelengths as long as 23 μm. Maps of far infrared continuum radiation at 60 μm and 100 μm of the very large nebulae NGC 6853 and NGC 7293 have been made using IRAS data (Zhang et al.; Leene et al.; in preparation). In these cases it is clear from a comparison with the ionized gas emission that the infrared originates in the ionized region.

IV PROPERTIES OF THE DUST

A. Dust mass and dust to gas ratio

The determination of the dust mass is not simple. The reason for this is that very little is known of the emission properties of the dust. Besides this, it is not known whether the dust properties, especially size and density, change in the course of evolution. If it assumes that they remain constant, and that the dust radiates as a blackbody with an efficiency Q_ν which is a function of wavelength, then it is found that the dust to gas ratio varies strongly with the nebular radius. A very high value, $M_D/M_G \simeq 10^{-1}$ is found for the very young nebulae, while a much lower value, 4×10^{-5} is found for large evolved nebulae.

The question arises as to whether this is a real effect, since it has been assumed in this analysis that dust and gas emission originate in the same region. If the far infrared radiation was produced in a neutral material surrounding the ionized gas a high ratio of M_D/Mg could be expected. Then as the evolution progresses and the fraction of ionized matter increases, the M_D/Mg ratio will decrease. But this possibility does not exist if the far infrared emission only comes from the ionized region. There is no information about the spatial extent of the infrared in Vy 2-2 and HB 12, but in BD + 30 3639, which has the 3rd highest ratio, the infrared radiation comes only from the ionized region. If this is true for all nebulae, the only way to interpret Fig. 6 appears to be the following. The matter first ejeced either contains at least 10% of the material in the form of dust, or this large amount of dust is formed within the first 100 years after the ejection. This dust is then constantly being destroyed, probably under the influence of the strong ultraviolet radiation field of the central star which will cause it to sputter and evaporate.

There is another argument that indicates that these dust to gas mass ratio's are both real and reliable. The abundances in the gas phase of Fe, Si, Mg and Al in the nebula BD + 30 3639 have recently been determined (Pwa et al., 1986). These elements are found to be substantially underabundant compared to the solar values. If they originally had the solar value in the pre-nebular stellar atmosphere and are depleted because they are now in the form of dust then the mass of these elements can be determined. Pwa et al. compute a mass fraction of 3.5×10^{-3} of the total nebular mass for these elements; if they are in the form of oxides or carbides it is likely that the mass fraction will increase to about 6 to 8×10^{-3}. The dust to gas mass ratio found from the far infrared emission using the above equation is 11×10^{-3}. This agreement is good considering the uncertainties involved and argues that the far infrared dust determination is probably reliable. It also argues that the dust probably contains substantial amounts of Mg and Fe compounds as well as Si and C.

In this respect HB 12 and Vy 2-2 form a problem because the high dust to gas ratio would require a substantially higher than solar abundance of these elements. Since this seems unlikely, either the dust to gas ratio has been overestimated, or the infrared emission is formed outside the ionized region in these two cases. The latter possibility is realistic, considering that these nebulae are extremely small.

B. Spectra between 8 μm and 23 μm

The very young nebulae have the highest dust temperature and the highest luminosity in the infrared. In these objects the infrared comes primarily from the region between 8 and 30 μm. It is therefore interesting to study the low resolution spectra from these nebulae.

Aitken et al. (1979) have studied the spectra of several young nebulae in part of this spectral range, between 8 and 13 μm. Many more have been observed by the IRAS LRS with a wavelength range extended to 23 μm. Some of these spectra, both young and more aged nebulae, are shown in Pottasch et al. (1986). Twelve of them are shown as Fig. 2. While some ionic emission lines can be seen in these spectra, especially the SIV line at λ 10.5 μm and the NeII line at λ 12.8 μm, the line emission does not dominate the infrared emission in these young nebulae as it may do in the later stages. General differences can be seen in these spectra. Almost all show broadband features which contain information about the chemical composition of the emitting dust. Three distinctly different types of spectra can be seen. In the first column of the figure, spectra of the nebulae Vy 2-2, HB 12, Pe 2-8 and CRL 618 are shown. They have a clear emission feature which is about 3 microns wide and is centered at about 10 μm. This feature is also seen in circumstellar shells of some oxygen rich stars and has been attributed to 'silicate' grains. The same features is usually seen in absorption in HII regions.

In the second column the spectra of Sw St 1, IC 418 and NGC 6572 are shown. They also show a broad feature but it is centered nearer 11.5 μm. This feature can sometimes be very strong as in NGC 6790 (spectrum illustrated by Pottasch et al., 1986 and Aitken et al., 1979). It is generally attributed to 'silicon carbide' grains.

The third type of spectra is shown in the last column. BD + 30 3639, He 2-113 and NGC 7027 are representatives of this type. Here there are much narrower features visible at 7.7, 8.6 and 11.3 μm which are attributable to 'polycyclic hydrocarbons'. Notice that these features appear together and apparently are not present in the first two types of spectra.

It is possible that these different types are related to the chemical composition of the nebulae or of the spectral type of the central star. To begin with the latter: there does not seem to be any relation. Representatives of all three types have Wolf Rayet central stars (HB 12, Sw St 1, BD + 30 3639, He 2-113) so that this does not seem to be an important factor. The chemical composition is likely to be more important but not enough is known of the composition in some of these young nebulae. It has been suggested that those showing the 'silicate' feature are oxygen rich nebulae but confirming evidence is sparse. The representatives of the other two types do seem to be carbon rich, however. This seems to be well established for IC 418, NGC 6572, NGC 7027 and BD + 30 3639.

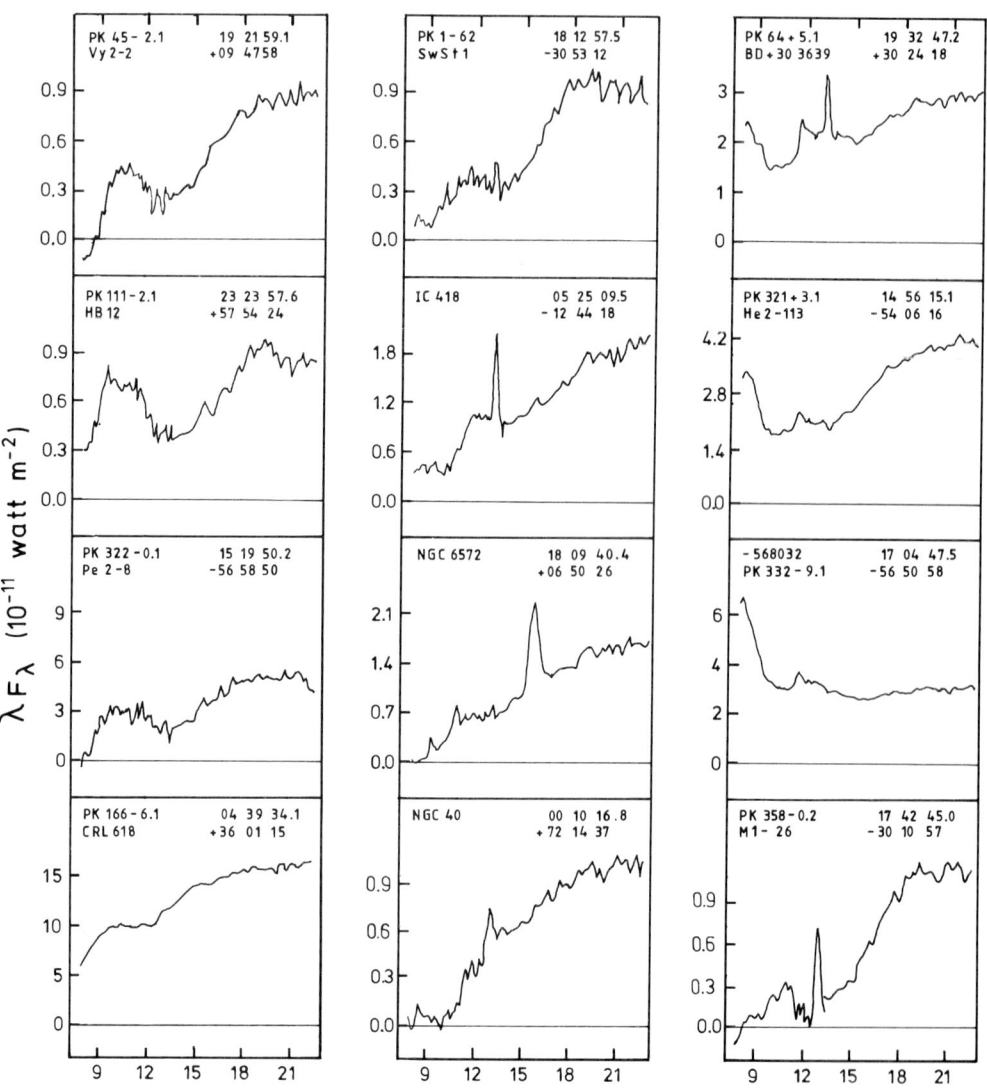

Fig. 2   The IRAS low resolution spectra of 12 young planetary nebulae.   The abscissa is the wavelength scale in microns, while the ordinate gives λF_λ in watts m^{-2}.

V DISTRIBUTION OF YOUNG PLANETARY NEBULAE NEAR THE GALACTIC CENTER

An attempt is being made to discover new nebulae near the galactic center using the IRAS data base. At present the point source catalogue has been used to find those objects within 15° of the center which have their peak flux density either at 25 μm or 60 μm. Further, the 25 μm flux density must be at least a factor of 2 higher than the 12 μm flux density, while the ratio 60 μm to 25 μm must not be greater than 5. The object need not be measured in all 3 bands, upper limits are often sufficient.

In a preliminary study about 1600 objects which meet the above criteria have been found. About 200 are apparently known planetary nebulae. Further study will be necessary to confirm the total number of actual nebulae. From the colors, it is very unlikely that many of these objects are galaxies or HII regions, although an occasional Seyfert galaxy may be in the sample. Some large planetary nebulae will be excluded, but if their properties are similar to nearby nebulae, this will not be more than 15%. The objects whose colors are so similar to PN that they can be confused with them are OH/IR stars and various stars with a large dust shell. These latter objects have only recently (since IRAS) been discovered and neither their frequency nor their evolutionary status is well known. It is possible that they like the OH/IR stars, are related to the early stages of planetary nebula formation.

It is expected that a majority of these objects are PN's, but they are not average nebulae. Their intrinsic infrared flux is much higher than is found for the nearby nebulae and compares with the intrinsically brightest nebulae such as NGC 7020, 6369 and many of the very young objects such as HB 12, He 2-113, Vy 2-2 and Sw St 1. It is expected that most of the nebulae seen by IRAS at the galactic center are therefore young PN's. A preliminary estimate of their spatial distribution is shown in Fig. 8. These objects are clearly seen to form a bulge at the galactic center, flattening into a disc further from the center. There are fainter older objects as well so that the total number may be considerably in excess of the value shown.

ACKNOWLEDGEMENT

The work described in section V is being done jointly C. Bignell, R. Olling, and K. Sahu.

REFERENCES

Aitken, D.K., Roche, P.F. 1983, Mon. Not. Roy. Astron. Soc. 202, 1233
Aitken, D.K., Roche, P.F., Spenzer, P.M. 1979, Astrophys. J. 233, 925
Bentley, A.F 1982, Astron. J. 87, 1810
Bentley, A.F., Hackwell, J.A., Grasdalen, G.L. Gehrz, R.D. 1984, Astrophys. J. 278, 665

Cohen, M., Barlow, M.J. 1974, Astrophys. J. 193, 401
Cohen, M., Barlow, M.J. 1980, Astrophys. J. 238, 585
Gathier, R., Pottasch, S.R., Goss, W.M. v. Gorkom, J.H. 1983, Astron. Astrophys. 128, 325
Gillet, F.C., Low, F.J., Stein, W.A. 1967, Astrophys. J. 149, L 97
Leene, A., Pottasch, S.R. 1986, IRAS Symp. ed. F. Israel (Reidel, Dordrecht)
Moseley, H. 1980, Astrophys. J. 238, 892
Mufson, S.L., Lyon, J., Marionni, P. 1975, Astrophys. J. 201, L 85
Pottasch, S.R. 1980, Astron. Astrophys. 89, 336
Pottasch, S.R. 1986, IRAS Symp. "Light on dark matter" p.131 ed. F.P. Israel
Pottasch, S.R. Baud, B., Beintema, D., Emerson, J. Habing, H.J. ed. F. Israel (Reidel, Dordrecht), Harris, S., Houck, J., Jennings, R., Marsden, P. 1984, Astron. Astrophys. 138, 10
Pottasch, S.R., Peite-Martinez, A., Olnon, F., Mo, J.E., Kingma, S. 1986, Astron. Astrophys. 161, 363
Preite-Martinez, A., Pottasch, S.R. 1983, Astron. Astrophys. 126, 31
Pwa, R., Pottasch, S.R., Mo J.E. 1986, Astron. Astrophys. 164, 184

DISCUSSION

Peimbert: The M_{dust}/M_{gas} ratio is proportional to the filling factor, ε, to the 1/2 power; since $M_{dust} \propto \varepsilon \, M_{dust}$ and $M_{gas} \propto (\varepsilon N_e^2)^{1/2}$; there is some evidence that indicates that ε becomes smaller for larger nebulae (e.g. Torres-Peimbert and Peimbert Rev. Mexicana Astron. Astrofis.), therefore this effect would go in the direction of reducing the M_{dust}/M_{gas} spread.

Pottasch: That is correct. But the filling factor in planetary nebulae usually has a value that is close to unity. For small nebulae it may be between 0.3 and 0.5, while for large nebulae it never goes below 0.1. Thus the spread in M_{dust}/M_{gas} would at most be reduced by a factor of 3 which is small compared to the factor 10^3 spread which is found.

Lenzuni: I noticed that in the log M_D/M_G - log R plot, there are two nebulae displaying impressively high values of M_D/M_G, near 10^{-1} or just below. Are these values physically meaningful? Whatever chemical composition you assume for condensed matter, you always need to have metal abundances much higher then solar, in order to form such a large amount of dust. How can this situation occur?

Pottasch: I believe in these two cases, Vy 2-2 and HB 12, an important part of the far infrared radiation is coming from outside the small region where most of the radio emission is concentrated. These are probably the only two nebulae of those I have studied where this is the case.

Roche: In the object like Hb 12 and Vy 2-2, if one takes the observed size of the ionized region as the extent of the dust emission region, the high fluxes at 25 and 60 μm imply that the dust emission is optically thick in the FIR. That gives good evidence that a substantial fraction of the FIR emission arises from outside the ionized region.

Monk: Point 1: One must be careful when judging temperatures from IRAS data as the line contribution in Band 1 can be considerable, and the corrected flux narrows the distribution severely for some nebulae. Point 2: Would you like to comment on the reliability of flux and continuum measurements form the LRS spectra.

de Muizon: The reliability of the LRS spectra is rather uneven, both in terms of line intensities and of continuum level. There are sources for which line fluxes vary by a factor up to 2 from one individual spectrum to another. Similar variations can also occur on the continuum. There is also a large number of strong and point sources for which the variations are smaller than 30%. One way to decide whether an LRS-catalogue spectrum is reliable and to estimate the error bars is to examine carefully all the individual spectra of that source and compare them to each other (line fluxes and continuum). By doing so, you may also "identify" some possible "features" that look real on the catalogue spectrum but which could in fact be a spike on one of the individual spectra.

Danziger: Is there evidence for more than one temperature distribution of dust in any given PN?

Pottasch: There is sometimes more emission in the 12 μm band than is consistent with the temperature derived from the longer wavelength radiation. This could be radiation from hotter dust, but it could also be atomic line emission contributing to the measured radiation. For those nebulae with Low Resolution Spectra where this problem can be resolved, there is no evidence for a higher temperature dust.

Preite-Martinez: If you refer only to IRAS data, then it is true that the distribution of survey fluxes is broader then a single temperature blackbody. In other energy ranges (near IR) it is clear from the observations of Perzi and myself that in some cases also hot dust must be present.

Clegg: Could some of the apparent correlation between nebular radius and dust-to-gas mass ratio be due to the contribution of emission lines to the IRAS 4-channel fluxes increasing with nebular size? In large PN the dust is cool and may only emit in Band 4, for example.

Pottasch: If a plot is made of only those objects where LRS spectra are available (and thus a correction for line emission can be made between 8 μm and 23 μm), and an attempt is made to subtract possible lines in the other bands, the correlation remains in approximately its present form. For the very largest nebulae careful studies of NGC 7293 and 6853 confirm low dust to gas mass ratios.

Danziger: Where do compact HII regions lie in 2-colour diagram?

Pottasch: The following is a color-color plot on which the position of HII regions are given. These are all 'compact' in the sense that they appear in the IRAS point source catalogue with meaningful fluxes. It has been checked that they have diameters less or equal to 2 arc minutes. The larger HII regions probably are found in a similar position on the diagram.

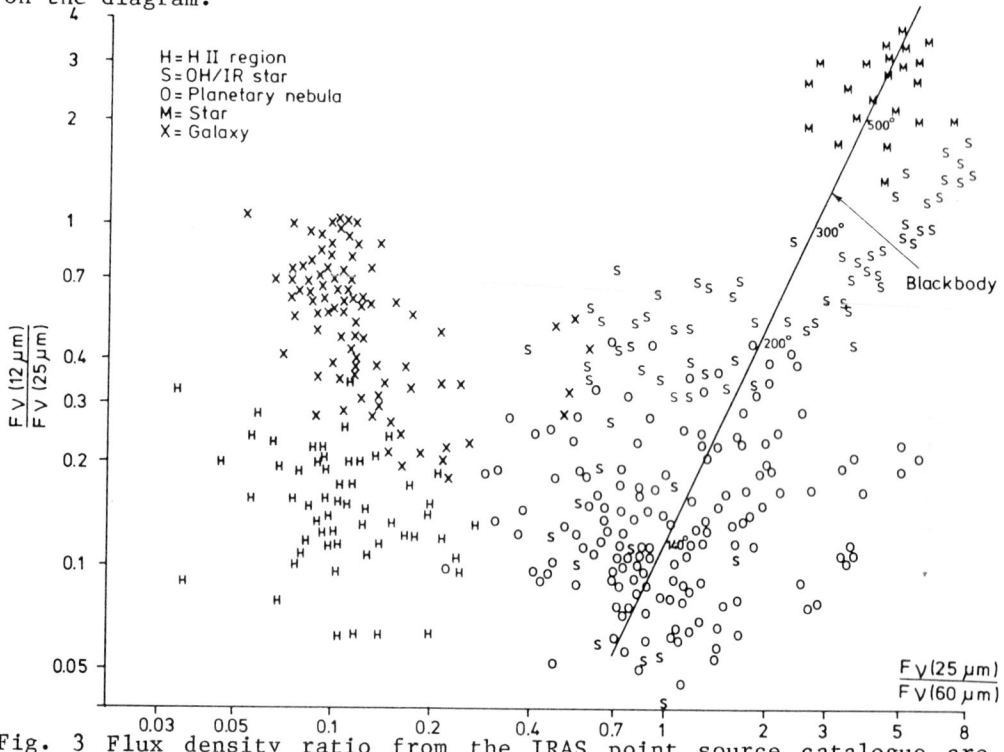

Fig. 3 Flux density ratio from the IRAS point source catalogue are plotted for various kinds of objects, as indicated on the diagram.

Preite-Martinez: Where is the location of Symbiotic stars in the colour-colour diagram you showed?

Pottasch: Symbiotic stars are usually weak infrared emitters except when they are part of a double star system in which the companion is a strong source, e.g. a Mira star. The other cases usually have 12 to 25 µm ratios greater or equal to one.

ADVANCES IN KNOWLEDGE OF PLANETARY NEBULAE FROM UV ASTRONOMY

M. Perinotto
Istituto di Astronomia, Università di Firenze
Largo Enrico Fermi 5
50125 Firenze, Italy

ABSTRACT. A summary is given of recent achievements of UV astronomy in planetary nebulae with emphasis on chemical abundances and comparison with predictions of stellar evolution theory and on properties of fast winds from central stars.

1. INTRODUCTION

Information on radiation from planetary nebula (PNe) shortwards of λ 3100 A has been marginal until the launch of the IUE satellite at beginning of 1978. The IUE satellite is a spectroscopy facility working between 1150 and 3200 A with resolutions of 6 and 0.15 A and two apertures (circular of 3" diameter and oval 10" x 20"). About 150 PNe have been observed with IUE at low resolution and about 40 at high resolution.

The spectra exhibit a continuum and lines. The continuum is dominated by radiation from the central star, unless this is too faint. Nebular continuum in the UV was first detected in a planetary nebula by Benvenuti and Perinotto (1980) in NGC 7662 with off-set spectra and shown to be essentially of atomic origin with little contribution from dust scattered light. This behaviour is opposite to the one in bright HII regions (Perinotto and Patriarchi, 1980), proving that the gas-to-dust ratio is quite higher in planetary nebulae than in HII regions, apart from improbable basic differences in dust properties in the two types of objects.

The stellar continuum provides a way of measuring a color temperature, or T_{eff} and gravity if a model atmosphere is used for the fit. This has been attempted by various authors, including Clegg and Seaton (1983), Adams and Barlow (1983), Cerruti-Sola et al. (1983). There are however problems, particularly for high temperature nuclei, in obtaining valuable temperatures for central stars. In general a black-body cannot fit the emission of high excitation objects in the wavelength range $\lambda\lambda$ 1200 - 1500 A, the observed emission being often higher than a black-body with infinite temperature. On the other hand realistic atmospheric models for hot nuclei of PN are not yet available, although very desirable if one wants to take full advantage of these observations. The line spectrum in the UV contains radiation formed: (i) in the (optical) nebula, (ii) at the star surface, (iii) in the first few stellar radii from the star, and (iv) in the interstellar medium. In this review I concentrate on (i) and on (ii) i.e., on the nebular lines and on P-Cygni like circumstellar lines.

2. NEBULAR LINES IN THE UV

The impact of IUE into our knowledge of nebular lines in PNe is evident from Tables published by Cerruti-Sola and Perinotto (1983) and Clegg (1985) where emission lines of several chemical elements observed in the optical and in the UV range are shown. A better diagnostics of physical conditions particularly in highly ionized regions of planetary nebulae (for details cf. Cerruti-Sola and Perinotto, 1983 and Clegg, 1985), is offered by IUE.

About chemical abundances, the very new information of IUE concerns carbon, whose optical lines are very faint and in addition still present problems with their interpretation as pure recombination lines (cf. Clegg, 1985). A valuable contribution comes even for nitrogen with observations of various ionization stages not seen in the optical. There is also useful additional information for oxygen and neon.

3. ABUNDANCES OF He, C, H, O, Ne

The reasons why accurate determination of the abundances of these elements is particularly important are well known: (i) they are the most abundant elements (after hydrogen), (ii) some of them are strongly modified by stellar nucleosynthesis, (iii) differently from e.g. Fe, Mg, Si, Ca, they should not be affected much by gas to dust depletion, therefore remaining indicative of the total elemental content.

An extended list of abundances in PNe has been recently published by Aller and Czyzak (1983), based on their own work, and a compilation of abundances which includes work by various authors by Zuckerman and Aller (1986). Comparison of the observed abundances in several PNe with theoretical predictions by Iben and Becker (1980) and Renzini and Voli (1981) has been made by various authors including Peimbert (1981) and most recently Pottasch (1984), Aller (1985) and Kaler (1985).

I felt justified in producing a new compilation not only to include more recent results but mainly to have a better appreciation of the quality of the adopted abundances. This of course depends on the weight given to the various independent determinations, which is based on the accuracy of the observations and on the technique used for their interpretation, whether: (i) simplified procedures with T_e and N_e constants in one or two nebular zones, (ii) detailed photoionization models or (iii) a combination of (i) and (ii).

I have considered all individual papers dealing with determination of chemical abundances in PNe using UV, optical or IR data, starting with the fundamental work of Torres-Peimbert and Peimbert (1977), up to June 1986. Thus all the modern photoelectric work is taken into account, as well as the many object studies made by Aller, Barker, French, Kaler etc. The total number of papers examined amounts to about 50. Even the relevant work made before my starting date is included because the authors have thereafter summarized their previous determinations. With the use of such a relatively recent starting date for the compilation, one has the advantage that also the atomic quantities used in the interpretation are quite modern. For the abundance of carbon, only determinations making use of UV lines have been considered except for quite few objects where very careful studies of optical lines have been performed, while the UV lines have not observed.

4. COMPARISON WITH THEORETICAL PREDICTIONS

The results are shown in Figs 1 and 2, where $\log(N/O)$ versus He/H and $\log(C/O)$ versus He/H, respectively, are compared with the theoretical predictions by Renzini and Voli (1981). (The behaviour of Neon is not presented here). The advantages of presenting abundances of heavy elements relative to oxygen instead of to hydrogen are evident: (i) errors in the determination of abundances of heavy elements (quite sensitive to T_e) are minimized, as are the (ii) effects of possible differences in (C+N+O)/H abundances due to vertical and to horizontal galactic chemical gradients. On the other hand oxygen is predicted to be little affected during the relevant stellar evolution, so that with the above presentation one really sees the variation of N and C abundances along with changes of He/H.

We simplify the matter of errors, indicating that He/H should be correct to within a 10% except for PNe of low excitation which must be corrected for neutral helium. To N/O and C/O an error of a 60% is attributed. Halo PNe are not included in Figs 1, 2 because they cannot be interpreted with the plotted theoretical predictions, valid for a solar initial chemical composition. Also not included are A30 and A78 whose very high He/H speaks in favour of an evolutionary stage very different from that of most of the observed PNe (cf. Iben and Renzini, 1983). In each figure two theoretical curves are plotted corresponding to extreme values of the parameter α describing the behaviour of the convection (α = mixing length/pressure scale height). The curves are labelled with values of the initial mass of the star, whose envelope reaches the chemical composition of the labelled point of the curve at the moment of the expulsion of the envelope itself.

Consider first Fig. 1, i.e. $\log(N/O)$ against He/H. Few objects of our compilation are not plotted because they are out of scale in He/H. Some of them do have He/H quite smaller than 0.08. It is likely that their He abundances are heavily underestimated, being PNe of very low excitation. Others do have high He/H and should be included in our discussion. They are: NGC 6563 (He/H = 0.15:, $\log(N/O)$ = 0.26), NGC 2346 (0.17,-0.38), and NGC 6302 (0.23, 0.23).

Observationally we see: (i) a clear correlation of N/O with He/H; (ii) the range of He/H extends from 0.08 to 0.23, i.e. $\Delta(He/H)$ = 0.15; (iii) the range of $\log(N/O)$ goes from -1.5 to 0.5, i.e. $\Delta(N/O)$ = 200.

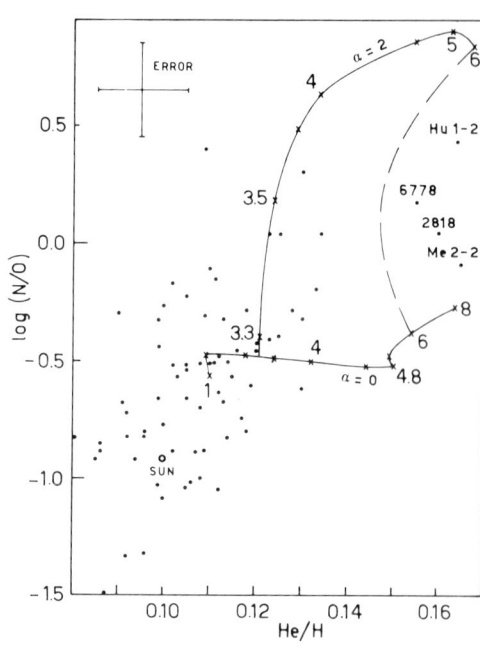

Figure 1.

Observed abundances of N/O versus He/H in planetary nebulae from a new compilation, are plotted against theoretical predictions by Renzini and Voli (1981) for progenitors with solar initial chemical abundances. Masses of progenitors (in solar units) are labelled along two theoretical curves corresponding to two extreme values of the "convection" parameter α.

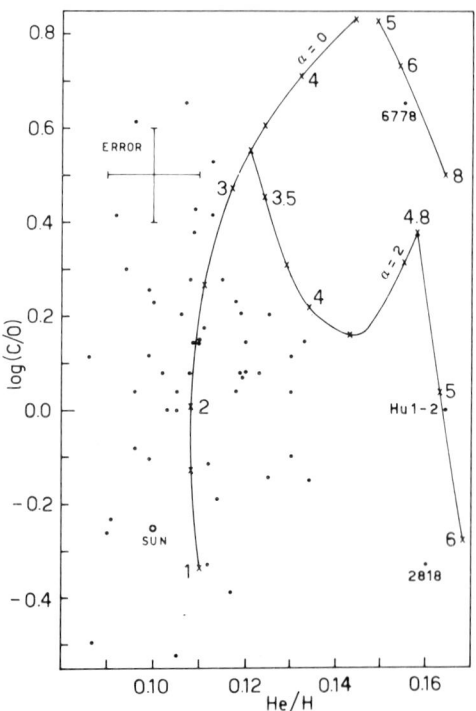

Figure 2.

Same as Figure 1, but for the abundance ratios C/O versus He/H.

Theoretically we expect: (iv) a minimum He/H of 0.11 and log(N/O) = = - 0.5 if the smallest progenitors of PNe do have main sequence masses of 1 solar mass; (v) no correlation of N/O with He/H in the range of progenitors masses: $1 \div 3$ $M_\odot$, which on evident grounds should be the most populated one; (vi) a strong correlation of N/O versus He/H for initial stellar masses between 3 and 8 $M_\odot$.

Expectation (iv) is not in agreement with observations. Part of the discrepancy in He/H might be removed with a proper correction for unseen He° in low excitation objects, but the He/H discrepancy (as well as the one in N/O: observed values smaller than the predicted ones) regards even well observed high excitation objects as NGC 7662. It is not clear how the N/O discrepancy can be removed except by admission that the initial N/O is lower than solar in many progenitors. In conclusion these discrepancies remain to be understood. Expectation (v) is not inconsistent with the observations. Expectation (vi) is consistent with observations. A bit strange is the fact that we see no one PNe with progenitor mass from 4 to 6 $M_\odot$, while we have 4 plotted points (plus NGC 2346, 6302, 6563), i.e. 7 objects with progenitor masses above 6 $M_\odot$.

The suspicion here arises that these objects do not form a continuous sequence with the other PNe in the diagram, but rather that they are of a somewhat different nature. This may have to do with the bipolar morphology of various of them.

Figure 2 shows log(C/O) agains He/H. Again few objects are not plotted being out of scale in He/H because of low He/H. NGC 6302 is out of scale for high He/H. In addition NGC 6302 and also M 1-14 are out of scale because of low (-0.70) and high (1.04) log C/O, respectively.

In Figure 2 we see facts in agreement with the theory: (vii) no correlation of C/O with He/H for most objects, as predicted by theory for progenitors with masses 1 to 3 $M_\odot$, but correlation for the same objects that in Fig. 1 exhibit high (N/O): NGC 2818, 6778, Hu 1-2 (Me 2-2 has no measured C/O). A bit extreme appears the carbon abundance of 6302 (log C/O =- 0.70), but still consistent with the theoretical predictions; (viii) the observed range of log C/O from -0.4 to +0.7 is consistent with the one predicted from theory for progenitors with masses 1 to $M_\odot$.

Instead one must explain why: (ix) there is a much higher concentration of PNe with predicted progenitors of masses $2 \div 3$ $M_\odot$ than in the interval $1 \div 2$ $M_\odot$, contrary to expectation; (x) again, as in Fig. 1, we have a complete lack of observed objects with progenitors of masses 4 to 6 $M_\odot$, which strengthens the above suspicion.

The discrepancy (iv) has already been suggested (cf. Kaler, 1985) to indicate a theoretical overestimate of abundance of nitrogen. The point made in (ix) may seem, at a first glance, to mimic an observational selection effect in the sense that we might observe less objects with smaller progenitors because they probably produce less massive PNe. Of course, the reality of this effect could be investigated by examining the properties of all the observed PNe in the progenitors range 1 to 3 $M_\odot$. There is however the possibility that (ix) indicates a theoretical underestimate of carbon.

If this is the case, an improvement on the theory able to produce less N/O in the first dredge-up would imply less C/O produced in the third dredge-up. The discrepancy (ix) on carbon would become worse, and would require other effects to overcompensate this reduction of C/O.

5. FAST WINDS FROM CENTRAL STARS

Although the existence of fast winds from central stars of PNe was known before the IUE satellite for those central stars exhibiting WR spectra, it was with IUE that the importance of the phenomenon was fully appreciated. In addition to the importance per se, fast winds may be relevant to the genesis, the dynamics and the excitation of the nebula, as well as for accelerating the evolution of the nucleus. After the first detection (Heap et al., 1978), several works have considered essentially individual objects while one study (Cerruti-Sola and Perinotto, 1985; CP) has dealt with the statistical properties of fast winds in many objects. I first summarize here the general properties of fast winds in PNe and then briefly discuss the status of art for the determination of the associated mass loss rate.

5.a General properties

Low resolution IUE spectra are quite adequate to reveal the phenomenon, which is outstanding in the UV lines: CIV 1549, NV 1240, OIV 1342, OV 1371, SiIV 1400, NIV 1719 A. They are also adequate for measuring the edge velocity, and useful for a first approximation derivation of the mass loss rate, which of course is better studied using high resolution IUE spectra.

According to CP, the phenomenon is quite common. P Cygni profiles have been detected in 62% of the studied objects, but the number of central stars with fast winds is probably higher than that because, as discussed by CP, cases of blue shifted absorption lines with little emission escape detection with low resolution IUE spectra. Various objects where P Cygni profiles are not seen, are high excitation objects with almost a continuum stellar spectrum. Regarding the behaviour in the HR diagram, a tendency is found for central stars with gravities higher than log g = 5.2 (cgs) to show less frequently the phenomenon of fast winds. The edge velocities range between 1400 and 5000 km s^{-1}. The ratio of edge to escape velocity v_∞/v_{esc} is between 1.5 and 5.5, with an average of 3.3, similar to population I OB stars. This seems to indicate that the mechanism responsible for the production of the wind is similar to the one accepted as able to fully explain the winds in population I OB stars, i.e., the line radiation driven wind mechanism (cf. Abbott and Lucy, 1985).

The velocity field of the expanding one-fluid gas is usually expressed by the velocity normalized to the edge velocity, as function of the radial distance normalized to the stellar radius ($x = r/R_*$) with

$$w(x) = v/v_\infty \cong (1 - \frac{1}{x})^\beta \quad . \tag{1}$$

In population I OB stars it appears that winds in almost all objects follow equation (1) with $\beta \cong 1.0$.

Instead studies of various PNe central stars indicate extremely different values of β, ranging from 0.5 to 4. The reason for such a wide difference in the character of the velocity field is not yet clear since small values of β are found both for high and low excitation objects and the same is true for high values of β.

The behaviour of the ionization across the wind can be inferred from the OIV, OV lines and even from the comparison of the CIV and NV profiles. The ionization is seen to increase outwards as expected from an outwards decrease in electron density, if the ionization is radiative.

5b. Mass loss rate

The procedures to determine mass loss rates from P Cygni profiles go back to the theory of line formation in expanding atmospheres (Sobolev, 1957; Lucy and Solomon, 1970; Castor, 1970; Lucy, 1971). Castor and Lamers (1979, CL) have produced an atlas of theoretical profiles in terms of three parameters: the exponent β of the velocity law in equation (1), the exponent γ of the opacity law:

$$\tau(w) \cong T(1-w)^\gamma \tag{2}$$

where T is the total optical depth in the line expressed by

$$T = \int_{w_0}^{1} \tau(w) \, dw = \frac{\pi e^2}{mc} f \lambda_0 v_\infty^{-1} N_i \tag{3}$$

where w_0 is the velocity at the base of the wind, f the oscillator strength of the line whose laboratory wavelength is λ_0 and N_i (cm^{-2}) the total number of absorbers in the line.

After, β, γ, T are obtained by fitting the observed with the theoretical profile of CL, one gets readily the mass loss rate $\dot{M}$ if the stellar radius, the chemical abundance of the element and the number of absorbers populating the lower level of the observed P Cygni line are known. The last implies a knowledge of the ionization fraction of the ion (q_{ion}) in one point or across a region of the wind, a quantity difficult to determine. This technique requires the use of high resolution profiles and has been used for few central stars.

Another useful procedure, under Sobolev approximation, has been worked out by Castor, Lutz and Seaton (1981, CLS) particularly for use with low resolution IUE data. This requires measuring the first moment of the flux distribution

$$w_1 \propto (F_\lambda - F_c)(\lambda - \lambda_0)/F_c \, d\lambda \, , \tag{4}$$

a quantity which can be obtained from low resolution IUE data. This technique assumes the CL parametrization of the velocity and opacity laws with $\beta = \gamma = 1$. The authors, to overcome the problem of the determination of the ionization fraction, propose the reasonable approximation $q(OIV)+q(OV) = 1$. This procedure has been extensively used by CP,

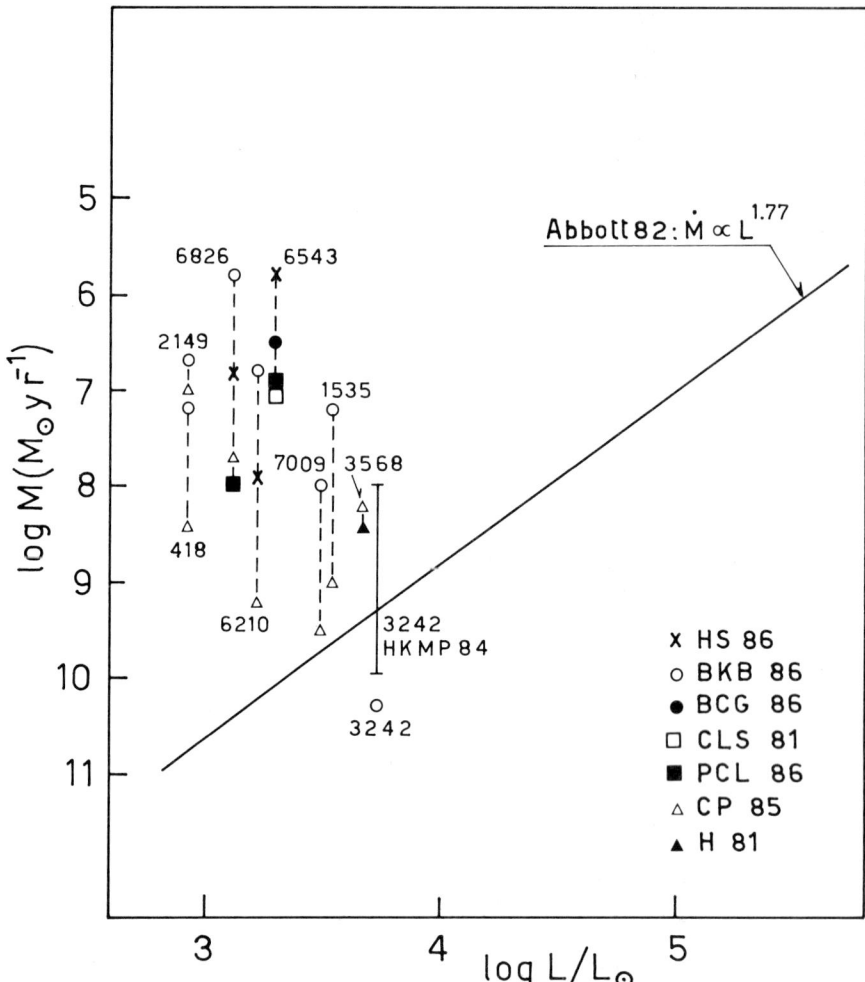

Figure 3. Observed mass loss rates from fast winds against luminosity in central stars of planetary nebulae are compared with the extrapolation to low luminosities of the Abbott (1982) empirical law for population I OB stars. The objects are indicated with their NGC or IC numbers. Only objects measured by at least two different authors are shown. A dashed line joins measurements of the same object made by the different authors. Their initials are attached to a specific symbol and the full names are easily recognized in the References. For the sake of clarity, differences in luminosity attributed to the same object by different authors are here disregarded.

who have derived $\dot{M}$ for 17 central stars. Surdey (1982) has further investigated the CLS method showing that w_1 is indeed independent of the velocity and opacity laws, and therefore the method is quite powerful but only for very optically thin lines. Otherwise with the CLS method one underestimates $\dot{M}$ (Hutsemekers and Surdej, 1986).

A more powerful technique (no Sobolev approximation) implies the full solution of the transfer equation in the fluid frame (Mihalas et al., 1975; Hamann, 1980). It has been applied so far to only one planetary nebula central star.

Finally Lamers, Cerruti-Sola and Perinotto (1986) have developed a technique called the SEI method intermediate between that of Hamann (1980) and the one used by CL. This procedure gives results very close to those of Hamann, includes collisions in the source function, what is not done by Hamann, and is much faster to use on a small size computer. It appears quite promising to provide valuable information even from lines not so thin as needed for an exact application of the CLS method. Clearly the problem of the ionization remains with us.

In Figure 3 the determinations of $\dot{M}$ in objects where values by different authors have been obtained, are plotted: they are compared with the extrapolation to low luminosities of the $\dot{M} \propto L^{1.77}$ law found by Abbott (1982) to describe closely the behaviour of population I OB stars. I do not enter here into details about the individual values and the possible reasons of the differences among them.

The main conclusion of CP concerning the behaviour of $\dot{M}$ in central stars of planetary nebulae, being higher than predicted by the mentioned extrapolation, remains valid, although uncertainties up to more than an order of magnitude are present on individual stars. By now, values of $\dot{M}$ between 10^{-10} to 10^{-6} $M_\odot$ yr^{-1} are suggested for different central stars, which means that significant effects on the nebula and on the evolution of the nucleus are possible at least for some planetary nebulae. Part of the discrepancy with the extrapolated line may be due to the chemical abundances in the winds being higher than those we have used. It is however unlikely the discrepancy is entirely due to this reason, because abundances higher by a factor of ten or more should be present in almost all the objects, what seems unreasonable.

REFERENCES

Abbott, D.C.: 1982, *Astrophys. J.* 259, 282.
Abbott, D.C. and Lucy, L.B.: 1985, *Astrophys. J.* 288, 679.
Adams, S. and Barlow, M.J.: 1983, in 'Planetary Nebulae', IAU Symp. No. 103, Ed. D.R. Flower, Dordrecht: Reidel, p. 537.
Aller, L.H. and Czyzak, S.J.: 1983, *Astrophys. J. Suppl.* 51, 211.
Aller, L.H.: 1984, 'Physics of Thermal Gaseous Nebulae', Dordrecht: Reidel, pp. 299.
Benvenuti, P. and Perinotto, M.: 1981, *Astron. Astrophys.* 95, 127.
Becker, S.A. and Iben, I.Jr.: 1980, *Astrophys. J.* 237, 111.
Bianchi, L., Cerrato, S. and Grewing, M.: 1986, preprint.
Bombek, G., Köppen, J. and Bastian, U.: 1986, preprint.
Cerruti-Sola, M. and Perinotto, M.: 1983, *Mem. S.A.It.* 54, 511.
Cerruti-Sola, M. and Perinotto, M.: 1985, *Astrophys. J.* 291, 237.

Cerruti-Sola, M., Perinotto, M., Cacciari, C. and Patriarchi, P.: 1983, 'Planetary Nebulae', IAU Symp. n. 103, Ed. D.R. Flower, Dordrecht: Reidel, p. 535.
Clegg, R.: 1985, 'Production and distribution of C,N,O elements', ESO Workshop, Garching, p. 261.
Clegg, R. and Seaton, M.J.: 1983, in 'Planetary Nebulae', IAU Symp. n. 103, Ed. D.R. Flower, Dordrecht: Reidel, p. 536.
Castor, J.I.: 1970, *Monthly Notices Roy. Astron. Soc.* 149, 111.
Castor, J.I. and Lamers, H.J.G.L.M.: 1979, *Astrophys. J. Suppl.* 39, 481.
Castor, J.I., Lutz, J.H. and Seaton, M.J.: 1981, *Monthly Notices Roy. Astron. Soc.* 194, 547.
Hamann, W.-R.: 1980, *Astron. Astrophys.* 84, 342.
Hamann, W.-R., Kudrizki, R.P., Méndez, R.H. and Pottasch, S.R.: 1984, *Astron. Astrophys.* 139, 459.
Harrington, J.P.: 1982, 'Advances in UV Astronomy, Four Years of IUE Research', NASA CP- 2238, p. 610.
Hutsemekers, D. and Surdej, J.: 1986, *Astron. Astrophys.*, in press.
Heap, S.R. et al.: 1978, *Nature* 275, 385.
Iben, I.Jr. and Renzini, A.: 1983, *Ann. Rev. Astron. Astrophys.* 21, 271.
Kaler, J.B.: 1985, *Ann. Rev. Astron. Astrophys.* 23, 89.
Lamers, H.J.G.L.M., Cerruti-Sola, M. and Perinotto, M.: 1986, *Astrophys. J.*, in press.
Lucy, L.B.: 1971, *Astrophys. J.* 163, 95.
Lucy, L.B. and Solomon, P.M.: 1970, *Astrophys. J.* 159, 879.
Mihalas, D., Kunasz, P.B. and Hummer, D.G.: 1975, *Astrophys. J.* 202, 465.
Peimbert, M.: 1981, "Physical Processes in Red Giants", Ed. I. Iben and A. Renzini, Dordrecht: Reidel, p. 409.
Perinotto, M., Cerruti-Sola, M. and Lamers, H.J.G.L.M.: 1987, 'Late Stages of stellar evolution', Dordrecht: Reidel, p. 387.
Perinotto, M. and Patriarchi, P.: 1980, *Astrophys. J.* 238, 614.
Pottasch, S.R.: 1984, 'Planetary Nebulae', Dordrecht: Reidel, pp. 233.
Renzini, A. and Voli, M.: 1981: *Astron. Astrophys.* 94, 175.
Sobolev, V.V.: 1957, *Astron. Zh.* 34, 694 (Trans. Soviet Astron. 1, 678).
Surdej, J.: 1982, *Astrophys. Space Sci.* 88, 31.
Torres-Peimbert, S. and Peimbert, M.: 1977, *Rev. Mex. Astron. Astrophys.* 4, 341.
Zuckerman, B. and Aller, L.H.: 1986, *Astrophys. J.* 301, 772.

DISCUSSION

M. PEIMBERT: Some of the differences between the predicted He/H, N/H values and the observed ones could be due to a range of about a factor of two in the Z values of the observed objects; while the theoretical track is for a single initial chemical composition. Even if there is scatter in the observational He/H, N/H diagram, there is a very strong correlation present in the sense that the higher the N/H the higher the He/H ratio.

I wish also to add that: a) since only about 10 to 20% of the ionizing stars of PN are hydrogen poor (Wolf-Rayet type), it follows that the mass loss rate has to be smaller than 10^{-8} to 10^{-9} $M_\odot$ per year for most objects; b) from the galactic distribution and from kinematical arguments it is expected that the number of PN in the 1-2 $M_\odot$ range should be larger than that in the 2-3 $M_\odot$ range. Therefore the C abundance results seem to give the wrong mass distribution, in addition to a higher efficiency of the dredge ups than that considered by Renzini and Voli (1981). The stellar wind phenomenon goes in the right direction increasing the C abundance in the shells of PN with central stars of the WR type.

N. PANAGIA: In your estimates of the mass loss rates from UV P Cyg line profiles, what do you assume for the chemical composition of the wind? The reason for asking this question is that since the central stars of PNe may have chemical abundances (especially for carbon) which are vastly different from each other, great care should exercised in adopting the most appropriate one. In fact, I guess that this effect may account for part of the spread and/or some of the exceedingly high values of $\dot{M}$ which are shown.

M. PERINOTTO: The abundances used are those measured in the individual nebulae. Unfortunately there is not yet information on the abundances in the stellar photosphere of the various objects, which clearly would be more appropriate to use. Do you have suggestions for a better choice?

N. PANAGIA: Unfortunately no, but I would like to strongly encourage everyone to think on this problem and contribute to solve it.

P. PERSI: The relationship $M \propto L^{1.77}$ given by Abbott 1982 is derived mainly using ratio data. You compare the mass-loss rate derived from UV from PNe with this relationship. Sometimes the two different methods in deriving M disagree. What is your opinion?

P. PERINOTTO: To my knowledge for the best studied population I OB stars, the two methods give close results for the mass loss rate.

J. KOPPEN: The mass loss rate of 10^{-6} $M_\odot$ yr^{-1} for NGC 6826 was found assuming $T_* = 35\,000$ K. If one does the analysis with $T_* = 40\,000$ K, in particular the excitation of the subordinate lines, the rate goes down by about a factor of ten.

M. PERINOTTO: Right. As I have mentioned, I did not wish to enter here into details about the possible reasons of the differences found in the same objects by different authors.

L. BIANCHI: About your plot of $\dot{M} \div L$: the various authors that you quoted also found different L values (not only different $\dot{M}$).

M. PERINOTTO: Right. In this simplified diagram I have attached only one luminosity to the same object for the sake of clarity, because we were concerned mostly with differences in $\dot{M}$. Anyhow the differences in luminosities of the different authors for most of the objects are not such to bring the points in much better agreement with the Abbott's empirical law shown in the diagram.

I.J. DANZIGER: To what extent are the large differences in the mass loss rates by different methods dependent on differences in the assumed (or calculated) mass fraction of the ion used for modelling?

P. PERINOTTO: Procedures used by the different authors are quite different in this respect and I do not have an immediate answer at hand. The authors who have used the assumption $q(OIV)+q(OV) = 1$ should be relatively safe from this point of view, because the assumption is probably good. In the case of IC 3568, one determination is made by ourselves with the above assumption and the other by Pat Harrington with another procedure. The two $\dot{M}$ are close, but this can be due to a favorable combination of the various quantities entering into the two determinations. Thus I do not have by now a precise answer.

STATUS OF THE SPECTROSCOPIC SURVEY OF PLANETARY NEBULAE*

Björn Stenholm Agnes Acker
Lund Observatory Observatoire de Strasbourg
Box 43 11, rue de l'Université
S-221 00 Lund, Sweden F-67000 Strasbourg, France

ABSTRACT. The status in August 1986 of the spectroscopic survey in the visual domain of planetary nebulas is given along with a short discussion of spectral classification problems of emission-line objects.

1. INTRODUCTION

In the Catalogue of Galactic Planetary Nebulae (CGPN), Perek and Kohoutek (1967), 1036 planetary nebulas were presented. The latest list from Centre de Données de Strasbourg contains 1563 objects. It was already stated by the authors of the CGPN, that many objects in the catalogue would come out as other types of objects than PN:s when sufficiently detailed observations had been made. This is true for the contents of the CGPN and it is of course also true for the more than 500 objects in this category found between 1967 and 1985. The reason for this is that the indication of a PN on e.g. an objective-prism plate may be very vague, but still the object will go into the catalogue as a possible or probable PN. Thus it is well known that the population of about 1600 PN:s is inhomogeneous. Lists of misclassified objects have also appeared. Sanduleak (1976) gives results from observations of his own on objective-prism plates, and Kohoutek (1978, 1983) quotes the recent literature.
 In order to increase the homogeneity of the known PN population and to improve our knowledge of the faint majority of the PN:s a spectroscopic project was started. The project, described by Stenholm and Lundström (1984), aims to observe all known PN:s and PN candidates with a slit spectrograph equipped with a modern digital detector. The programme was started at European Southern Observatory in 1984 and is still running. Since July 1986, Observatoire Haute Provence in France is also contributing. A previous status report giving the situation in February 1986 is found in Stenholm (1986), while this paper gives the situation in September 1986.

---

* Based on observations collected at European Southern Observatory, La Silla, Chile and Observatoire Haute Provence, France.

2. THE INSTRUMENTATION

In the following table some instrumental and observational parameters are summarised:

Table 1. Instrument configurations

	European Southern Observatory (ESO)	Observatoire Haute Provence (OHP)
Telescope	1.52 m	1.93 m
Spectrograph	Boller & Chivens	CARELEC
Detector	IDS	CCD
Number of useful pixels	2053	511
Aperture	4x4"	2.5x3"
Appr. wavelength range	400-740 nm	385-740 nm
Dispersion	17 nm/mm	26 nm/mm
Appr. resolution	1 nm	2: nm
Normal exposure time	10 min	10 min

Some of the content in the table deserves special comments. The detectors are both giving data in digital form, and they have high sensitivity, which is necessary for the project. However, the sensitivity goes down very quickly below 400 nm. This is most important for the IDS but the CCD behaves in a similar way. This means, that the astrophysically important line of [O II] at 372.7 nm is not recorded.

The resolution, given as approximate numbers above, is illustrated in figures 1 and 2 on spectra from ESO, which show the blends of [N II] lines with Hα and the sulphur doublet around 672 nm. The difference in wavelength between the latter components is 1.4 nm and the intensities are of about the same order of magnitude and can easily be resolved. The difference between Hα and the two nitrogen components is 1.5 and 2.0 nm respectively. As is shown in figure 2, these lines can easily be resolved, given that the intensities are not too different. For high-excitation nebulas, where the Hα line is very dominant, the nitrogen contribution is difficult or impossible to establish, and the Gaussian fitting procedure is not converging.

The apertures used are very small. Most objects are, however, point-like with no or little angular extension beside the seeing disc. This majority of objects is also the less observed part of the population. On the other hand, extended objects, particularly those of low area brightness, are difficult to observe in this manner.

Depending on the large number of objects, the exposure time for each of them must be short, in order to execute the programme during a reasonable period of time. A standard exposure time of 10 minutes has been chosen. The signal-to-noise ratio then gives indirectly a measure of the faintness of the object. When dictated on practical grounds, however, in some cases a longer exposure time has been used.

STATUS OF THE SPECTROSCOPIC SURVEY OF PLANETARY NEBULAE

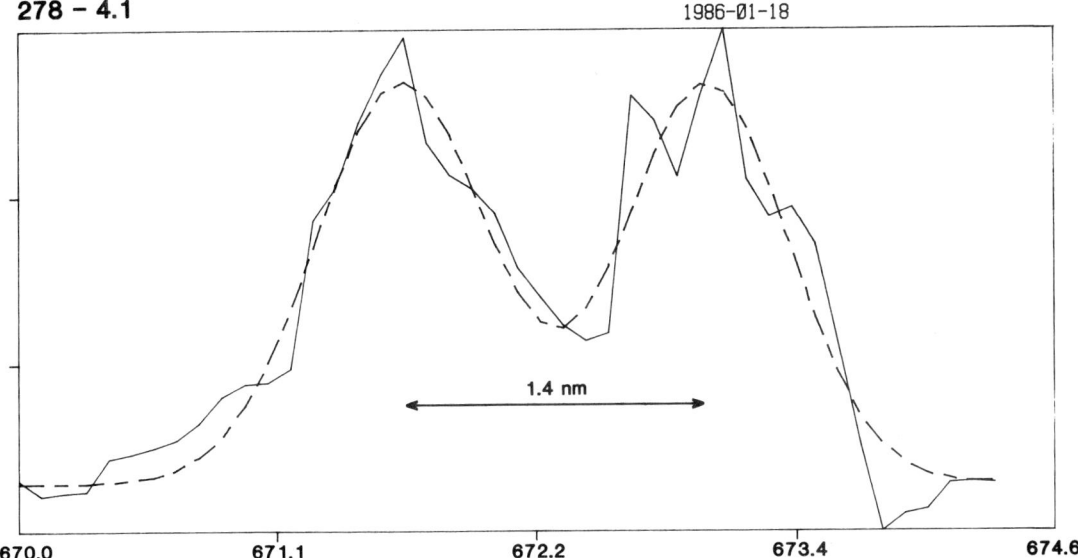

Fig. 1. Observed doublet of sulphur at 671.7 and 673.1 nm in the object 278-4.1 along with a Gaussian fitting (dashed line). ESO, La Silla.

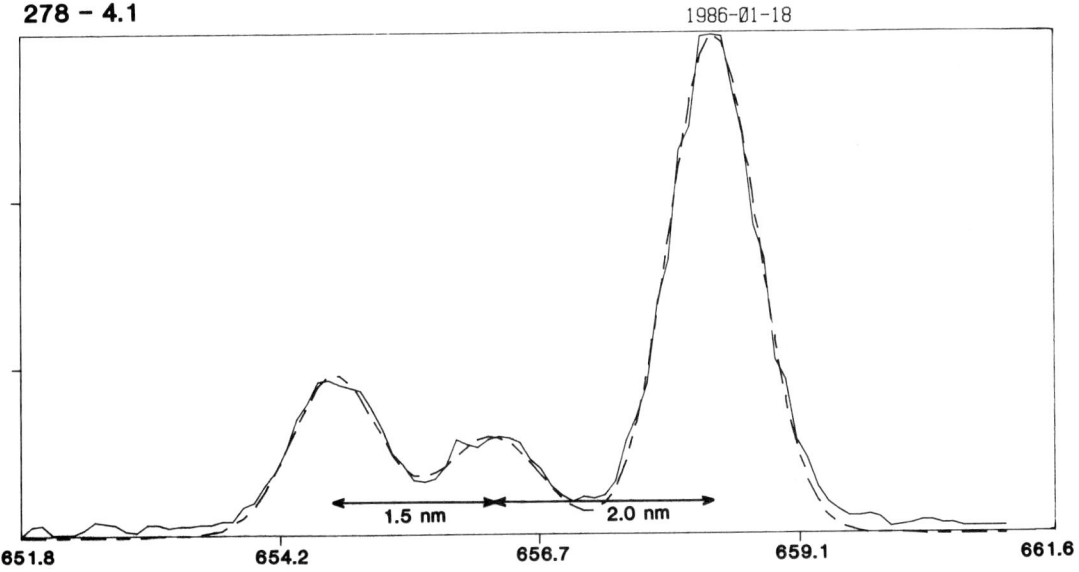

Fig. 2. Observed Hα line at 656.3 and nitrogen doublet at 654.8 and 658.3 in the object 278-4.1 along with a Gaussian fitting (dashed line). ESO, La Silla.

3. QUANTITATIVE RESULTS UP TO AUGUST 1986

Up to now four sessions have been executed at ESO, La Silla, namely April 1984, July-August 1985, January 1986 and July 1986. For the northern part of the project, the programme at OHP has just started with the first session in July-August 1986. The PN population is heavily concentrated to southern declinations which motivates the earlier start in the south. Figure 3 shows the distribution of planetary nebulas in right ascension according to the most recent list, 1563 objects. Of them are now 718 observed within this programme, 656 at ESO and 74 at OHP having 12 objects in common. An example of the observations made at La Silla is in figure 4, giving a high-cut spectrogram of the whole spectral range for the little known planetary nebula 3-17.1.

4. QUALITATIVE RESULTS

Up to the present day the main efforts have been put upon data collection and reduction and only little stress has been put on evaluation. However, some preliminary results are given in Stenholm (1986). The majority of the objects in the tables by Sanduleak (1976) about misclassified and faint planetary nebulas has been scrutinised in a separate paper, Stenholm and Acker (1986).

To get an idea about the material compiled so far we show in the following table the number of objects not considered as "true" planetary nebulas by us or following comments in the literature:

Table 2

Type	Number
Symbiotic stars	74
H II regions	14
Galaxies	18
Reflection nebulas	4
SNR knot	1
Plate faults	13
Undetected or uncertain status	37
Total	161

An extrapolation of this gives that about 20% of the present catalogue content is not true planetary nebulas.

5. PROBLEMS OF SPECTRAL CLASSIFICATION

A planetary nebula is the consequence of stellar evolution for solar-type stars. A red giant (or Mira variable) sheds its atmosphere into

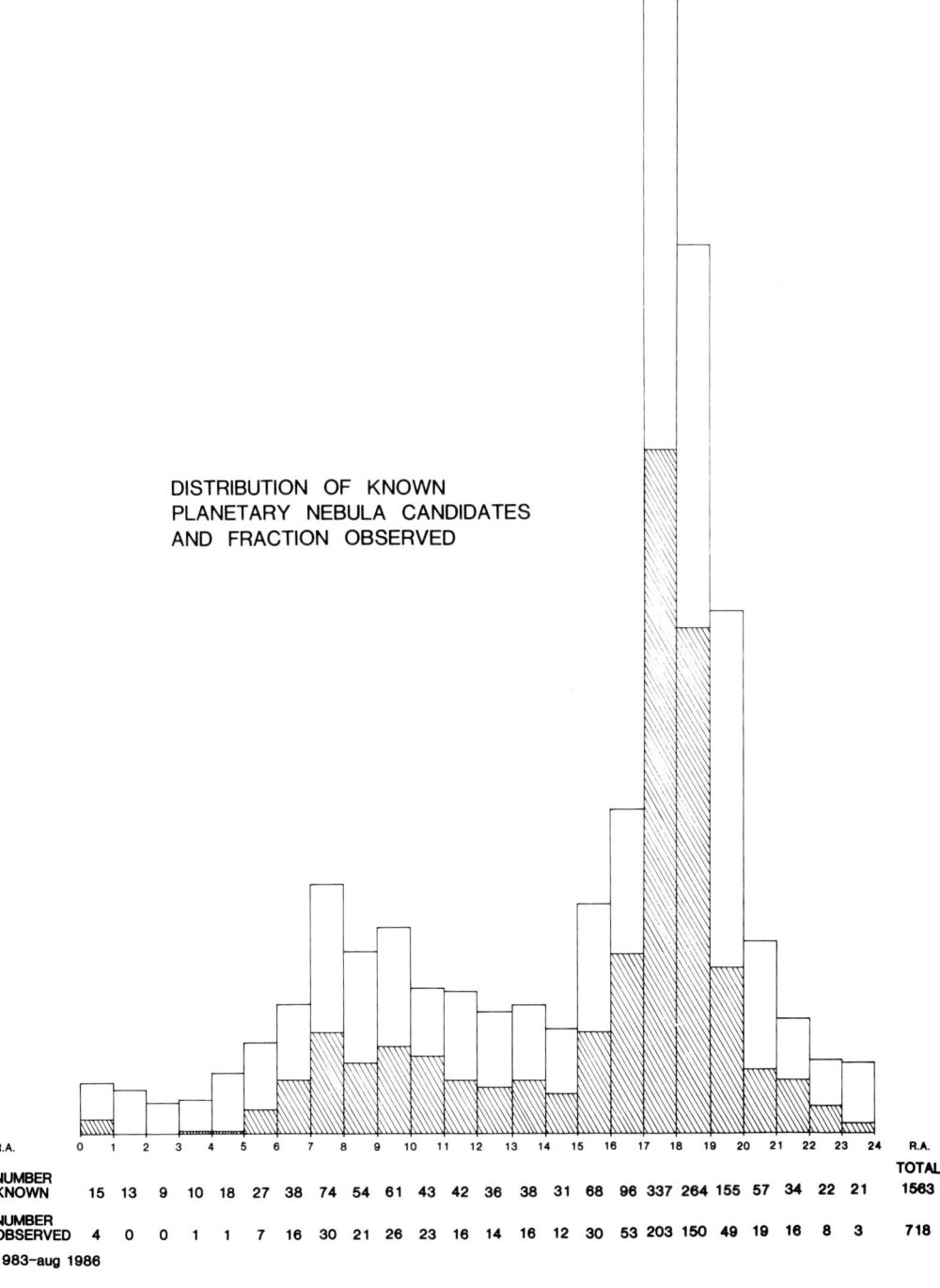

Fig. 3. Distribution of planetary nebulas in right ascension. The hatched area shows the fraction observed until August 1986.

circumstellar space and its core contracts to a hot white dwarf, exciting the nebular shell. This process gives rise to certain observational parameters, morphology, spectrum, expansion etc. If all these parameters were accurately observed for a certain object, it would be easy to discern PN:s from other objects. In most cases, most observational parameters are missing, maybe only a faint trace of spectrum has been observed on e.g. an objective-prism plate. However, in the present programme we collect some spectral information in the visual domain. According to the above, this is not always sufficient to sort out PN:s from other emission-line objects. Let us therefore discuss this matter.

The largest number of the misclassified planetary nebulas above belongs to the symbiotic stars. They are usually known as binaries with mass transfer from one component to the other and their emission-line spectrum is formed in the mass-transfer process. Thus, the symbiotic stars are stars, and normally the spectrum from a symbiotic star is quite different from that of a PN. See figure 5 for an example. Many of the 74 objects above are already in the catalogue of symbiotic stars by Allen (1984). The emission lines on a spectrum of a symbiotic star normally rest on a background continuum of a late-type star, one of the component, giving the saw-tooth pattern in the red end of the spectrum. This is generally easily recognised. Moreover, the symbiotic stars have a different set of emission lines (indlucing high-excitation lines of e.g. He II, [Fe II], [Fe III], [Fe VI], [Fe VII], [Ca VII], when e.g. the nebular lines of [O III] and [N II] in planetary nebulas are depressed or absent. Symbiotic stars usually also have two lines of unknown origin at 683 and 709 nm and they should also be seen. Generally, symbiotic stars are easily discerned from planetary nebulas. However, in recent years there has been a discussion about a possible evolutionary connection between planetary nebulas and symbiotic stars, and there exist individual objects which spectroscopically show both properties. See e.g. the discussion about "Protoplanetary nebulae and symbiotic stars" in Pottasch (1984).

H II regions and particularly small, compact H II regions are difficult to discern from planetary nebulas of very low excitation (VLE) objects, especially when only spectroscopic data exist, which is the case for the majority of the objects here. What we have when we examine H II regions and planetary nebulas, is clouds of gas, excited by nearby, hot stars, which then will give rise to identical spectra. The cloud does not know how it was formed. As the exciting stars of H II regions are cooler than PN nuclei, high excitation is in favour of planetary nebulas, low excitation means H II regions. This does not work in the individual case, however. But there may be other parameters helping to discriminate between the types, such as morphology and kinematics. Such data are, however, missing for most of our objects.

Some types of emission-line galaxies, Seyfert galaxies and star-burst galaxies, may show a spectrum resembling that of a planetary nebula. A distinct redshift will normally sort out these objects.

Results and discussions of the above mentioned material are given in a paper by Stenholm and Acker (1986) and more will be given in a paper by Acker, Chopinet, Pottasch and Stenholm (in preparation)

STATUS OF THE SPECTROSCOPIC SURVEY OF PLANETARY NEBULAE

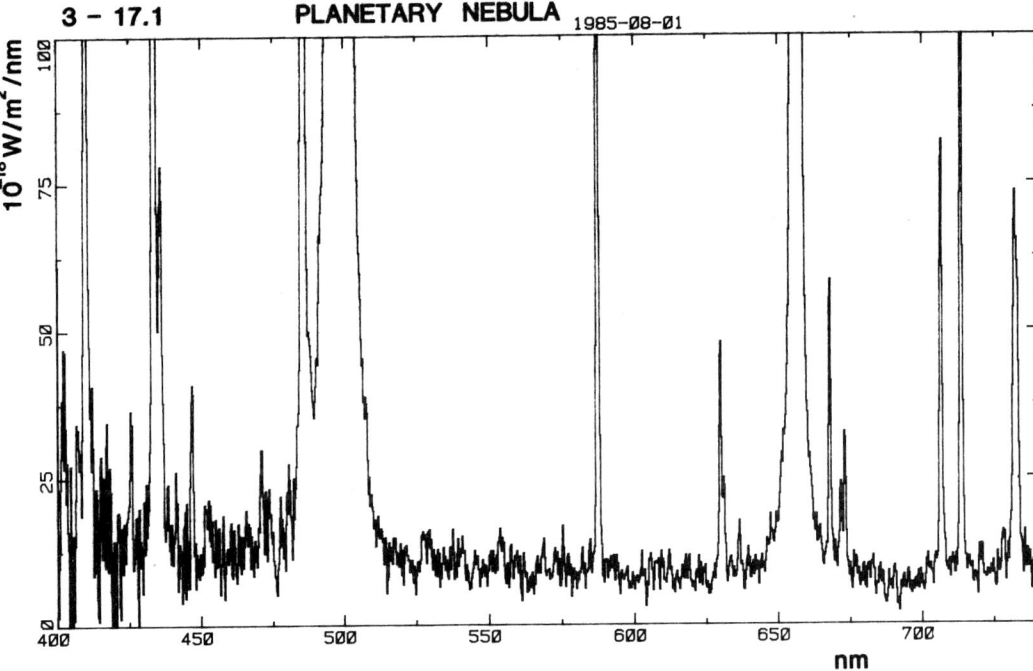

Fig. 4. IDS spectrogram from ESO of the little known planetary nebula 3-17.1. The flux is high-cut at $100 \cdot 10^{-18}$ W/m²/nm.

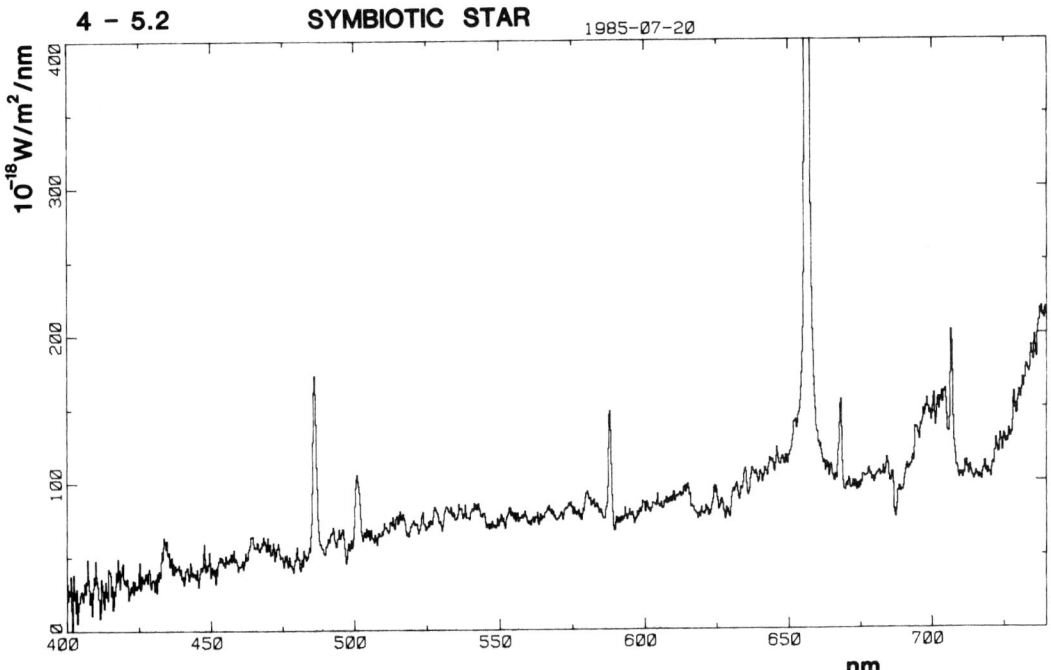

Fig. 5. IDS spectrogram from ESO of the object 4-5.2, a symbiotic star according to this investigation. High-cut at $400 \cdot 10^{-18}$ W/m²/nm. Cf. fig. 4.

together with a compilation of current literature. A detailed review of some classification problems is given by Acker (1986).

6. GENERAL PROPERTIES OF THE POPULATION OF PLANETARY NEBULAS

Along with the observations, measurements of line intensities are also going on, which eventually will give relative line intensities for all lines observed in each spectrogram. Hopefully, this will yield a lot of new data which can be used e.g. for the following purposes:

For individual nebulas the line ratios can be used for determinations of electron temperature, density and to a lesser degree also of ionic abundances. For the population in general, correlation studies can be made of line strengths vs spectral type of the central star, morphology and position in the Galaxy.

7. THE STRASBOURG-ESO CATALOGUE OF GALACTIC PLANETARY NEBULAS

The spectroscopic survey described above is part of a bigger project, a new catalogue of planetary nebulas. This catalogue is now being prepared by A. Acker in Strasbourg and her colleagues and is planned to be issued as a co-production between Observatoire de Strasbourg and European Southern Observatory. The catalogue will appear in two parts. The first will be issued as soon as the observations are finished, which is planned for June 1988. It will contain all basic data (including selected line ratios) and all literature published since the catalogue of Perek and Kohoutek (1967). The second part will contain finding charts, spectral tracings, and monochromatic images. It is foreseen that the second part will be issued in 1989. Comments on the project are greatly appreciated by the authors.

REFERENCES

Acker, A., 1986, Comptes rendus de la 8e Journée de Strasbourg, Obs. Strasbourg
Allen, D.A., 1984, Proc. Astron. Soc. Australia 5 (3), 369
Kohoutek, L., 1978, in *Planetary Nebulae*, IAU Symp. No. 76, Y. Terzian (ed.), p. 47
Kohoutek, L., 1983, in *Planetary Nebulae*, IAU Symp. No. 103, D.R. Flower (ed.), p. 17
Perek, L., Kohoutek, L., 1967, *Catalogue of Galactic Planetary Nebulae*, Academia, Praha
Pottasch, S.R., 1984, *Planetary nebulae*, D.Reidel Publ. Comp., Dordrecht/ Boston/Lancaster, p. 260
Sanduleak, N., 1976, Publ. Warner and Swasey Obs. $\underline{2}$, No. 3
Stenholm, B., 1986, Comptes rendus de la 8e Journée de Strasbourg, Obs. Strasbourg
Stenholm, B., Acker, A., 1986, Astron. Astrophys. (in press)
Stenholm, B., Lundström, I., 1984, Report Observatory and Astrophysics Laboratory Helsinki, No. 6/84, 97

DISCUSSION

Grewing: What precautions have been taken to allow the relative emission intensities to be converted to absolute fluxes? Is there any plan to determine in a parallel program absolute H-beta fluxes for the targets in your list?

Stenholm: For extended objects the position of the aperture relative to the appearance of the nebula will be given. We have no plan of calibrating the relative line fluxes into absolute values, but we feel that such a calibration would add great value to our observations.

Preite-Martinez: Did you find any Be star previously classified as planetary nebula?

Stenholm: Yes, there are some Be stars, Be pec., and other types of emission line stars.

D'Antona: Did you find any Nova shell among the 37 unidentified objects?

Stenholm: I am not shure, but there might be such objects in that group.

ANNOUNCEMENT:
THE STRASBOURG-ESO CATALOGUE OF GALACTIC PLANETARY NEBULAE

A. Acker (^)
Observatoire Astronomique de Strasbourg
11, rue de l'Universite'
67000 Strasbourg, France

1. INTRODUCTION

In the framework of the VIIth Strasbourg Day, held on February 6th 1986 at the Strasbourg Observatory and devoted to Planetary Nebulae (PN), a general discussion was devoted to fixing the criteria and main conditions of the production of a catalogue, which would be useful to the astronomical community, according to the recommendations of the members and Council of Commission 34 which met at the IAU General Assembly in New Delhi.

The main catalogue will consist of two parts, the first one containing the data with bibliographical references, the second one containing the finding charts. A number of annexes is also in preparation.

The new Catalogue should appear on the occasion of the I.A.U. Symposium on "Planetary Nebulae" to be held in Mexico in November '87.

2. THE DATA AND BIBLIOGRAPHICAL REFERENCES.

Available as a book, this part should also be machine readable (available on magnetic tape).

2.1. Objects and names

The Strasbourg - ESO Catalogue (SEC) concerns true and possible PN, as well as related objects, such as proto PN. Objects which are called "PN" and which should not be included in the SEC, are discussed in a paper to be published in the Supplements of Astronomy and Astrophysics (1987), in accordance with the selection made by Kohoutek in 1978, 1983. The objects are listed in the order of their galactic longitude (Perek - Kohoutek number).

All known names of the nebulae should be given, including the "IRAS" number.

(^) Contribution presented by A. Preite Martinez.

2.2. Coordinates

Equatorial coordinates are given for two equinoxes. For the first values corresponding to 1950.0, the most precise published determinations were chosen, and the authors cited in reference. The "IRAS" coordinates will be given for all objects. The positions 2000.0 have been calculated from α, δ 1950.0, and are given with an uniform precision, which may, however, not always be significant. Annual precession refers to 1950.0.

The reference noted "disc" corresponds to the discovery; if the discovery was made before 1965, the PN appears in the Perek-Kohoutek Catalogue, and the reference "PK 67" is given.

Proper motions are from the Catalogue of the Central Stars of true and possible PN (Strasbourg, 1982-3-4: AG82).

2.3. Radial and expansion velocities

The mean value of the radial velocity is taken from Schneider et al. (1983); if this value does not exists, the most precise more recent individual value is given, if available.

The value of the expansion velocity for the [OIII] lines by Sabbadin (1984) is given; if other values are published for other lines, the comment "+ other ions" will appear. If the PN being considered is not included in Sabbadin's work, the most precise more recent individual value is given, if available.

2.4. The spectrum

Bibliographical references are given, according to the studied spectral range. For the UV range, we also give the number of the IUE spectra concerning the PN studied.

Selected values of the flux are given, with corresponding references. In the optical range, the majority of the line ratios should be taken from the spectroscopic survey done by Stenholm and Acker since 1984, using uniform orientation and size of diaphragm. If other ratios or fluxes are given, the corresponding characteristics should be looked for in the paper referred to.

2.5. Dimension

The mean diameter of the main envelope is given in arcsec, as well as the wavelength and the reference of the observation.

2.6. Central star

Data are taken from the AG82 catalogue. All known names of the central star should be given. If available, V, B-V, U-V and E(B-V) values are given, with the reference of the corresponding observations; if these are not available, other magnitudes should be given. The most reliable spectral type is given, as well as the references of the corresponding observations.

2.7. Notes

Peculiarities are noted here, with the corresponding references, concerning : (i) the status of the object: "true PN" or "Possible", "doubtful", or "proto-PN", etc.; (ii) the star: "very hot nucleus", or "binary star", etc.; (iii) the nebula: "Type I", "bipolar", or "high polarization", etc. Distances are also given, specifying the method used ("spectroscopic parallax", "absorption", etc.).

2.8. Bibliographic references

The reference of the two main catalogues PK67 and/or AG82 is given, if the PN is to be found there. In order to be consistent with the PK67 catalogue, the references of all papers published since 1965 are given. We give the author's name, the name of the magazine or book, volume and page number, and the title of the paper. This is followed by the CDS's number, which begins with two numbers denoting the year of publication.

2.9. Discovery list

The discovery lists of all objects are given at the end of Part I of the Catalogue, with the corresponding bibliographical references.

3. FINDING CHARTS

Finding charts (only available as a book) are given with the following scale, corresponding to the resolution of the ESO survey plates: 25 mm = 2'.5; field : $\pm(7'.5)$. If the dimension of the PN is greater than the 7'.5 field, the scale of the finding chart will be adapted to all such objects.
    For each PN, a monochromatic image with good resolution will be given, if available. The chosen wavelength, as well as the bibliographical reference will be indicated.

4. ANNEXES

An Atlas of selected objects is also in preparation. It will contain (i) <u>monochromatic images</u> in chosen wavelengths in the optical and radio ranges, and (ii) <u>spectrograms</u>.
Spectrographic recordings of $\phi(\lambda)$ are collected in the optical range for most PN (and possibly in the UV for the PN observed with IUE).

References

Acker,A., Gleizes,F., Chopinet,M., Marcont,J., Ochsenbein,F., Roques, J.M., 1982, "Catalogue of the central stars of true and possible planetary nebulae", Publ. Sp. du C.D.S. n.3, Strasbourg; Complements I (1982), II (1983), III (1984). (AG82)

Kohoutek,L., 1978, "Planetary Nebulae", I.A.U. Symp. n.76, p.47
Kohoutek,L., 1983, "Planetary Nebulae", I.A.U. Symp. n.103, p.25
Perek,L., Kohoutek,L., 1967, "Catalogue of galactic planetary nebulae",
   Academ. publish. House of the Czechoslovak Acad. of Sciences
Sabbadin,F., 1984, Astron.Astrophys.Suppl.Ser. $\underline{58}$,273
Schneider,S.E., Terzian,Y., Purgathofer,A., Perinotto,M., 1983,
   Astrophys.J.Suppl.Ser. $\underline{52}$,399

Discussion

   Panagia: How will the size of an object be defined for the catalogue?

   Grewing: The diameter of PN from monochromatic images is clearly a matter of definition. One can easily define e.g. an isophote which corresponds to a certain fraction of the peak nebular intensity. For absolutely calibrated images one would instead chose an isophote which corresponds to a certain flux per unit surface area. In either case the results will depend in general on the particular ion under study.

   Renzini: In the first Acker's Catalogue of PN there appear objects with WC and WN nuclei in roughly equal number. This is quite surprising, as other sources maintain that all PN nuclei of Wolf-Rayet type belong to the WC subclass. So, does anybody here know if there is a "bona fide" planetary nebula with WN spectrum? The point is that theorists may find easy to make WC's, but rather hard to make WN's.

IRAS ADDITIONAL OBSERVATIONS OF PLANETARY NEBULAE

A. Leene, C. Y. Zhang, S. R. Pottasch
Kapteyn Astronomical Institute
Postbus 800
9700 AV Groningen
The Netherlands

ABSTRACT. The IRAS additional observations of planetary nebulae are described. Extended sources can have very wrong integrated flux densities, when compared with the PSC or SSC. Only 7 of the 33 sources observed turn out to be extended. Maps of NGC 6853 and A 21 are presented.

1. INTRODUCTION

During the IRAS mission some 60% of the observing time was devoted to the survey observations, which have resulted in the Point Source Catalog (PSC), the Small Structure Catalog (SSC) and the Skyflux products. The other 40% was used for special pointed observations (AO's). Two different instruments could be used for the AO observations: the Chopped Photometric Channel (CPC) or the Survey Array (SA). The advantage of the CPC is the circular beam, whereas the SA beam is rectangular. For both instruments several observing programs (macro's) were defined. The relevant macro's here are the DSD01A, the CPC09A, the CPPF0A and CPC03A macro's. The DSD macro was carried out with the small edge detectors of the SA and resulted in a 9×9 arcmin^2 map. The other macro's made use of the CPC and resulted in a 9×9, 3×3 or 3×1 arcmin^2 map for the CPC09A, CPC03A and CPPF0A, respectively. The latter two macro's were not officially published due to gain problems of the detectors.

Table 1 Summary of observations			
macro type	number of observations	no. of objects detected	not detected
DSD	12	10	1
09A	66	31	14
F0A	30	13	1
03A	47	24	13

The main goal of the program was to obtain improved integrated flux densities and to determine whether an object was extended or not. It was expected that the CPC would be more sensitive than the Survey Instrument. But due to a too low focal plane temperature this was not the case. Thus half way the mission the DSD01A macro was defined and the observing program was slightly adapted. The sources were mainly selected on their optical size and optical brightness. This resulted in a sample of 65 objects, of which 10 are proto planetary nebulae and 5 are young planetary nebulae. A summary of the observations is given in Table 1.

2. INTEGRATED FLUX DENSITIES

As the main goal of the program was to get improved flux densities a comparison has been made between the PSC, SSC and the AO flux densities. This also gives an indication of the accuracy of the different flux densities. For this purpose the AO and PSC flux densities have been compared. Within the errors both flux densities are similar. The errors are comparable to those quoted in the Explanatory Supplement (1985) and are entirely due to the repeatability of IRAS. Only one source, NGC 6853, has a much larger AO flux density than found in the PSC. The flux density listed in the SSC is more comparable to the AO flux density. It is however clear that both the PSC and SSC have not found all the flux for this object, which implies that the source is extended with respect to the IRAS beams.

For the CPC the errors found are also equal to the those quoted in the CPC Explanatory Supplement (1985). These errors are much larger than those of the PSC, which is due to a worse calibration. There is no source, except NGC 6853, which stand out as really extended. NGC 650/1 seems to be somewhat extended. This could however not be comfirmed on a second observation. Probably NGC 3587 has a weak halo at 50 µm. Table 2 gives a summary of the flux density comparison. The mean and error are quoted as a (PSC - AO) / AO flux density ratio.

Table 2 Summary of PSC - AO fluxdensity comparison				
macro type	wavelength	AO-PSC difference		
		mean	error	No. Obs.
DSD	12	-0.09	0.27	6
DSD	25	0.04	0.08	7
DSD	60	0.02	0.13	7
DSD	100	-0.06	0.13	6
CPC	50	0.21	0.85	25
CPC	100	0.51	0.59	22

It is clear that the AO's can provide improved integrated flux densities. For this purpose only the DSD macro's can be used, as the CPC is not accurate enough. The basic limitation to the accuracy is the repeatability of IRAS.

3. THE SOURCE SIZE

The deciding test to determine whether a source is extended or not, is to look at the source shape. One of the goals was to make an unbiased test, which could decide whether a source is extended, has a halo, etc. For this purpose the beam shape must be very well known. No beam shapes have been published for the survey AO's and the only way to find the beam shape is to look at observations of asteroids like Ceres. For the CPC09A macro this results in a FWHM size of 88 and 100 arcsec for the 50 and 100 μm band, respectively. Although these observations give an indication of the beam shape, the actual error and a possible dependence on the source brightness is unknown.

In order to determine these parameters the source sizes of the planetary nebula sample have been measured at the 5, 10, 20 and 50% level of the peak brightness in the in-scan and cross-scan direction. From this it is possible to define the beam shape and its error. No dependence on the source brightness could be found. The beam determined in this way is comparable to a simple theoretical beam and to the observations of Ceres. The differences with the theoretical beam can be explained by hysteresis effects and by a non uniform cross scan responsivity of the detectors.

The requirement for a source to be extended at some level, is that its size must be larger than $\bar{x} + 3\sigma$ for that percentage level. The result of this method is that NGC 6853 is extended at all wavelengths. Furthermore A21 is extended at 25 and 60 μm. These two sources are the most obvious extended sources. NGC 3195 seems to have an halo below the 10% level at 12 and 25 μm. The same is true for NGC 2346, which seems to have an halo at all wavelengths. This is confirmed by the PSC correlation coefficients.

The CPC the results are less good. NGC 650/1 might be extended at 50 and 100 μm. NGC 6720 is extended at 100 μm and marginally at 50 μm. And possibly NGC 7008 is extended at 50 μm.

In summary some 7 sources have been found to be extended out of the 33, which could be analysed. No attempt has been made to deconvolve the beam for the true source size. The results are not accurate enough to warrant such an analysis. Fig. 1 and 2 show the results for NGC 6853 and A 21, respectively.

4. FUTURE PROSPECTS

It is clear from these results that it is difficult to find extended sources. The analysis of the CPC03A and CPPFOA macro still needs to be done and this might enlarge the sample of extended objects. It is however as yet unclear how one can extract the information from such small maps. It will certainly be impossible to derive improved flux densities. All the observations need to be processed again from the raw data in order to use the better gain correction procedure, which is now available.

The only new method to analyse the data will be by co-adding all the data and by keeping the original resolution. Such a project is currently underway in Groningen, the GEISHA project. This project will be able to deliver full resolution maps of the entire sky at the end of 1986. The gain in sensitivity will be at least a factor of two, depending on the position in the sky. Such a resolution and sensitivity can already be obtained by using the raw calibrated data (CRDD). An example of such an analysis is the map is NGC 7293 (see elsewhere this volume). It is the intention to make similar maps for A 7, A 21, A 31, A 35 and NGC 1360.

5. REFERENCES

- Explanatory Supplement: 1985, eds. Beichman, C.A. et al., US Government Printing Office.
- CPC Explanatory Supplement: 1985, Wesselius et al., Internal Report ROG, Groningen.

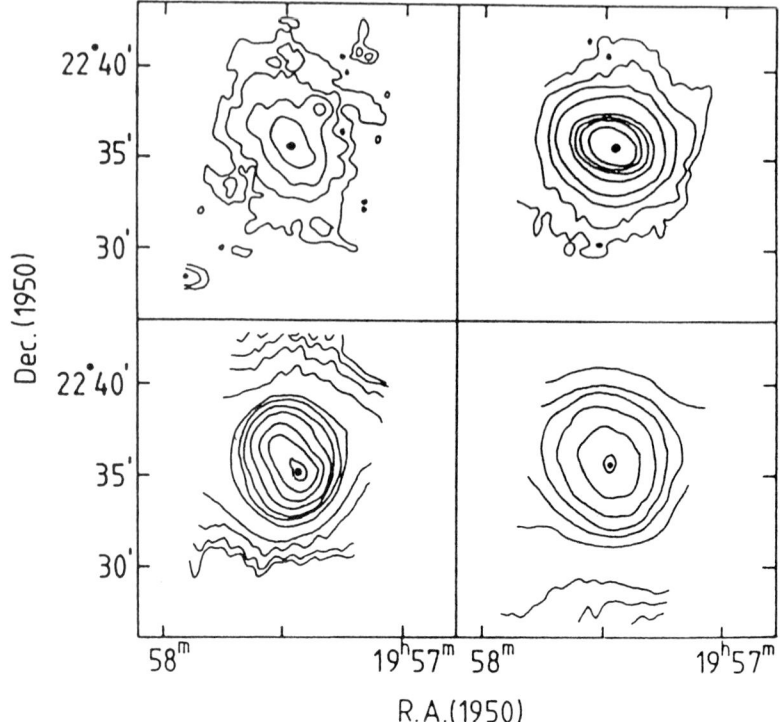

Fig. 1: DSD observations of NGC 6853 at 12, 25, 60 and 100 μm (starting top left, clockwise). The dots indicate the position of the central star. The contours lie at 0.4, 0.8, 1.6, 3, 10, 15, 20, 30, 50, 70 and 90 MJy ster^{-1}.

IRAS ADDITIONAL OBSERVATIONS OF PLANETARY NEBULAE

6. DISCUSSION

Persi: could your source be affected by the presence of "cirrus"?

Leene: in general the objects are very distinct, they really stand out on the maps. This makes it easy to define the background. Only at 100 μm there is sometimes a problem in the form of a background gradient. This can however in most be removed by a higher order polynomial fit.

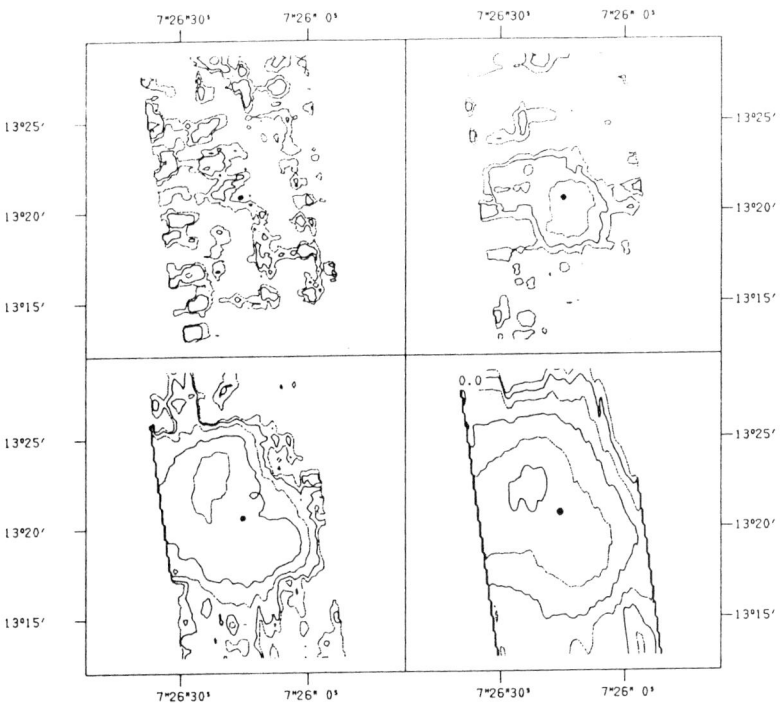

Fig. 2: DSD observations of A 21 at 12, 25, 60 and 100 μm (starting top left, clockwise). The dots indicate the position of the central star. The contours lie at 0.2, 0.4, 0.8, 1.6, 3.2 and 4.8 MJy ster^{-1}.

SPECTROSCOPY AT 8-13 μm AS A PROBE OF THE CARBON TO OXYGEN RATIO IN PLANETARY NEBULAE.

P.F. Roche
Department of Physics & Astronomy,
University College London,
Gower Street, London WC1E 6BT.

ABSTRACT. The spectral properties of planetary nebulae near 10 μm are summarized. Those nebulae with strong dust emission near 10 μm can often be classified as oxygen- or carbon- rich by the detection of spectral signatures of dust grains whose formation depends upon the C/O ratio of the material in which they condense. The classification of pn as C or O rich according to their 8-13 μm spectra can be accomplished quickly and simply without recourse to detailed models, and provides a means of investigating abundances in objects which are difficult to study by other techniques. Some preliminary results on the properties of pn located near the Galactic centre are presented.

1. Introduction

The mid-infrared is a rich spectral region providing a wealth of information on the gas and dust in planetary nebulae (pn); see Barlow (1983) for an excellent review. Collisionally-excited fine-structure transitions produce bright ionic line emission, and measurements of these lines, in conjunction with other observational data, can provide estimates of abundances of a number ions. The bright emission lines of [AIII], [SIV] and [NeII] at 8.99, 10.52 and 12.81 μm which can be measured readily with ground-based instruments, have been detected in many pn and several other important lines, which lie outside the atmospheric windows, have been measured by the LRS on IRAS (see the contribution by Pottasch in these proceedings). In addition, many weaker lines, due to recombination as well as forbidden transitions, lie throughout the infrared. Increased instrumental sensitivity through better detectors and long detector arrays and eventually through space missions such as ISO, will permit the use of these transitions for detailed abundance and excitation analyses.

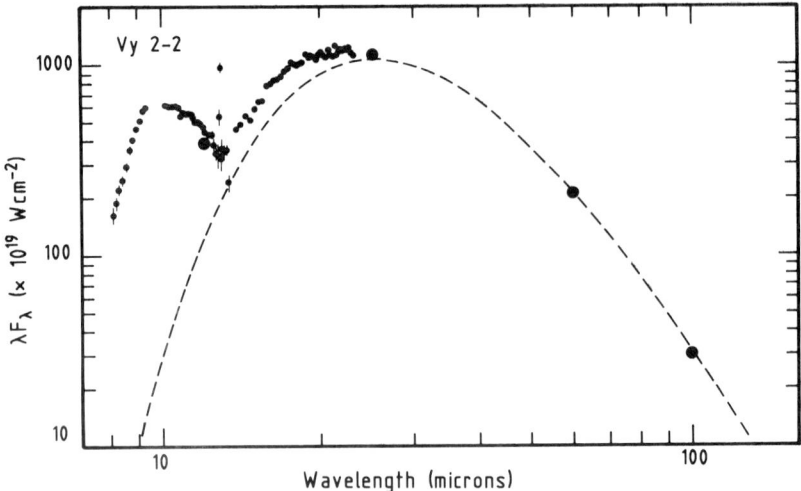

Fig 1. The IR spectrum of Vy 2-2. The 8-13 μm spectrum is taken from AR whilst the 13-23 μm spectrum is from the IRAS LRS. Open circles are the IRAS Point Source Catalogue data and the dashed curve represents emission from grains with $Q=1/\lambda^2$ at 95K.

In order to maintain reasonable instrument sensitivity in the face of the high thermal background at ground-based telescopes, most observations of pn at mid-infrared wavelengths are carried out using relatively small instrument apertures (<10 arcsec). This has lead to a bias towards observations of compact objects which often, though by no means always, have emission from dust at higher temperatures than the more evolved nebulae. Observations of compact nebulae from the ground have recently been supplemented by low resolution 8-23 μm spectra from IRAS (Pottasch et al 1986) although the spectra achieve a sufficient signal - to - noise ratio in the continuum to distinguish the various dust emission bands in only a few objects. An exception is NGC 6578 which had not been observed spectroscopically from the ground and which shows evidence of emission from silicate dust in the LRS spectrum.

2. Observed properties of dust in Planetary Nebulae.

The continuum emission from pn in the near-infrared (1 - 3 μm) is generally weak and close to the level expected from free-free and bound-free transitions in ionized gas when extrapolated from observations in the visible and radio. However, emission from dust grains becomes more important with increasing wavelength and, in many pn, dominates at wavelengths longer than ~8 μm. The chemistry of the emitting dust reflects the abundances in the pn ejecta and this can be recognised through the presence of emission features in the infrared spectra. Emission in dust features at wavelengths shorter than 13 μm requires relatively warm grains with T >200K. This is substantially warmer than the grains responsible for the bulk of the IR emission which

A PROBE OF THE CARBON TO OXYGEN RATIO IN PLANETARY NEBULAE

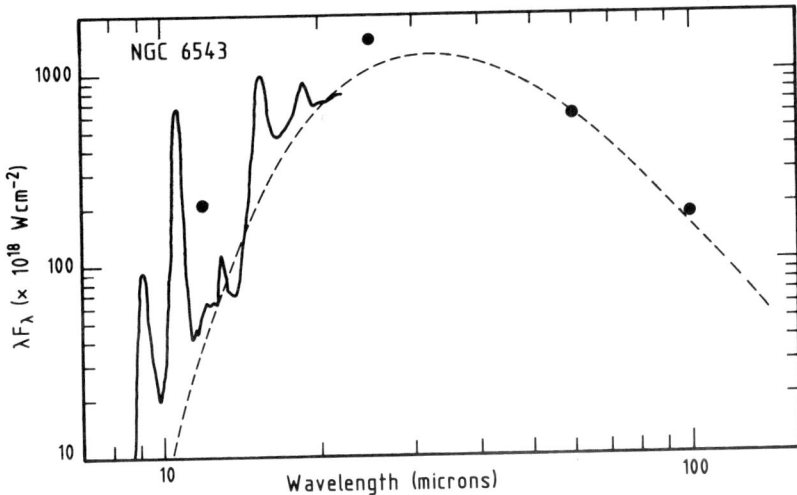

Fig 2. The IR spectrum of NGC 6543. The 8-23 μm spectrum is the combined LRS data from Pottasch et al (1986) and the dashed curve represents dust with $Q=1/\lambda$ at 90K.

peaks near 30 μm and comes from grains at temperatures near 90K. Some pn contain little warm dust and so have very weak emission from dust at wavelengths near 10 μm (compare the spectra of Vy 2-2 and NGC 6543 in figures 1 and 2); in these objects the mid-IR flux arises mainly from line emission. The continuum emission at 10 μm varies over a huge range from about 2 - 1000 times greater than the free-free emission expected from the ionized gas.

The physical location of dust in the planetary nebulae is not yet clear. Observations of ultraviolet resonance lines (e.g. for NGC 7662, Harrington et al 1982) show that there is significant absorption of C IV $\lambda1549$ by dust, so that some grains at least exist inside the ionized region. On the other hand, spatial scans across NGC 7027 have shown that the dust emission band at 11.25 μm peaks up just outside the ionized region and that the grains responsible for the emission lie in a thin neutral shell (Aitken & Roche 1983). Further evidence of dust in the neutral region in very compact nebulae comes from consideration of the brightness temperature of the far-infrared emission which would be uncomfortably close to the observed colour temperature at 30 μm if the FIR comes from within the ionized nebula. It may be that the warm dust emitting near 10 μm lies within the ionized zone, whilst much of the FIR emission arises from the neutral region. High spatial resolution observations are crucial in investigating this question.

To date, 10 μm spectra have been published for some 50 pn and they fall into 4 groups (see Aitken & Roche 1982 (AR) and Roche & Aitken 1986 and references therein). The different dust emission spectra seen in pn at 8-13 μm are illustrated in Fig 3 and described here:

(a) Oxygen-rich pn show emission in the 9.7 μm silicate band which is characteristic of the circumstellar dust shells of oxygen-rich late type stars. These objects, e.g. Vy 2-2 and IC 4997, tend to be compact and of low or moderate excitation. In addition, several bipolar pn (e.g. M 2-9 and Mz-3) have spectra which show apparent silicate absorption features. Because the extinction towards these nebulae is much less than that implied by the depth of the apparent silicate absorption, the 8-13 μm spectrum is interpreted as arising from radiative transfer effects in optically thick discs or rings of dust close to the exciting star; these dusty rings probably play an important role in the bipolar morphology of the nebulae. The presence of the silicate grain signature implies that the dust formed in an oxygen-rich environment so that we conclude that these pn are O-rich.

(b) A broad emission feature peaking near 11.2 μm is often seen in the circumstellar dust shells of carbon stars, and is attributed to a resonance in silicon carbide grains which are expected to form where the C/O ratio is $\geqslant 1$. A similar emission band is seen in a number of pn (e.g. IC 418, NGC 6790) providing evidence that they are carbon-rich.

(c) A third group of objects have spectra dominated by narrow but resolved emission bands at 11.3, 8.6 and 7.7 μm (e.g. NGC 7027, BD +30 3639). These bands are part of a family of emission features with other members at 6.2 and 3.3 μm which have been seen in a variety of astronomical sources but whose origin and excitation are still problematical. Current thinking favours the idea that they are produced by very small (~10A carbon-based grains and possibly the polycyclic aromatic hydrocarbons discussed by d'Hendecourt in this volume. Cohen et al (1986) have found a good correlation between the strength of the 7.7 μm emission band and the C/O ratio in a small number of pn, giving further credence to the idea that the grains giving rise to the emission features form in C-rich environments. These results confirm the suggestion of Barlow (1983) that the pn showing strong emission in the narrow emission bands have very high carbon abundances with C/O $\geqslant 2$.

(d) The fourth category of pn comprises those objects with continua too weak to permit classification from the presence of dust signatures (e.g. NGC 6543, NGC 7009). These objects appear to contain very little warm (T>200 K) dust although some are very compact with diameters < 0.05 pc, and contain similar amounts of cool dust to the other objects. There may be little dust close to the central star in these pn.

Of the 49 pn for which spectroscopic observations made from the ground have been published, 15 are dominated by emission in the family of narrow dust features and probably have high C/O ratios, 9 show SiC emission and have C/O >1, 7 plus a further 5 bipolar nebulae show evidence of silicate grains and have C/O <1 whilst the remaining 13 objects have continua too weak for calssification in this way. The assignment of pn as O or C rich on the basis of the dust chemistry derived from their 8-13 μm spectra has been confirmed in those objects for which detailed abundance analyses have been carried out on good quality optical and ultraviolet spectra (see Seaton 1983). The excellent agreement obtained between the UV/optical analyses and the classification by dust features demonstrates that infrared spectroscopy can be used to investigate the C/O ratio in pn and give reliable results.

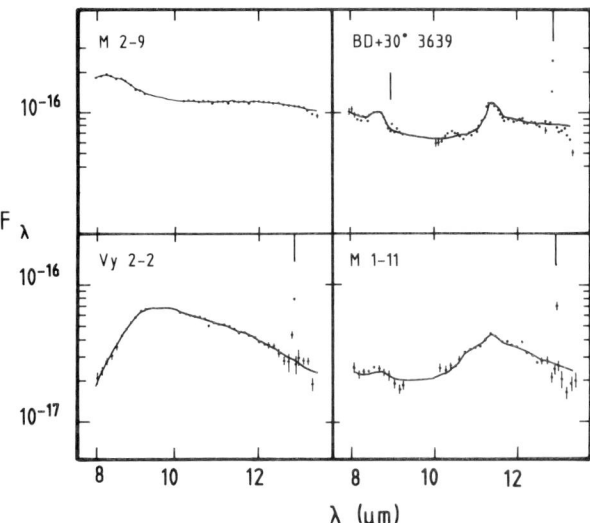

Fig 3. Representative 8-13 μm spectra of planetary nebulae taken from AR. Vy 2-2 displays a prominent 9.7 μm silicate emission feature whilst the bipolar nebula M 2-9 has a spectrum fitted by emission from optically thick silicates. M 1-11 shows the 11.2 μm silicon carbide feature together with weak emission from the family of narrow emission features. The spectrum of BD +303639 is dominated by emission from the unidentified emission bands; those at 11.25 and 8.65 μm are clearly seen together with the long wavelength wing of the 7.7 μm feature. The solid lines are fits to the continua using the various dust emissivity curves. Arrows indicate detected fine-structure emission lines at 9.0 and 12.8 μm due to [AIII] and [NeII].

3. Planetary Nebulae near the Galactic centre.

Because pn have almost exclusively been discovered in surveys at visible wavelengths, the known planetaries are mostly relatively nearby and optically bright objects. This means that the properties of Galctic pn refer mostly to objects within a few kpc of the sun.

It has been established (e.g. Shaver et al 1983) that there are abundance gradients in the Galaxy, with increasing fractions of heavy elements towards the Galactic centre. However, it is difficult to measure abundances accurately close to the centre because of the high interstellar extinction. There is evidence from infrared ionic line emission that the abundances of at least some elements are higher than the solar values (see Lacy 1982). Further evidence of enhanced metallicity comes from the discovery a super-metal-rich population of red giants in the Galactic bulge by Wood & Bessell (1983). These are relatively young long-period variable stars (>3M$_\odot$) which have increased metallicities by a factor of about 2.5 over the solar value. The obvious explanation for the higher fraction of heavy elements is that it results from high rates of star formation and nuclear processing.

A consequence of increased metallicity is that it becomes harder to produce sufficient processed material to form carbon stars and this is reflected in the relative numbers of C and M stars. For example the metal-poor

Magellanic clouds have a large fraction of carbon stars, the Galactic disk has a smaller number, but the Galactic bulge has very few indeed (e.g. Blanco et al 1978):

	SMC	LMC	Local	G.C.
C/M	25	2	0.01	0.002

We expect the properties of pn near the Galactic centre to follow the C/M star ratio.

Largely because of the limitations imposed by the high interstellar extinction between the sun and the centre of the Galaxy, abundance determinations from optical spectroscopy have been carried out for very few Galactic bulge pn. An exception is the study of H1-55 by Price (1981) who found a large overabundance of oxygen and enhanced abundances of N, S and A, but infrared techniques promise to allow a much more comprehensive study. The difficulty lies in choosing the sample because there are important selection effects in any optical survey of G.C. pn. The main effect is that the brightest pn will be picked out most easily. This could mean that chance alignments rather than true bulge pn would be selected. Fortunately, radial velocity measurements exist for many of the known pn towards the G.C. (Schneider et al 1983), and by careful selection the contamination from foreground objects can be reduced. A radio study using the VLA, in which 34 optically-selected pn were detected, has been published by Gathier et al (1983), and they estimate that about 90% of these are true bulge planetaries. This sample is biased towards the more luminous objects, as a consequence of initial selection at visible wavelengths.

Despite the obvious selection effects, the sample studied by Gathier et al is an attractive starting point for a follow-up investigation of the C/O ratio of G.C. pn. Planetaries near the Galactic centre are at distances which are known quite accurately and the measurements of radio continuum flux at 5GHz with the VLA allow calculations of ionic fractions from detected IR fine-structure lines. In addition, all the objects in this sample are compact with diameters <3 arcsec so that the whole nebula can be included in the entrance aperture of the spectrometer. Here, preliminary results on part of the Gathier et al sample are presented.

Spectra at 8-13 μm were obtained of 7 pn using the UCL spectrometer at the Anglo-Australian telescope in collaboration with Drs M. Barlow and D. Aitken. The spectra are shown in Figure 4.

It is immediately obvious that two of the objects, H 1-40 and M-2-23 show prominent emission from silicate grains, whilst the continua of the remaining objects have insufficient signal-to-noise to permit classification. The remaining objects do, however, have strong line emission in the [AIII], [SIV] or [NeII] lines, permitting calculation of the ionic fractions of these elements (Table 1). The ionic fractions are broadly similar to those found in local pn with no evidence of dramatic overabundances but, of course, only a single ionization state is available for each element.

Table 1. Ionic lines detected in Galactic centre pn.

Name	Line	I line[1]	$\frac{N_{ion}}{N\text{ element}}$[2]	S_{5GHz}[3]	d(pc)
H 1-40	[SIV]	3.9	0.05	56	0.031
	[NeII]	3.8	0.25		
M 3-44	[NeII]	8.1	0.86	35	0.092
M 4-6	[SIV]	5.5	0.09	42	0.048
	[AIII]	1.6	0.22		
M 2-27	[SIV]	3.5	0.05	50	0.057
M 1-37	[NeII]	3.9	0.96	15	0.065
NGC 6644	[SIV]	1.3:	0.01:	97	0.074
M 2-23		<2.5		41	0.017

Notes: 1. Line intensities in 10^{-19} Wcm^{-2}.
2. Ionic fraction assuming solar values for A, S and Ne.
3. 5GHz radio flux in mJy from Gathier et al., except for H 1-40 where the 14.7 GHz measurement of Milne and Aller was used.

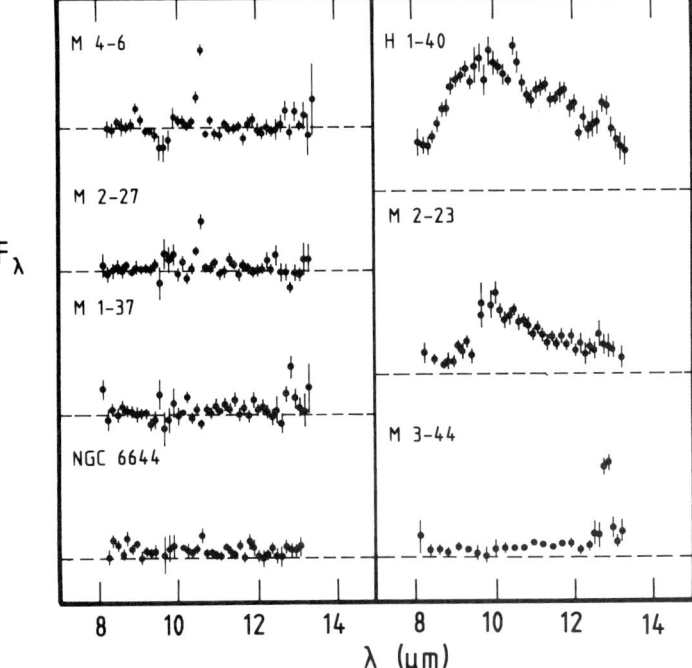

Fig 4. 8-13 μm spectra of 7 Galactic centre pn selected from the list of Gathier et al (1983). Note that the flux scale is linear.

The continuum emission is more intriguing. Although seven objects constitutes too small a sample for meaningful statistics, it is interesting that two are clearly oxygen-rich, and are the two most compact nebula. There is evidence to suggest that these two objects may be optically thick at 5 GHz as measurements at 14.7 GHz by Milne and Aller (1982) give fluxes about 1.5 times those at 5 GHz. H 1-40 and M 2-23 are probably very dense and presumably young objects, and very similar to the more nearby pn like IC 4997 and Hb 12 that show silicate emission. None of the seven objects shows evidence of emission from carbon-rich grains and indeed, none would be expected to if they are near the Galactic centre. The other five objects should also be oxygen-rich and the non-detection of silicate emission may indicate that the dust is mostly ouside the ionized region. Almost all of the pn which have been found to show silicate emission are very compact, with diameters < 0.05 pc, and so presumably the silicate signature becomes harder to detect as the nebula expands. It may be that many of the pn with weak continua at 10 μm are just these objects, and further observations of planetaries near the Galactic centre will be of great value in investigating the evolution of oxygen-rich objects.

I am grateful to Drs M. Barlow and D. Aitken for allowing me to use the data on the Galactic centre pn in advance of publication.

References:

Aitken, D.K. & Roche, P.F., 1982. M.N.R.A.S., 200, 217.
Aitken, D.K. & Roche, P.F., 1983. M.N.R.A.S., 202, 1233.
Barlow, M.J., 1983. IAU Symp 103 "Planetary Nebulae" ed D.R. Flower, p105.
Blanco, B.M., Blanco, V.M. & McCarthy, M.P. 1978. Nature, 271, 638.
Cohen, M., Allamandolla, L.J., Tielens, A.G.G.M., Bregman, J., Simpson, J., Witterborn, F., Wooden, D. & Rank, D., 1986. Ap.J., 302, 737.
Gathier, R., Pottasch, S.R., Goss, W.M. & van Gorkom, J.H., 1983. Astr. Ap. 128, 325.
Harrington, J.P., Seaton, M.J., Adams, S. & Lutz, J.H., 1982. M.N.R.A.S., 199, 517.
Lacy, J.H., 1982. "The Galactic Center." p 53. Amer. Inst. Phys. New York.
Milne, D.K. & Aller, L.H., 1982. Astr. Ap. Suppl, 50, 209.
Pottasch, S.R., Preite-Martinez, A., Olnon, F.M., Jing-Er, Mo & Kingma, S., 1986. Astr. Ap., 161, 363.
Price, C.M., 1981. Ap.J. 247, 540.
Roche, P.F. & Aitken, D.K. 1986. M.N.R.A.S., 221, 63.
Schneider, S.E., Terzian, Y., Purgathofer, A. & Perinotto, M., 1983. Ap.J. Suppl., 52, 399.
Shaver, P.A., McGee, R.X., Newton, L.M., Danks, A.C. & Pottasch, S.R., 1983. M.N.R.A.S., 204, 53.
Seaton, M.J., 1983. IAU Symp 103 "Planetary Nebulae" ed D.R. Flower p129.
Wood, P.R. & Bessell, M.S., 1983. Ap.J., 265, 748.

Discussion

Persi: Have you studied a possible correlation between the unidentified features at 11.3 and 3.3 μm?

Roche: It is difficult to do properly because of the different beam sizes used for the observations. Certainly if you see one, you expect to see the other but quantitative results are less certain. There probably are real differences in the flux ratios in the two bands by factors of 3 or 4 in different objects.

Persi: What is the maximum resolution of the IR grating spectrometer?

Roche: A resolving power of 300 using only 5 detectors and 100 with 25.

Clegg: Are any of the set of unidentified features detected from N-type stars?

Roche: No, presumably because there are no UV photons to excite the grains.

Leene: Is there a colour separation of the four pn samples in the IRAS colour-colour plots?

Roche: Yes, a clear separation is visible in the 12/25:25/60 plot. Where $Q = 1/\lambda^n$ O-rich pn with emission from silicates lie in the region requiring grain emissivities that fall off steeply with wavelength with $1<n<2$, pn that show SiC emission lie near the $n \sim 0.5$ line and the other objects lie closer to the blackbody line; the 25/60:60/100 plot does not show any clear separation (see Roche & Aitken 1986). The IRAS data on most of the pn near the Galactic centre are badly affected by confusion, so that the colour-colour diagrams cannot be used for these objects.

van der Veen: Can you give an explanation for the fact that you see 10 μm silicate emission and absorption in your sources?

Roche: The sources which require silicates in emission and absorption to fit their 10 μm spectra are all bipolar nebulae. We believe that the emission arises from a thin disk around the exciting star which is optically thick in the plane of the disk. Apparent silicate absorption will then appear because of optical depth effects and temperature gradients.

PROTOPLANETARY NEBULAE

Luis F. Rodríguez
Instituto de Astronomía, UNAM
Apdo. Postal 70-264
04510 México, D.F., MEXICO

ABSTRACT. Protoplanetary nebulae are objects that precede closely in time the well-studied stage of planetary nebula. It is now firmly established that the most likely progenitors of planetary nebulae are stars at the tip of the asymptotic giant branch that are losing mass in a superwind phase ($\dot{M}_* \geq 10^{-5}$ $M_\odot$ yr^{-1}). The situation is less clear in what refers to the identification of "transition" objects, that is, those that are moving from the tip of the asymptotic giant branch to the left in the H-R diagram. I discuss some recent results related to the most probable transition objects, CRL618, Vy2-2 and CRL2688.

1. INTRODUCTION

While preparing this review, I faced a problem in what regards to the definition of protoplanetary nebulae. It is clear that a protoplanetary nebula is a stage that precedes closely in time that of the planetary nebula. However, during the last decade the term protoplanetary nebula has been used in different senses. In his review paper, Zuckerman (1978) used the term to describe those evolved objects that are losing mass at a rate $\gtrsim 10^{-5}$ $M_\odot$ yr^{-1}, and which will become planetary nebulae themselves within $\lesssim 10^5$ yr. Pottasch (1984) has a broader definition and includes not only the evolved stars undergoing large mass loss rates, but also objects that are in a transition from the tip of the asymptotic giant branch to the left in the H-R diagram. Finally, Kwok et al. (1986) use the term protoplanetary nebula associated with stars that have passed through the asymptotic giant branch evolution and are evolving to the left of the H-R diagram. These "transition" objects are very important to study because they may represent a direct observational confirmation of a major stage of stellar evolution. Perhaps one of the goals of this Workshop could be to select one of these definitions or to come with a new one.

In any case, I decided to adopt Pottasch's definition, that includes objects at the tip of the asymptotic giant branch as well as those evolving to the left in the H-R diagram. As a result of this adoption, I review in §2 the situation with regard to evolved stars of low and intermediate mass undergoing mass loss rates in excess of 10^{-5} $M_\odot$ yr^{-1}. In §3, I discuss some of the theoretical expectations on what should a transition object look like. In § 4, I give a list of possible transition objects and discuss their pecularities in some detail. Finally, I summarize the conclusions in §5. I must note that this review emphasizes the radio results and that the references should be consulted to gain a broader perspective of the problem.

2. THE PROGENITORS OF PLANETARY NEBULAE

Shklovskii (1956) was the first to suggest that red giant stars are the progenitors of planetary nebulae. His main argument was that if one imagines the evolution of a planetary nebula going backwards in time, at an earlier epoch its physical structure would resemble the envelope of a red giant. Ten years later, Abell and Goldreich (1966) gave a more ample discussion on the issue, presenting a variety of arguments and reaching again the conclusion that red giant stars are the precursors of planetary nebulae. However, as it was emphasized by Zuckerman (1978), the typical red giant star could not be the inmediate precursor to the planetary nebula stage. These stars have a mass loss rate in the range of $\dot{M} \sim 10^{-8}$ to 10^{-6} $M_\odot$ yr^{-1} and a wind velocity of $V \sim 10$ km s^{-1} (Goldberg 1985). Taking as an average value $\dot{M} \sim 10^{-7} M_\odot$ yr^{-1}, a typical red giant would require about 10^6 yr to produce an envelope with a mass of 0.1 $M_\odot$. However, this envelope would be much larger (~ 10 pc) and less dense (a few particles cm^{-1} at 0.1 pc from the star) than compact planetary nebulae. Clearly, the formation of a planetary nebula must be preceded by a large mass loss rate, $\dot{M} \sim 10^{-5}$ $M_\odot$ yr^{-1}. Zuckerman (1978) also noted that if the dust to gas ratio in these massive winds was similar to the interstellar value, the progenitor stars will be heavily obscured ($A_V > 100$) and their search and study would be better achieved in the radio and infrared wavelengths.

Renzini (1981) proposed that the required very efficient mass loss, that he called the superwind, could appear for low- and intermediate-mass stars at the tip of the asymptotic giant branch. In his picture, once the star reaches a critical luminosity, most of the residual hydrogen-rich envelope is ejected on a time scale very short compared with the previous asymptotic giant branch lifetime (Iben and Renzini 1983).

Kwok (1982) has proposed an alternative model to explain the high densities observed in planetary nebulae. In his interacting stellar winds model, the slow wind of a red giant is compressed by the fast, hot wind that develops once the core is exposed. The observations made with the International Ultraviolet Explorer satellite revealed that these hot, fast winds are commonly found in central stars of planetary nebulae (Perinotto 1983), and have $\dot{M} \sim 10^{-7}$ to 10^{-9} $M_\odot$ yr^{-1} and $V_{exp} \sim 1000$ to 3000 km s^{-1}. However, the observed mass loss rate and momentum transfer generally fall short of the values required in the interacting stellar winds model. Furthermore, Sabbadin et al. (1984) argue that the radius-expansion velocity diagram for planetary nebulae are accounted better by the superwind hypothesis. Finally,

Bedogni and D'Ercole (1986) note that the interacting stellar winds model requires that the ratio of mass loss rate of the fast wind over mass loss rate of the slow wind, $\dot{M}_{fast}/\dot{M}_{slow}$, should be in a very narrow range, 1 to 4×10^{-2}, a restriction that appears to limit the applicability of the model.

Of course, the best support for the superwind phase is the growing number of observed cool stellar objects with $\dot{M} \gtrsim 10^{-5}\ M_\odot\ yr^{-1}$. These objects are better detected with the low level rotational transitions of carbon monoxide (CO; see Figure 1) in the millimeter range and with the 1612 MHz maser emission of hydroxil (OH). Using the CO lines, Knapp and Morris (1985) report the detection of mass loss from 50 evolved stars. Eighteen of these stars have $\dot{M} \geq 10^{-5}\ M_\odot\ yr^{-1}$; among them the highest value corresponds to the OH/IR star IRC+10420 with $\dot{M} \simeq 2.6 \times 10^{-4}\ M_\odot\ yr^{-1}$. Zuckerman and Dyck (1986) reported the CO detection of 39 new objects from the InfraRed Astronomy Satellite (IRAS) and the Revised Air Force Geophysics Laboratory (RAFGL) catalogs. These authors estimate the current total of evolved stars detected in the CO lines in about 130, a good percentage of them having $\dot{M} \gtrsim 10^{-5}\ M_\odot\ yr^{-1}$ (Knapp 1987; Figure 2).

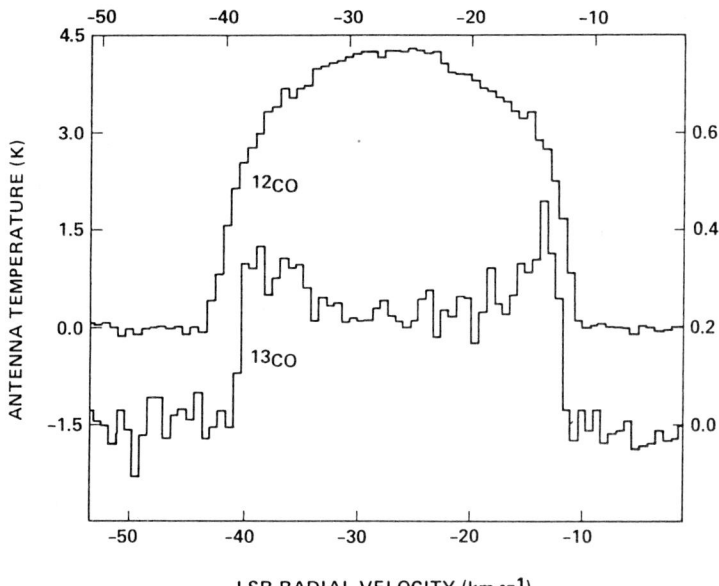

Figure 1. Spectra of the ^{12}CO and ^{13}CO, J=1-0 line emission from the carbon-rich star IRC+10216 (Morris, Jura and Stark 1985). The ^{12}CO line has the characteristic parabolic shape of an optically thick transition from an expanding envelope, while the ^{13}CO line exhibits the flat-topped profile expected for an optically thin transition.

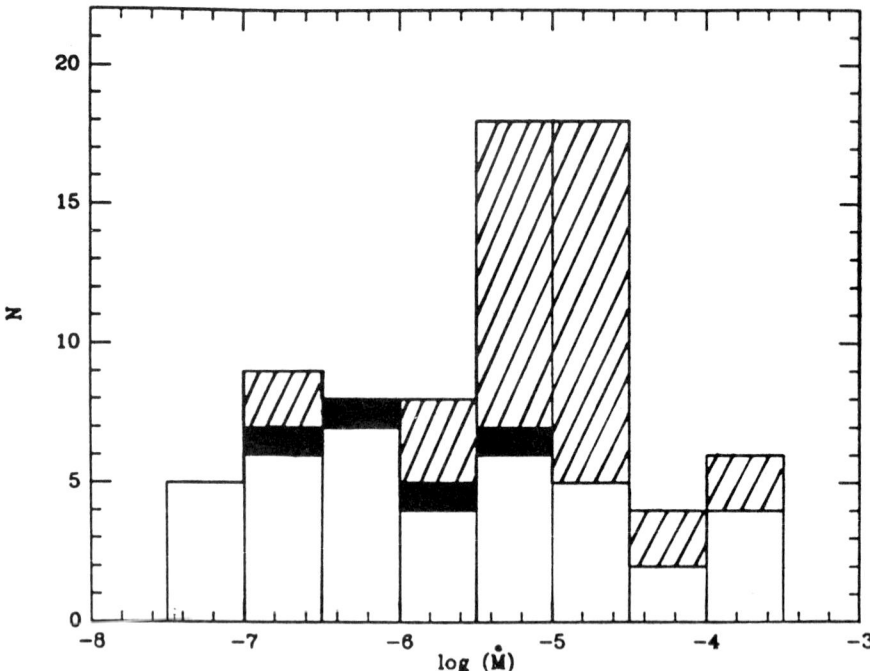

Figure 2. Histograms of mass loss rates determined from CO observations as compiled by Knapp (1986). The white areas correspond to oxygen stars, the black ones to S stars, and the hatched ones to carbon stars. About 1/3 of the determinations give $\dot{M} \geq 10^{-5}$ $M_\odot$ yr^{-1}.

When the star is oxygen rich, oxygen-bearing molecules can form in their envelopes. In particular, hydroxyl (OH), water-vapor (H_2O), and silicon monoxide (SiO) can exist and emit as masers. Usually, a star exhibits only one or two of these maser emissions, but there are cases where all three are present. It is known that the emitting regions for these three molecules are not coincident in space. On the contrary, they exist at different radii (Reid and Moran 1981). SiO maser emission originates at 10^{14} to 10^{15} cm, very close to the stellar photosphere. H_2O maser emission comes from an intermediate region, 10^{15} to 10^{16} cm from the star. Finally, the OH maser emission is produced in a shell at 10^{16} to 10^{17} cm from the star (Figure 3). The 1612 MHz maser emission has proved to be extremely useful to study very long period variables with large mass loss rates. These infrared-bright stars with 1612 MHz maser emission are usually referred to as OH/IR stars.

As shown by Herman and Habing (1985) in a thorough review, one can use the 1612 MHz emission to determine the terminal wind velocity, the distance to the star, and to estimate the mass loss rate. Again, $\dot{M} \gtrsim 10^{-5}$ $M_\odot$ yr^{-1} is frequently found for OH/IR stars.

It appears to be observationally established that these IR stars

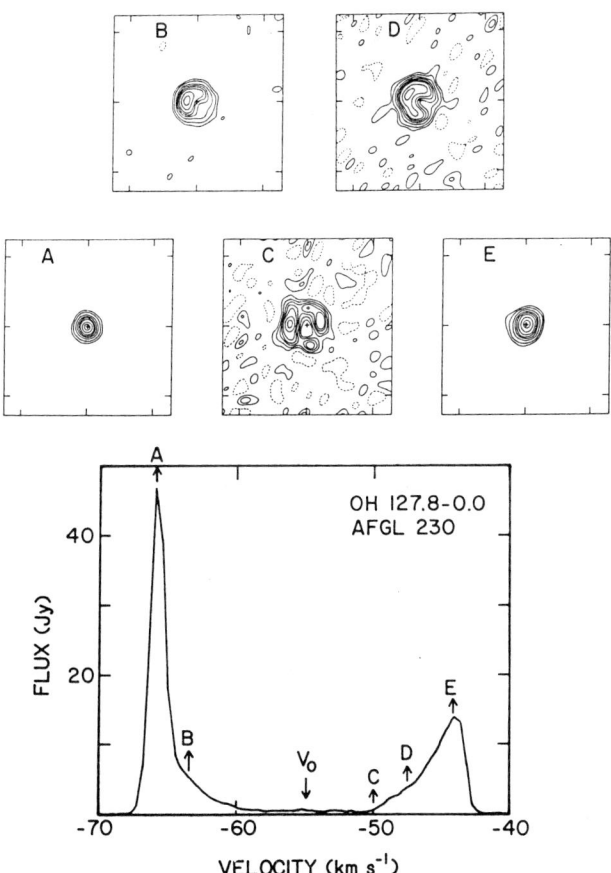

Figure 3. Spectral line profile and VLA maps of the 1612-MHz OH emission for OH127.8-0.0 (Bowers, Johnston and Spencer 1983). The maps are made at the velocities indicated by each letter (A to E) and marked in the OH spectrum. The tick marks in the maps are separated by 5". The cross on each map indicates the inferred position of the star. Emission from the OH peaks (A and E) originates from the front and back of an expanding shell, while the emission at intermediate velocities originates from the edge of the shell.

undergoing large mass loss rates are the progenitors of planetary nebulae. At present only about 300 OH/IR stars are known (Jewell et al. 1985). However, it has been shown that OH/IR stars can be selected with good confidence ($\sim 80\%$) from the large IRAS data base (Hrivnak, Kwok and Boreiko 1986; Herman, Burger and Penninx 1986). This promises that the number of known OH/IR stars will be increased by a large factor in the near future.

3. SOME CONSIDERATIONS ON THE OBSERVATIONAL CHARACTERISTICS OF TRANSITION OBJECTS

At the end of the superwind regime, the expelled matter keeps expanding, while the remnant star continues its evolution toward high temperatures, although approximately maintaining its luminosity. When the stellar temperature reaches about 30,000 K, photons with energy above 13.6 eV begin to be produced significantly by the star and the inner parts of the expanding envelope become ionized. The time required by the star to evolve from the end of the superwind phase to a temperature of 30,000 K depends sensitively on the mass of the remnant at the end of the envelope ejection, M_c. Thus for $M_c \sim 0.565\ M_\odot$, $\Delta t \sim$ a few 10^3 years, but for $M_c \sim 0.546\ M_\odot$, $\Delta t \sim 10^5$ years (Schöenberger 1983). In the last case the envelope will disperse before significant photoionization takes place. On the other extreme, stellar remnants with $M_c \sim 1\ M_\odot$ will have $\Delta t \sim$ a few decades (Paczynski 1978).

The different times required to start producing photoionization suggest that we can roughly divide the transition objects into two types: those where ionized gas appears after the superwind envelope has expanded significantly, and those where ionized gas appears before the superwind envelope has changed significantly.

I will speculate on the appearance of the first type of objects. Since most planetary nebulae are believed to have low mass cores, it is expected that the majority of objects will fall in this category. After about 2,000 yr the envelope will have an inner cavity of 10^{17} cm. Probably the fast wind from the nucleus will stop the fill-in of this cavity (Kwok 1983; Kahn 1983). It may be possible to detect this cavity with interferometer maps of the CO emission or with high angular resolution (a few arc sec) maps of the far-IR emission. The OH, H_2O and SiO maser emissions will have disappeared since they exist in the slow wind only for $R \sim 10^{17}$ cm. The optical extinction will have decreased significantly and it may be possible to observe in the optical the central star, which should have an intermediate temperature (5,000 to 20,000 K). Perhaps CRL2688 is in this stage (see below). Also the non-variable OH/IR stars discussed by van der Veen (1986) may be an example.

The situation becomes more interesting when we consider the case of objects with a more massive nucleus. Here, ionization will begin before the superwind envelope expands significantly. The time evolution of the resulting ionized zone has been considered by Spergel, Giuliani and Knapp (1983), Rodríguez, Gómez and García-Barreto (1985), and Okorokov et al. (1985). This evolution depends strongly on the envelope mass and the ionizing flux of the stellar nuclei. The results of a model with $M=0.1\ M_\odot$ in the envelope and a core star with mass $1.2\ M_\odot$ is shown in Figure 4. Full ionization of the envelope takes place in about 500 years. There is a period of about 100 years when the ionized volume is internal to the region where the OH maser emission originates. This implies that one expects some OH/IR stars to have a small ionized zone that could be detected in the radio continuum. Since the expected cm fluxes are in the mJy level, the search can be attempted only with the Very Large Array (VLA). Such a program was carried out by Herman, Baud

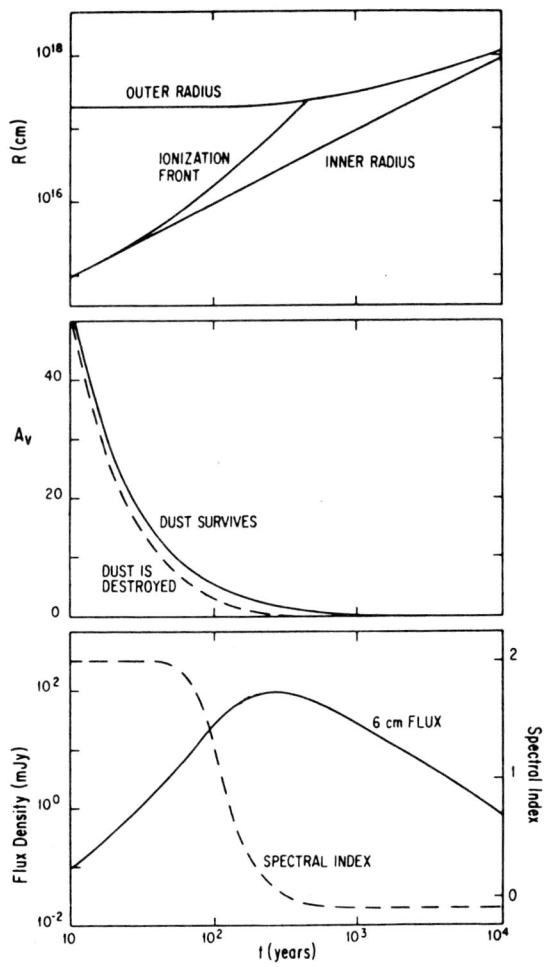

Figure 4. Observational parameters as a function of time for a model planetary nebula with core mass of 1.2 M$_\odot$ and envelope mass of 0.1 M$_\odot$ (Rodríguez et al. 1985). Top: inner radius, position of the ionization front and outer radius. Full ionization of the envelope is obtained at t$\sim$ 500 years. Middle: extinction in the visible for a line of sight going from the outer surface of the envelope to the central star. The solid line is for the case of dust survival within the ionized gas and the dashed line is for instantaneous destruction of dust within the ionized volume. Bottom: flux density at 6-cm (solid line) and spectral index, as derived from 6 and 2 cm measurements (dashed line).

and Habing (1985) toward 12 OH/IR stars with negative results. Rodríguez et al. (1985) and Knapp and Bowers (1986) noted that if an inner ionized zone is present it will be optically thick at 1612 MHz and will produce an OH maser profile with the blueshifted peak much stronger than the redshifted one. Nevertheless, the searches toward OH/IR stars with these characteristics have not proved successful in finding radio continuum sources associated with OH/IR stars.

Apparently, the detection of a radio continuum source in the inner parts of an OH/IR envelope requires the observation of a large sample of objects. Assuming that the ionized zone and the OH maser emission coexist for about 100 years (Rodríguez et al. 1985) and that the superwind phase lasts about 10^4 years (Herman 1985), it is clear that to have a significant probability of detection one should observe several hundred OH/IR stars. Things are even worst than this, since most OH/IR stars will not pass by a phase where OH and ionized gas coexist. This means that perhaps as much as 10^3 sources should be observed to find one transition case where ionized gas and OH maser emission are simultaneously present. Although the amount of VLA time required for this project is large, the detection and monitoring of one of these radio continuum sources could provide definitive observational proof of the evolution of OH/IR stars to planetary nebulae.

Another observational test for this transition could be the disappearance of SiO maser emission in an evolved giant star. As the ionization of the envelope proceeds, the region producing the SiO maser emission will be ionized and dissociated in a few years. Unfortunately, this test is hampered by the intrinsic variability of SiO masers (Reid and Moran 1981), and by the fact that one would have to monitor a few thousand sources over a decade. Unfortunately, only about 130 SiO masers have been reported (Jewell et al. 1985).

4. OBSERVATIONS OF POSSIBLE TRANSITION OBJECTS.

What do we expect for the observational appearance of a transition object? Since the outer molecular envelope is still unaffected by what is happening closer to the star, we expect to detect extended (>0.1 pc) CO emission. Possibly 1612-MHz OH maser emission will be present if the envelope has an oxygen-rich chemistry. The object should also be a relatively strong far-infrared source with dust temperatures in the 100-200 K range. If fast stellar winds have appeared one could hope to see H_2 infrared line emission, usually considered to be a tracer of shocked molecular gas.

If we are dealing with a massive core, photoionization of the envelope should be detected via free-free emission; furthermore, this emission could be time variable (with a time scale of years or decades).

Quite a number of sources have been alleged to be objects in the transition between the tip of the asymptotic giant branch and the planetary nebula stage. I have tried to produce a list of the objects that, in my impression, are the best cases. I have not included in the list objects like the carbon-rich star IRC+10216 (=CW Leo) or the variable OH/IR stars since they most probably still are in the asymptotic giant branch. On the other end of the transition there are young planetary nebulae like NGC7027, IC418, NGC6302, NGC6790 and IC4997, that have the outer parts of their envelopes neutral (Mufson, Lyon, and Marionni 1975; Knapp et al. 1982, Rodríguez et al. 1985; Gathier et al. 1986; Altschuler et al. 1986), but that possess well-developed ionized regions. Although we do not consider these objects as transition examples they are obviously an important observational evidence that planetary nebulae envelopes where once cool and neutral. One important point in relation to the presence of neutral gas in planetary nebulae is that it does not necessarily guarantee that we are dealing with a young object. This has been shown by Huggins and Healy (1986a; 1986b), who detected CO in the evolved planetary nebulae NGC2346 and NGC7293 (the Helix Nebula).

I also excluded from the list objects that probably are symbiotic stars, such as HM Sge and V1016 Cyg. These emission-line candidates have underlying M giant characteristics that appear to imply a binary nature (Kaler 1985).

I considered also other sources, such as MWC349 (White and Becker 1985), the Red Rectangle (Daintly et al. 1985), and M2-9 (Kwok et al. 1986) but decided not to include them since they apparently lack evidence of a molecular envelope, one of the key ingredients of our selection criteria.

I list in Table 1 the objects that seem to be the best candidates for a transition-object status. There are only three objects in the list. In what follows we will discuss some new results on the sources given in Table 1. Recent papers with more discussion are those of Kwok and Bignell (1984) for CRL618, Seaquist and Davis (1983) for Vy 2-2, and Kawabe et al. (1986) for CRL2688.

4.1. CRL618

This is probably the object in the sky that more closely fits our preconception of a transition object. It complies with all our criteria, except that it does not have 1612-MHz OH maser emission. This could be due to the envelope being carbon-rich and thus not favoring the formation of O-bearing molecules. The high angular resolution VLA maps of Kwok and Bignell (1984) show the compact radio continuum source to be elongated (0".4x0".1) and to be surrounded by a halo of several arc sec. Kwok and Bignell argue that the compact radio source is an ionized ring or toroid around the central star. This could explain the bipolar morphology of the object observed in the optical.

The expansion velocity of the object is 18 km s^{-1} for the molecular gas, but 80 km s^{-1} for the ionized component. The high velocity ionized gas component could be due to acceleration by the fast stellar wind.

4.2. Vy 2-2

In the radio continuum this objects has a compact, shell-like structure (Seaquist and Davis 1983). The evidence for an outer molecular envelope comes from the OH maser emission. There is no observed CO, but deeper searches could reveal its presence.

The OH emission is asymmetric, that is, we only can see the blueshifted component of the typical OH spectrum. This is expected because the redshifted component arises from behind the ionized core, which is optically thick at 1612-MHz. One of the puzzling aspects of the OH emission is that it is not centered on top of the free-free source (Bignell 1983), as expected from symmetry arguments, and that its flux is much weaker than it should be.

4.3. CRL2688

This possible transition object is observable in the visible as a bipolar reflection nebula and lacks detectable ionized gas. Kawabe et al. (1986) proposed a model for this object that is similar to that proposed by Kwok and Bignell (1984) for CRL618. Kawabe et al. detect in

TABLE 1
POSSIBLE TRANSITION OBJECTS

Name	Chemistry	CO Envelope?	1612-MHz OH?	H_2?	Strong Far-IR Source?	Radio Continuum?	Radio Continuum Variability?	Optical Appearance?	T_*(K)	Expansion velocity (km s^{-1})
CRL618	N[a]	Yes	No	Yes	Yes	Yes	Yes	Bipolar	30,000	18/80[b]
Vy 2-2	O	No	Yes	No	Yes	Yes	No?	Stellar	50,000	10
CRL2688	C	Yes	No	Yes	Yes	No	—	Bipolar	7,000	20/40[c]

(a) There is no information on the C/O ratio, but Calvet and Peimbert (1983) found that CRL 618 is nitrogen-rich.

(b) The first number refers to the molecular gas, the second to the ionized gas.

(c) This source has two molecular outflow velocities.

the CO lines evidence for a bipolar cavity and for a high velocity
($\sim$40 km s^{-1}) molecular flow superposed on the bulk molecular flow that
has $v_{exp} \simeq$ 20 km s^{-1}.

5. CONCLUSIONS

The observational and theoretical study of protoplanetary nebulae has
advanced significantly during the last years. The detection of
evolved stars (OH/IR as well as carbon stars) with mass loss rates in
excess of 10^{-5} M$_\odot$ yr^{-1} strongly suggests that these objects are the
progenitors of planetary nebulae. However, our knowledge on transition
objects (those moving from the tip of the asymptotic giant branch to
the left in the H-R diagram) is still relatively poor. Only a few
objects (CRL618, Vy2-2 and CRL2688) have enough observational evidence
to be reliably considered transition objects. It is necessary to model
more adequately this evolutionary phase. To achieve this one requires
better information on the evolution of the cores and to consider
bipolar geometries. This modeling will enable observers to know better
what to search for. Finally, the developments in mm and IR facilities
should permit in the near future the detection and detailed study of
many more transition objects.

REFERENCES

Abell, G.O. and Goldreich, P. 1966, P.A.S.P., 78, 232.
Altschuler, D.R., Schneider, S.E., Giovanardi, C. and Silverglate, P.R. 1986, Ap. J. (Letters), 305, L85.
Bedogni, R. and D'Ercole, A. 1986, A.A., 157, 101.
Bignell, R.C. 1983 in IAU Symp. 103 Planetary Nebulae, ed. D.R. Flower (Dordrecht: Reidel), p. 69.
Bowers, P.F., Johnston, K.J. and Spencer, J.H. 1983, Ap. J., 274, 733.
Calvet, N. and Peimbert, M. 1983, Rev. Mexicana Astr. Astrof., 5, 319.
Dainty, J.C., Pipher, J.L., Lacasse, M.G. and Ridgway, S.T. 1985, Ap. J. 293, 530.
Gathier, R., Pottasch, S.R. and Goss, W.M. 1986, A.A., 157, 191.
Goldberg, L. 1985, in Mass Loss from Red Giants, Astrophysics and Space Science Library, Vol. 117, p. 21.
Herman, J., Baud, B. and Habing, H.J. 1985, A.A., 144, 514.
Herman, J. 1985 in Mass Loss from Red Giants, Astrophysics and Space Science Library, Vol. 117, p. 215.
Herman, J. and Habing, H.J. 1985, Phys. Reports, 124, 255.
Herman, J., Burger, J.H. and Penninx, W.H. 1986, to appear in A.A.
Hrivnak, B.H., Kwok, S. and Boreiko, R.T. 1986, Ap. J. (Letters), 294, L113.
Huggins, P.J. and Healy, A.P. 1986a, M.N.R.A.S., 220, 33p.
Huggins, P.J. and Healy, A.P. 1986b, Ap. J. (Letters), 305, L29.
Iben, I. and Renzini, A. 1983, A.R.A.A., 21, 271.
Jewell, P.R., Walmsley, C.M., Wilson, T.L. and Snyder, L.E. 1985, Ap. J. (Letters), 298, L55.
Kahn, F.D. 1983, in IAU Symp. 103, Planetary nebulae, ed. D.R. Flower

(Reidel: Dordrecht), p. 305.
Kaler, J.B. 1985, A.R.A.A., 23, 89.
Kawabe, R. et al. 1986, submitted to Ap. J.
Knapp, G.R. and Morris, M. 1985, Ap. J., 292, 640.
Knapp, G.R. 1987, in the Proceedings of the Workshop on the Late Stages of Stellar Evolution, Calgary. (Reidel: Dordrecht), p. 103
Knapp, G.R. and Bowers, P.F. 1986, in preparation.
Knapp, G.R., Phillips, T.G., Leighton, R.B., Lo, K.Y., Wannier, P.G., Wootten, H.A. and Huggins, P.J. 1982, Ap. J., 252, 616.
Kwok, S. 1982, Ap. J., 258, 280.
Kwok, S., Hrivnak, B.J., Milone, E.F. and Boreiko, R.T. 1986, preprint.
Kwok, S. 1983, in IAU Symp. 103, Planetary Nebulae, ed. D.R. Flower (Reidel: Dordrecht), p. 293.
Kwok, S. and Bignell, R.C. 1984, Ap. J., 276, 544.
Kwok, S., Purton, C.R., Matthews, H.E. and Spoelstra, T.A. 1986, to appear in A.A.
Morris, M., Jura, M. and Stark, A A. 1985, in preparation.
Mufson, S.L., Lyon, J. and Marionni, P.A. 1975, Ap. J. (Letters), 201, L85.
Okorokov, V.A., Shustov, B.M., Tutukov, A.V. and Yorke, H.W. 1985, A.A., 142, 441.
Paczynski, B. 1978, in IAU Symp. 76, Planetary Nebulae, ed. Y. Terzian (Reidel: Dordrecht), p. 201.
Perinotto, M. 1983, in IAU Symp. 103 Planetary Nebulae, ed. D.R. Flower (Dordrecht: Reidel), p. 323.
Pottasch, S.R. 1984, Planetary Nebulae, Astrophysics and Space Science Library, Vol. 107.
Reid, M.J. and Moran, J.M. 1981, A.R.A.A., 19, 231.
Renzini, A. 1981, in Physical Processes in Red Giants, ed. I. Iben Jr. and A. Renzini (Dordrecht: Reidel), p. 165.
Rodríguez, L.F., Gómez, Y. and García-Barreto, J.A. 1985, Rev. Mexicana Astron. Astrof., 11, 139.
Rodríguez, L.F. et al. 1985, M.N.R.A.S., 215, 353.
Sabbadin, F., Gratton, R.G., Bianchini, A., Ortolani, S. 1984, A.A., 136, 181.
Schoenberger, D. 1983, Ap. J., 272, 708.
Seaquist, E.R. and Davis, L.E. 1983, Ap. J., 274, 659.
Spergel, D.N., Giuliani, J.L. and Knapp, G.R. 1983, Ap. J., 275, 330.
Shklovskii, I.S. 1956, Astr. Zhurnal, 33, 315.
Van der Veen, W.E.J.C. 1986, preprint.
White, R.L. and Becker, R.H. 1985, Ap. J., 297, 686.
Zuckerman, B. 1978, in IAU Symp. 76, Planetary Nebulae, ed. Y. Terzian (Reidel: Dordrecht), p. 305.
Zuckerman, B. and Dyck, H.M. 1986, Ap. J. (Letters), 304, 394.

DISCUSSION.

<u>Danziger</u>: How does the space distribution of OH/IR stars compare with that of planetary nebulae, and what is known about either near or far IR spectroscopy of OH/IR stars?

<u>Rodriguez</u>: OH/IR stars have a space distribution similar to Peimbert's Type I PN, suggesting massive progenitors. As for IR spectroscopy, very little is kown: most of the IR work made is photometric.

<u>D'Antona</u>: What do you know about the luminosity of these three transition masses, to infer their core masses?
If they are $\sim 10^4 L_O$, they have massive cores and in the evolutionary scheme you gave they should not be there. I will present a possible explanation for this behaviour this afternoon.

<u>Weidemann</u>: The galactic bulge OH/IR stars have average luminosities corresponding to 0.6 M_O core masses. The low-mass transition objects could be completely hidden in the kind of non-variable OH/IR stars which Kwok's group has recently shown to exist, without any optical counterpart (less bipolarity for larger mass objects).

<u>Rodriguez</u>: The non-variable OH/IR stars could well be transition objects, but this has to be proved observationally.

<u>van der Veen</u>: I like to comment on your viewgraph about Vy2-2. You find OH emission only at a specific part of the shell. This also seems to occur in very young Mira's that have just started their mass-loss. Because Vy2-2 is a planetary that is at the end of its mass-loss phase the explanation can be that both in the case of young Mira's and in Vy2-2, mass-loss occurs in "blobs". Only if the mass-loss increases to 10^{-7} M_O or larger all "blobs" seem to form a dust shell. The idea of a dust shell consisting of "blobs" was also used by Alcock in its recent article to explain the OH maser emission in OH/IR stars.

<u>Rodriguez</u>: I agree that this is a possible explanation.

THE EVOLUTION FROM MIRAS TO PLANETARY NEBULAE AS DERIVED FROM OBSERVATIONS OF MIRAS AND OH/IR STARS

W.E.C.J. van der Veen, H.J. Habing
Leiden Observatory.
T. Geballe,
UKIRT and the Netherlands Foundation for Astronomy (ASTRON)

ABSTRACT. The evolution of oxygen-rich Miras and OH/IR stars at the top of the asymptotic giant branch (AGB) is discussed. We assume that all AGB stars with main sequence masses larger than about 0.8 $M_\odot$ start to pulsate at a certain moment. The pulsation occurs simultaneously with, and probably causes a mass loss at a rate orders of magnitude larger than earlier in the stars evolution. From this moment on, the mass loss rate increases continuously until the total stellar envelope is lost. IRAS observations of these Miras and OH/IR stars show a continuous sequence in the $F_{25\mu m}/F_{12\mu m}$ ratio that can be interpreted as due to an increase in mass loss. A semi-empirical equation for the mass loss rate is given. Main sequence masses of these stars can be calculated by integrating the mass loss rate over time and by adding the core mass, determined from the stellar luminosity.
When there is only a small fraction of hydrogen left in the shell surrounding the stellar core, the star will stop to pulsate and as a consequence will stop losing mass. Such objects, known as 'non-variable' OH/IR stars, are good candidates for proto planetary nebulae. Following a suggestion by Bedijn we are lead to a simple model for such non-variable OH/IR stars; it consists of a stellar core with an expanding gas/dust shell in which a cave is formed between the stellar core and the shell. This model is supported by recent ground based observations at near IR wavelengths ($\lambda < 12\mu m$).

1. INTRODUCTION

When the first infrared survey was made at 2μm at the end of the sixties it became clear that the sky was crowded with IR objects; see the infrared catalogue (IRC) of Neugebauer and Leighton (1969). Most of these IRC sources correspond to optically identified red giants, but they have stronger IR radiation than expected. This is explained by the presence of dust close to the star. Many of the sources also emit OH maser emission. Maps of the OH maser emission in the stronger sources (always at 1612MHz) obtained with Merlin and with the VLA show a ring-type structure corresponding to a circumstellar shell. The double peaked maser line profile proves the shell to be expanding. Thus arises the picture of an expanding shell of dust and gas which is fed continuous by ejection of matter by the central star. The dust absorbs the stellar radiation at around 1μm and reradiates it at much longer wavelengths. The central star that is shining through the dust shell varies in luminosity with a period between 100 and 600 days. Most stars were already known as optical Mira variables.

Similar objects were also discovered at much longer wavelengths than the IRC, and than often a visible counterpart was lacking. In the IRAS survey at 12, 25, 60 and 100μm few thousand such red objects were found, although the IRAS catalogue contains also several ten thousands objects similar to Mira variables. The OH maser emission found in such invisible objects, commonly refered to as OH/IR stars, was much stronger than the emission from the visible Miras, but nevertheless they showed the same ring-type structure and double peaked line profile. This suggests that the OH/IR stars are closely related to the Miras. Monitoring the OH radiation showed that the OH/IR stars also vary in luminosity, but with much longer periods, between 500 and 3000 days. As the OH/IR stars have no visible counterpart the amount of dust around the central star must be much larger than in the case of the Miras. This suggests that the OH/IR stars have mass loss rates much larger than those of Miras.

Because the OH/IR stars radiate all their energy between 2 and 60μm, the IRAS fluxes combined with ground based observations between 2 and 18μm yield the total flux of these stars. For some OH/IR stars distances could be determined using the so called 'phase lag method' (Herman and Habing, 1985), and luminosities between 2,000 and 50,000 $L_\odot$ were found. The combination of the facts that the objects are pulsating and that they undergo heavy mass loss suggests that they are situated at the top of the AGB and are the precursors of the planetary nebulae.

IRAS also detected some of the few (OH discovered) OH/IR stars that have no significant luminosity variations. They appeared to be even redder than the variable OH/IR stars (Olnon et al, 1984). Bedijn (priv. comm.) proposed that these 'non-variable' OH/IR stars have stopped their pulsations and therefore are no longer losing mass. These objects must then be more closely related to planetary nebulae than to variable OH/IR stars.

A recent review on Miras and OH/IR stars is given by Herman and Habing (1985). In the next sections we will present a general picture for the evolution from Miras to planetary nebulae as derived from the observations of Miras and OH/IR stars.

2. AN EQUATION FOR THE MASS LOSS RATE OF MIRAS AND OH/IR STARS

During the first part of the AGB evolution the stellar luminosity rises very steeply until pulsation sets in. Probably at the same time, the star begins to lose mass at a rate much larger than earlier in its evolution. The time scale for the loss of all the envelope (10^5yr) is shorter than the nuclear time scale on which the star increases its luminosity (10^6yr). Hence during this last phase the stellar luminosity will increase by less than a factor of two. Mass loss rates increase from $10^{-7} - 10^{-6} M_\odot/yr$, when the star is visible as a Mira variable, to $10^{-5} - 10^{-4} M_\odot/yr$ at the end of the AGB. This 'superwind phase' lasts only a few thousand years and thus the stellar luminosity remains approximately constant; i.e has reached its maximum luminosity. Stars in these last thousand years will be referred to as dust/gas envelope stars (DES); those for which OH maser emission has been detected might be called OH/IR stars. Both are cool giants (T_{eff} = 2000 - 3000 K) surrounded by an optically thick dust shell. IRAS has found a few ten thousands of them concentrated to the galactic plane and to the bulge (Habing, 1987).

The mass loss rate, $\dot{M}$, for these stars may be written as a function of the stellar luminosity, L_*, the expansion velocity of the dust shell, v_e, and the optical depth in the dust, τ_d, as given by Salpeter (1974)

$$\dot{M} \sim v_e^{-1} L_* \tau_d . \qquad (1)$$

For our stars we expect L_* and v_e to remain approximately constant during the 'superwind phase', whereas the mass loss rate increases considerably. Hence the increase in mass loss must be given by τ_d.

The color - color diagram (Figure 1) as derived from the IRAS data of known Miras and OH/IR stars shows that there is a smooth sequence from the lower left to the upper right corresponding with an increase in the F_{25}/F_{12} and the F_{60}/F_{25} ratio, where F_{12}, F_{25} and F_{60} are the IRAS flux densities at 12, 25 and 60µm respectively. Model calculations of circumstellar dust shells with increasing mass loss rates yielded IR colors that reproduce the observed sequence in Figure 1 (Bedijn, 1986).

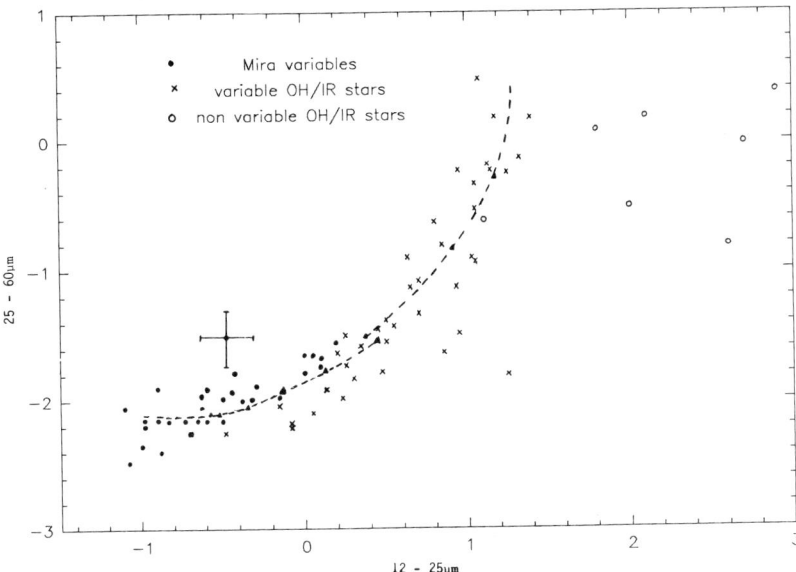

Figure 1. Color - color diagram of known Miras and OH/IR stars. The colors are derived from the IRAS data. The dashed line represents model calculations made by Bedijn (1986).

This suggests that the observed F_{25}/F_{12} ratio of Miras and OH/IR stars can be used to estimate the optical depth. We will assume a power law

$$\tau_d \sim (F_{25}/F_{12})^\alpha . \qquad (2)$$

Combining equations (1) and (2) yields

$$\frac{\dot{M}}{M_\odot yr^{-1}} = C_1 \left(\frac{v_e}{15 kms^{-1}}\right)^{-1} \left(\frac{L_*}{10^4 L_\odot}\right) \left(\frac{F_{25}}{F_{12}}\right)^\alpha , \qquad (3)$$

where C_1 and α can be determined from observations of stars with known mass loss rates.

Using stars with well determined distances for which Baud and Habing (1983) and Knapp (1985) derived mass loss rates, best values for C_1 and α are $7\ 10^{-6}$ and 3 respectively (Figure 2).

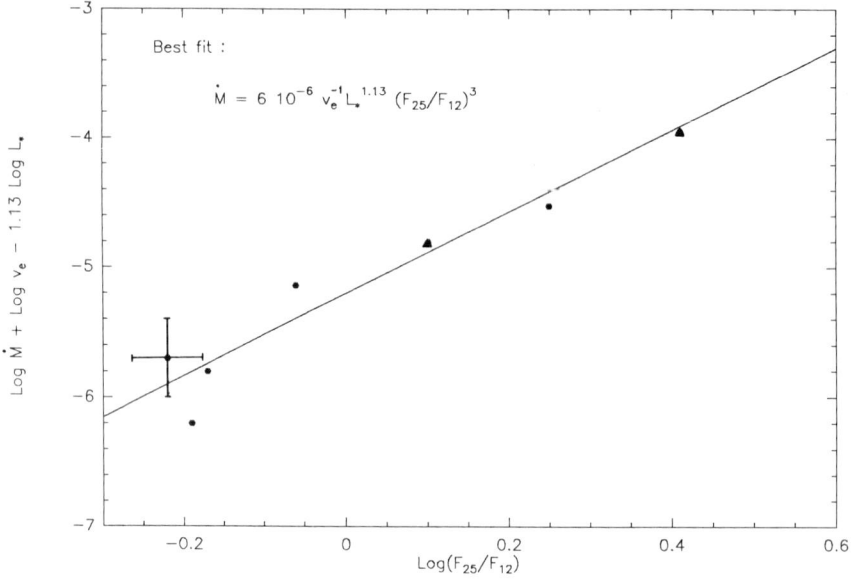

Figure 2. $\dot{M} v_e / L_*$ as a function of the ratio F_{25}/F_{12}. Dots are from CO observations by Knapp (1985), triangles are from OH observations by Baud and Habing (1983). An error estimate is indicated.

For the objects in Figure 2, only geometrical distances as obtained by Herman and Habing (1985) were used. It is clear that the F_{25}/F_{12} ratio is a good measure for the mass loss rate. The scaling factor L_*/v_e is important; without this factor the fit is much worse. A more elaborate derivation and argumentation of the equations given above will be given by Van der Veen (1987, in prep).

3. DETERMINATION OF MAIN SEQUENCE MASSES FOR OH/IR STARS

A star on the AGB is predicted to consist of two parts: a degenerate carbon-oxygen core of mass M_{co} and a hydrogen envelope of mass M_e. Since we assume that the star ejects all of its envelope, the envelope mass may be estimated by integrating $\dot{M}$ over time, which leads to

$$M_e = C_2 \left(\frac{v_e}{15 \text{kms}^{-1}}\right)^{-1} \left(\frac{L_*}{10^4 L_\odot}\right), \qquad (4)$$

where C_2 is a integration constant that will assumed to be constant for stars with different mass and chemical composition. We have also assumed that the luminosity and the expansion velocity are constant in time. The first follows from the fact that the time scale for the loss of all the envelope is much smaller than the nuclear time scale (10^5 versus 10^6 yr); the latter follows from the distribution of the expansion velocities (Winnberg et al., 1985), which shows a random distribution around v_e = 15 km/s. If the expansion velocity would evolve one should expect a distribution that is heavily skewed or even a monotonically increasing or decreasing function of the expansion velocity.

The core mass at the start of the TP-AGB can be found from the initial mass – core mass relation as given by Iben and Truran (1978). If we assume the the total mass lost before the TP-AGB equals 20% of the initial mass, the initial main sequence mass, M_i, is given by

$$M_i = 1.38 \, C_2 \left(\frac{v_e}{15 \text{km/s}}\right)^{-1} \left(\frac{L_*}{10^4 L_\odot}\right) + 0.59. \qquad (5)$$

Note that the initial main sequence mass, the chemical composition and the age of the star determine the final luminosity, the expansion velocity and the mass loss rate. In equation (5) however we take an diametrically opposed viewpoint and use the expansion velocity and luminosity to derive the initial main sequence mass. The gas to dust ratio and hence the metallicity can also be derived from expansion velocity and luminosity as will be shown by Van der Veen (1987, in prep.).

Because the observed luminosity yields the final mass, M_f, of the star via the Paczynski relation between luminosity and core mass (Paczynski, 1970), we are able to construct curves in the initial - final mass diagram (Figure 3). The data points are from Weidemann and Koester (1983, their Figure 1). Final masses are derived from the temperatures (black dots) and surface gravities (open squares) of white dwarfs in open clusters; the initial masses are obtained by comparing the cluster age minus the WD-cooling age with pre WD-ages as a function of M_i.

The foregoing can be used to find the main sequence masses of the progenitors of the OH/IR stars. For the calculations we will take C_2 = 2. The lowest luminosity found for OH/IR stars is about 2,000 $L_\odot$. In combination with an expansion velocity of 15 km/s - an average value (Winnberg et al, 1985) - this yields main sequence masses of about 1 $M_\odot$ and a resulting white dwarf mass of 0.53 $M_\odot$. Equation (3) predicts mass loss rates ranging from 5 10^{-8} $M_\odot$/yr, when the pulsation starts to 10^{-4} $M_\odot$/yr at the end of the AGB. For a luminous OH/IR star of 20,000 $L_\odot$ and with an expansion velocity of 15 km/s a main sequence mass of about 6 $M_\odot$ is found and a resulting white dwarf mass of 0.84 $M_\odot$. For this star mass loss rates will range from 5 10^{-7} to 10^{-3} $M_\odot$/yr.

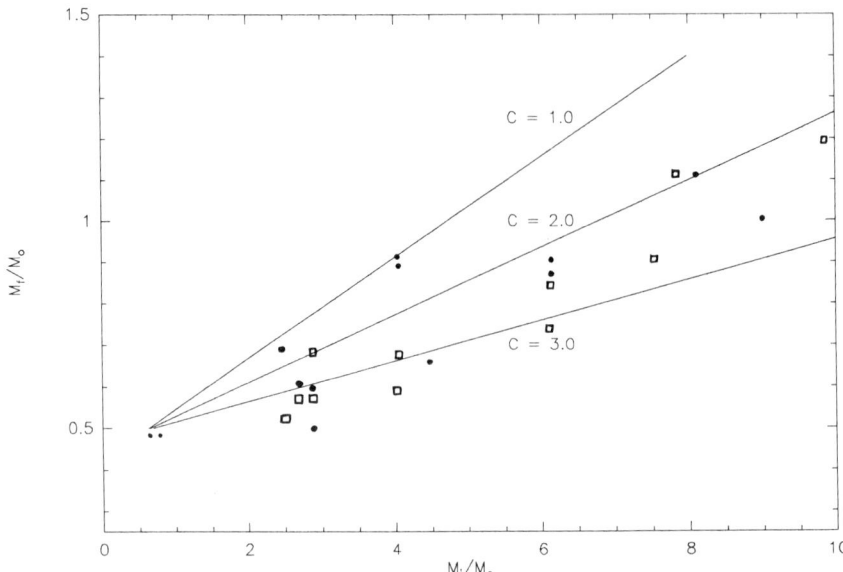

Figure 3. Initial - final mass diagram. The labeled curves are for constant $C = C_2/v_e$. Data points from Weideman and Koester (1983, their figure 1) : squares indicate data points derived from log g; dots indicate data points derived from the effective temperatures.

The same calculations made for stars with expansion velocities of 20 km/s give slightly lower main sequence masses. Because the distribution of expansion velocities is rather wide ($8 < v_e < 24$ km/s; Winnberg et al. (1985)) this can explain the scatter around the line $C = 2$ in the initial - final mass diagram (Figure 3). This wide distribution of expansion velocities is probably due to metallicity effects (Van der Veen, 1987, in prep).

4. THE EVOLUTION FROM AGB TO PLANETARY NEBULAE: PROTO PLANETARY NEBULAE

The duration of the super wind phase with mass losses in excess of 10^{-5} $M_\odot$/yr can last for only a few times 10^4 years. Then the hydrogen shell around the stellar core will be so thin that the pulsations will stop. Because these pulsations are thought to be the driving force of the mass loss, the mass loss rate will decrease enormously. Consider (following a suggestion by Bedijn) what will happen next. The dust shell produced during the 'superwind phase' will expand. The spectrum of the expanding envelope rapidly becomes much redder and its optical depth decreases. Therefore in the spectrum a 'shoulder' grows up at short wavelengths, when the remnant star begins to shine through the shell. Hydrogen burning in a thin shell will keep the star on the luminosity it had at the end of the AGB. The effective temperature however, will increase which causes the star to move to the left in the HR-diagram. The shift towards the red of the full spectrum is caused by cooling of the outward moving dust shell; the shoulder is caused by the expansion of a dust shell, to which no fresh material is added at the inner radius, causing a decrease in the optical depth. A spectrum as descibed above was indeed observed for the most luminous 'non⇾variable' OH/IR star OH17.7-2.0 (Figure 4). For comparison we give in this figure also the spectrum of a 'variable' OH/IR star OH26.5+0.6.

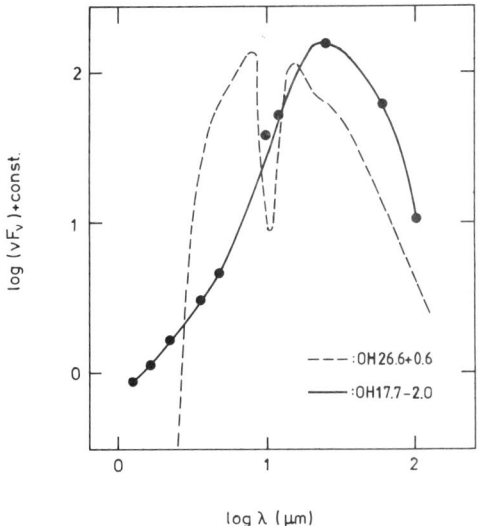

Figure 4. Low resolution spectrum of a variable OH/IR star (OH26.5+0.6) and a 'non⇾variable' OH/IR star (OH17.7⇾2.0). The data for OH26.6+0.6 have been taken from Figure 9 in Keinmann et al, 1981; for OH17.7-2.0 they are from Herman et al. (1984) and from Le Bertre et al. (1984).

In spite of the different IR spectra the OH maser emission of the variable and 'non⇾variable' OH/IR stars have the same characteristics. This is because the maser emission originates in the outer parts of the dust shell that are not yet affected by the decrease in mass loss of the central star. The scenario described above is exactly what one expect for a proto planetary nebula: a star that has suddenly decreased very much its mass loss but has not yet ionized its surrounding dust shell.

5. RECENT NEAR INFRA RED OBSERVATIONS OF NON⇾VARIABLE OH/IR STARS

In order to test whether OH17.7⇾2.0 is an unique case, new near IR observations were made for a sample of non variable OH/IR stars with the 3.6M ESO telescope at La Silla (july 1986) and the UKIRT telescope at Mauna Kea (august 1986). Six objects were observed between 1 and 13μm and nine between 1 and 5μm. In all objects a shoulder at the near IR wavelengths was found, wheras the energy distribution peaks at about

25μm. The form of the shoulders differ from object to object. Four different forms can be distinguished (Figure 5).

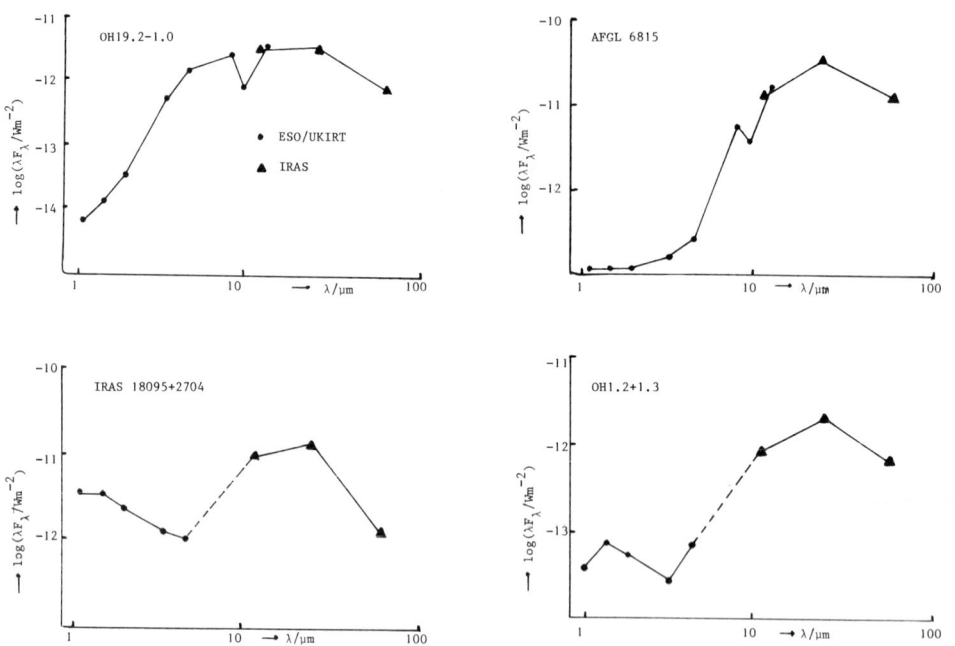

Figure 5. The form of the near IR spectrum of non-variable OH/IR stars found from recent observations at ESO and UKIRT.

For each category we have at least three examples. Of course one has to be very careful with these low resolution spectra; in particulary in planetary and proto planetary nebulae many lines may be present within the broadband IR filters. For this reason spectroscopy is necessary. However, the global form of the spectrum expected from the scenario for the formation of planetary nebulae (section 4), seems to be confirmed. Note that the planetary nebulae show near infra red spectra similar to those of the 'non-variable' OH/IR stars.

Figure 6 shows the evolution from Miras to planetary nebulae for both the stellar and the dust component. The horizontal axis of this color – color diagram describes the evolution of the dust (12 – 25μm); the vertical axis the evolution of the central star (3 – 4μm). For the Mira variables (lower left) the dust gets cooler when mass loss increase but the mass loss rate is still low and hence the central star is still visible. When mass loss increases the central star disappears. This is the case for the OH/IR stars (mid upper part). After the termination of the mass loss phase the dust moves outwards and become cooler, but also become optically thin at short wavelengths, which causes the reappearance of the central star and a bluer 3 – 4μm color. This is the case for the non-variable OH/IR stars and the planetary nebulae (lower right of the diagram).

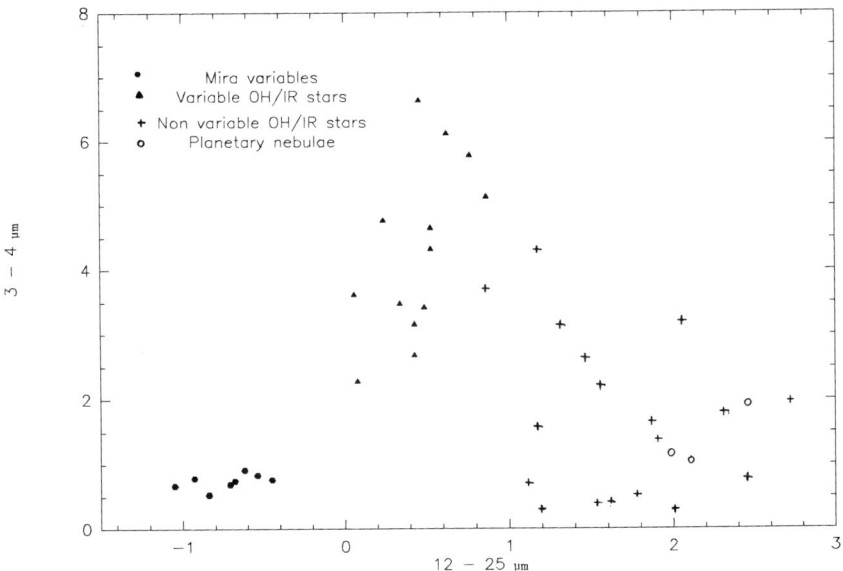

Figure 6. Color - color diagram obtained from IRAS data and ground based observations, showing the evolution of both the stellar component (vertical axis) and the dust component (horizontal axis).

6. CONCLUSIONS

An equation is presented for the mass loss rate as a function of luminosity, the expansion velocity of the dust shell and the ratio between the 12 and 25μm IRAS flux densities. A rather good fit exists to the mass loss rates obtained from CO and OH measurements. Stars with the same F_{25}/F_{12} ratio can have different mass loss rates depending on the stellar luminosity and the expansion velocity of the dust shell. The relation between initial mass, luminosity and expansion velocity was fitted to the data in initial - final mass diagram of Weideman and Koester (1983). The scatter in the diagram can be explained by differences in expansion velocities of the circumstellar dust shell which is probably a metallicity effect. For OH/IR stars with luminosities between 2,000 and 20,000 $L_\odot$, main sequence masses between 1 and 6 $M_\odot$ are found. These values depend on the value for the expansion velocity of the dust shell: larger expansion velocities result in smaller main sequence masses. After the termination of the mass loss phase the stars move to the left in the HR-diagram. When the central stars are hot enough (30,000 K) they will ionize the circumstellar dust shell. Between this moment and the moment the stars have stopped their mass loss we have to do with proto planetary nebulae. Recent observations in the near infrared show that the 'non-variable' OH/IR stars have all the characteristics of such proto planetary nebulae.

ACKNOWLEDGEMENTS

This work is supported by a grant from the Netherlands Organization for the Advancement of Pure Research (ZWO). W. van der Veen acknowledges travel support from the Leids Kerkhoven Bosscha Fonds.

REFERENCES

Baud, B.; Habing, H.J.: 1983, Astron. Astroph. **127**, 73
Bedijn, P.J., 1986, in "Light from dark matter", ed. F.P. Israël, Dordrecht, Reidel Publ., P. 119
Habing, H.J., 1987, in "Late stages of stellar evolution", ed. S. Kwok, Dordrecht, Reidel Publ.
Herman, J.; Habing, H.J.: 1985, Physics Reports, Vol 124, No 4, pp 255-314
Iben, I.; Truran, J.W.: Astroph. J. **220**, 980
Kleinmann, S.G.; Gillett, F.C.; Joyce, R.R.: 1981, Ann. Rev. Astron. Astrophys. **19**, 411
Knapp, G.R.: 1985, Astroph. J., **293**, 273
Le Bertre, T.; Epchtein, N.; Rieu Ngnyen-Q.: 1984, Astron. Astrophys. **138**, 353
Olnon, F.M.; Baud, B.; Habing, H.J.; de Jong, T.; Harris, S.; Pottasch, S.R., 1984, Astroph. J. **278**, L41.
Paczynski, B.: 1970, Acta Astron. **20**, 47
Salpeter, E.E.: 1974, Astroph. J. **193**, 585
Weidemann, V., Koester, D.: 1983, Astron. Astroph. **121**, 77
Winnberg, A.; Baud, B.; Matthews, H.E.; Habing, H.J.; Olnon, F.M.: 1985, Astrophys. J. **291**, L45-L50

QUESTIONS

L. Rodriguez: Your proposition of 'non-variable' OH/IR stars as transition objects is very interesting. Have you checked that, as expected, these objects do <u>not</u> have SiO and H_2O maser emission associated?

W. van der Veen: We did not check this yet, this will be done in the near future.

V. Weidemann: You showed Bedijn's fit (dashed line in Figure 1) which was calculated under the assumption of a constant luminosity. Which value did he take for the fit?

W. van der Veen: Bedijn used a luminosity of 10,000 L_o. But as I have showed the luminosity is only a scaling factor for the mass loss rate and has almost no influence on its time evolution; i.e. the chosen value of the luminosity has no influence on the fit in the color ⇸ color diagram.

THE POSITION OF PN NUCLEI IN THE H-R DIAGRAM: PRESENT STATUS

S.R. Pottasch
University of Groningen
the Netherlands

ABSTRACT

Recent improvements in the determination of the distances to various planetary nebulae are discussed. This increases the accuracy of the luminosities to a selected group of nebulae. The central star temperature determination has also been substantially improved for a large group of very faint hot objects. This involves especially new measurements of the central star magnitudes. The results are combined in giving an improved H-R diagram. A comparison between model calculations and observed positions on the H-R diagram gives a value for the age of the central star since it left the AGB. Another value for this age can be obtained from the size of the nebula and the observed expansion velocity. A discrepancy between these two ages seems to exist.

1. INTRODUCTION

The H-R diagram is very useful for discussing the evolution of many different kinds of stellar objects, and this is true for planetary nebulae central stars as well. This diagram is a plot of the stellar temperature against the stellar luminosity. Each of these will be discussed in turn.

A. Central star luminosity

This can be determined in several ways. The most direct way is to determine the effective temperature (T_s) (in a manner to be discussed presently) and the radius (R_s) separately, and combine these as $4\pi\sigma R_s^2 T_s^4$. The angular radius, R_s/d, where d is the distance, can be determined reasonably accurately when the visual magnitude is known (the magnitude determination will also be discussed presently). This leaves the distance of the nebula as the principal unknown factor in determining the luminosity.

Another way of determining the luminosity is to measure all the line and continuum flux which is emitted from the nebula. If the nebula is optically deep in hydrogen Lyman continuum radiation it converts all the radiation shortward of 912Å to radiation longward of 912Å, which in principle is measurable. But this method is just as sensitive as the previous one to the precise value of the distance.

For many nebulae ultraviolet and far infrared measurements are available, but for many more nebulae they have not been made. In those cases the total flux longward of 912Å can be approximated as some factor multiplied by the Hβ flux. This have been discussed by Pottasch (1984) who uses the factor 100 and by Gathier (1984) who gives the factor 110. It is clear that the main uncertainty comes in the distance determination.

B. Central star temperature

There are several ways of determining the temperature. A most useful method is that proposed by Zanstra sixty years ago. This involves the realization that Balmer line and continuum photon emitted by the nebula represents one photon emitted by the star capable of ionizing hydrogen. A comparison of this number of ionizing photons with the visual magnitude then yields a stellar temperature. This assumes that the nebula is optically deep to ionizing photons and is independent of distance. The optical HeII line spectrum can similarly be used to determine the number of photons capable of ionizing He$^+$ and combine with the visual magnitude, can give the so-called HeII Zanstra temperature.

Another method of determining the stellar temperature is the Energy Balance method. This make use of the ratio of the total flux in collisionally excited nebular lines to the Hβ flux. This can give a temperature because the Hβ flux represents the number of ionizing photons and the collisionally excited lines the energy put into the nebula by the ionization. The ratio therefore represents the average excess energy per ionizing photon, which is a measure of the temperature. It may be noted that this method can be used even when the nebula is optically thin to ionizing radiation (see Preite-Martinez and Pottasch, 1983). The most important disadvantage of this method is that measurements in the ultraviolet or infrared are necessary to determine the total collisionally excited flux. The IUE, for example, does not measure the OVI lines, which may be an important source of energy loss in nebulae with very hot central stars.

Other possibilities of measuring the temperature are less reliable. Sometimes use is made of measurements of the continuum stellar energy distribution: comparison to either a blackbody or a model atmosphere would give a temperature. But for hot stars even ultraviolet wavelengths measureable with the IUE are on the Rayleigh-Jeans part of the Planck curve and are therefore not very sensitive temperature indicators. Often the calibration is not better known than 15 to 20% (IUE) and possible corrections for interstellar or nebular extinction can be very uncertain. Kaler (1983) for example, found higher slopes than would be predicted by blackbodies of infinite temperature. Thus this method is considered less reliable and emphasis should be placed on the Zanstra and Energy Balance methods in determining the central star temperature.

II. DISTANCE DETERMINATION

A summary of the various methods of determining distances to planetary nebulae has been given by Pottasch (1984). The problem is difficult since no trigonometric parallaxes are known (but the Hipparcos satellite may change this) and only for a very few binary stars are spectroscopic parallaxes available.

Statistical distances are not acceptable for discussions concerning the evolution of nebulae. This is because statistical methods assume that all nebulae have the same properties: either the same ionized mass, the same size or the same Hβ flux. Any of these assumptions will introduce systematic deviations and selections, which could be interpreted as an evolutionary effect, while in reality it may be simply the result of the assumption made. It is therefore important to use individual distances determined independently of any assumed nebular property. A recent discussion of this is given by Gathier (1987).

For the purposes of the present discussion we have made use of a selection of about 30 nebulae whose distances have been reasonably well determined by independent methods. Those nebulae whose distances have been determined by the spectroscopic method or the angular expansion rate are summarized by Pottasch (1984). A recent addition to the angular expansion rate method is the result of Masson (1986) for NGC 7027 using the VLA at two epochs about 3 years apart. This interesting method may be applicable to more nebulae in the near future.

In addition to these methods, the results of two other methods have now become available, which double the number of reliable distances known. The first makes use of the relation between extinction and distance in a given direction on the sky. If many stars are measured in the direction of a nebula this relation can be determined with relatively little scatter. If, in addition, the extinction to the nebula is known, its distance follows directly. This method has been applied very carefully to 13 nebulae by Gathier et al. (1986a), resulting in quite reliable distances for these nebulae.

The second method makes use of the 21 cm H line absorption measured against the nebular continuum. This absorption is caused by material in a spiral arm between us and the nebula. Each spiral arm causes an individual absorption line whose velocity can be measured and is therefore identifiable. We know then that the nebula is either in or beyond the furthest of these spiral arms. Since the distance of the spiral arm is known from other consideration, this gives a good indication of the nebular distance. If, in addition, the 21 cm H line absorption is measured in an extragalactic source, which is close to the nebula in the plane of the sky, the spiral arms which are behind the nebula can also the ascertained. This places quite stringent limits of the nebular distance. Recent measurements of about 20 nebulae have been reported by Gathier et al. (1986b), who also give resultant distances. This method can only be used for nebulae which are within about 10° of the galactic plane.

Combining the various distances from the different methods, and including NGC 2818 whose distance can be found from its membership in an open cluster, a sample of nearly 30 nebulae with reasonably well determined distances is available.

III. MAGNITUDE DETERMINATION

The difficulty in determining the central star magnitude arises from the presence of strong nebular emission from around the star. Sometimes this emission is so strong that in a photograph of the nebula the star cannot be seen. This is true, for example, of 10% of the nebulae in the first catalogue of planetary nebulae (Curtis, 1918).

Curtis tried to obtain a magnitude by comparison of the central stars with other stars measured on the photographic plate. This involves an error of 0.2 mag or perhaps somewhat more because of the presence of nebular emission. Curtis could not see the central star of NGC 2440 and estimated it to be 19th mag or fainter.

In the past 15 years photoelectric photometry has been used to measure magnitudes. The use of a diaphragm causes a large amount of nebular emission to be measured at the same time. The nebular emission must be removed. This can partly be done with filters, which can remove the strong line emission. The nebular continuum emission must be removed by theoretically subtracting the expected emission from that observed. The method can work reasonably well if the star is bright compared to the nebula but can go very wrong when 90% or more of the

TABLE 1

Nebula	Recent Magnitudes Determinations		Previous
	Reay et al.	Gathier Pottasch	
N 2438	18.4		17.74
2440	18.9	18.8	14.3
2452	17.9	17.3	18:
2818	19.5		18.52
2867	16.5		14.4
3211	19.4	18.1	16.2
3918	16.7	15.8	14.6
5315	13.3		14.63
6302	No Det.	≤ 21.2	−
6369		16.6	14.66
6439	No Det.	20.2:	18.5
6445	18.9	19.0	

observed emission must be subtracted. An example of when it went wrong is the report of Kaler (1976) that the central star of NGC 2440 has a m_v= 14.3. In the large aperture used by Kaler (usually 26" or 40" diameter) the nebular continuum and the sky background add so much noise that the flux of the dark night sky appears as bright as a 14th magnitude star, while the nebular continuum is much brighter.

It is clearly necessary to return to that aspect of the photographic plate that allows the star to be seen and that its magnitude may be estimated by comparison with known stars. This can now be done (Reay et al., 1984) using CCD images of nebulae through narrow band filters which eliminate all known nebular lines. In this way the central star of NGC 2440 was found to have m_v= 18.9 (Atherton et al., 1986, Gathier and Pottasch, 1987). Magnitudes determined in this way for a selection of 12 nebulae, many of which from the sample previously defined, are shown in the 1st three columns of Table 1. A comparison with previous determinations, shown in the last column, indicate that large improvements can sometimes be made (and are being made).

IV. HR DIAGRAM

From the best available magnitudes, the Zanstra temperatures may be computed. These are shown for twelve nebulae in Table 2, for both hydrogen and helium. Many of these stars are very hot with temperature sometimes well in excess of 150.000K. For these hot stars an additional correction has to be made (Stasinsky and Tylenda, 1986). The reason for this is as follows. For T_z(HI) an overestimate of the number of ionizing photons is made, because some photons which initially ionize helium can be degraded into several photons each of which can ionize

TABLE 2

Nebula	Zanstra Temperatures T_z(HI)	T_z(HeII)
N 2438	1.7×10^5	1.6×10^5
2440	4.0 (3.5)	3.1 (3.4)
2452	0.89	1.2
2818	1.7	1.75
3211	2.6 (2.2)	2.2
3918	2.4 (2.1)	1.95
5315	0.49	0.61
6153	0.91	1.1
IC 2165	1.7	1.7
J 900	1.1	1.2
N 7027	7.6 (5.5)	4.3 (5.5)
7662	0.76	1.1

hydrogen. T_z(HeII) is underestimated because some of the photon capable of ionizing He^+ are absorbed instead by hydrogen. The corrected values of temperature following the calculations of Stasinska and Tylenda are shown in parenthesis in the table.

For the hot stars the (corrected) values of T_z(HI) and T_z(HeII) are in good agreement. For the stars with temperatures below 10^5K, the value of T_z(HeII) is greater than T_z(HI). This is a well known effect (see a discussion in Pottasch, 1984). It has been variously interpreted as (1) due to departures from blackbody radiation in the exciting stars, in which as T_z(HI) is probably correct, or (2) due to incomplete absorption of radiation in the nebula, in which case T_z(HeII) is probably correct. For the present purpose as average value has been used.

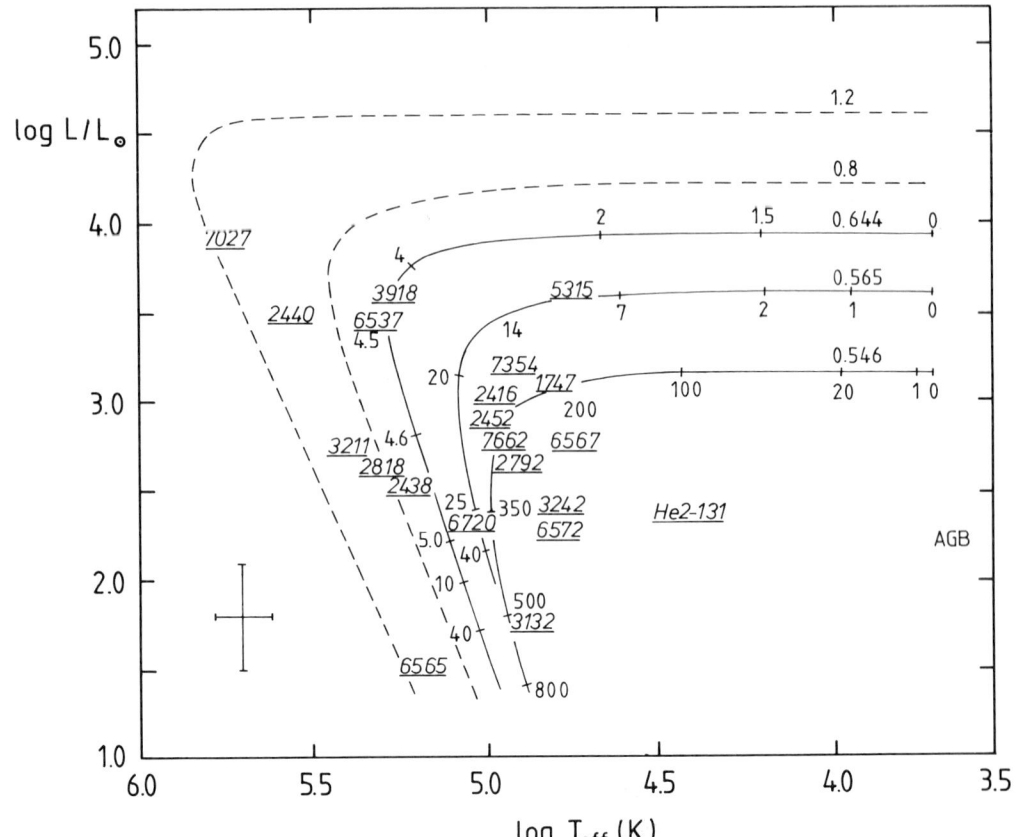

Fig. 1 The H-R diagram for a selected sample of planetary nebulae whose luminosity is reliably known. The theoretical tracks of Schönberner (1981, solid lines) and Paczynski (1971, dashed lines) are also shown. Evolutionary times (in 10^3 years) are indicated on the theoretical curves. Estimated average errors are indicated in the lower left hand corner.

Although this is wrong in principle, the uncertainty does not cause a large error in the position on the HR diagram.

This position is shown in Fig. 1, where each nebula is indicated by its NGC, IC, etc. number which is underlined. Also shown on the diagram are predicted evolutionary tracks. Those given by the solid line are taken from Schonberner (1981) while the evolution for the higher mass stars is taken from Paczynski (1971) and is shown by the dotted line. The core mass is labled on the right side of each track and the predicted evolutionary time in units of 10^3 years is given at various points on the track. The zero point for the measurement of the evolutionary time is arbitrary and is indicated in the figure. Typical error bars are indicated in the lower right hand corner of the figure.

From the figure, the age as predicted from these evolutionary theories can be determined. These are listed in the second column of Table 3 as 'prediction'. The third column of the table gives the observed age which is determined by dividing the radius of the nebula by

TABLE 3

Comparison of predicted lifetimes with observed nebular ages (in 10^3 years)

Nebula	Prediction	Observation
N 246	100	8
IC 1747	200	3
N 2440	0.5	8
N 2452	250	6
N 2792	300	3
N 2438	4	17
N 2818	3	22
N 3132	600	6
N 3211	2	4
N 3918	4	4
N 5315	9	1
Ne 2-131	> 1000	0.5
N 6537	≤ 4	4
N 6565	1	1
N 6567	300	2
N 6572	500	1
N 6720	20	4
N 7027	0.05	1
N 7354	50	3
N 7662	300	2

the expansion velocity. These two ages may now be compared. In making the comparison it must be born in mind that (a) the errors in luminosity and temperature may cause considerable error in the predicted age, and (b) the zero points in the time are not necessarily the same.

Even with this in mind, there appear to be large discrepancies between the two ages. Systematic effects are present as well. The very hot stars, especially in NGC 2440 and 7027 appear to be an order of magnitude older than predicted. This is true to a lesser extent for NGC 2438 and 2818 as well. On the other hand, the nebulae with central stars whose temperature is lower than 10^5 K appear to be systematically younger than the predicted age by at least an order of magnitude.

The tentative conclusion which can be drawn from this is that, while a rough quantitative agreement exists between the theoretical tracks and the position of the central star on the HR diagram, the details still are not well understood. Especially the predicted evolutionary times are not in agreement with the observed nebular age. Furthermore the observations indicate the existence of young, low temperature (T $\simeq$ 40000K) central stars with luminosities less than 10^3 $L_\odot$. The existance of these objects is not predicted by present theories.
A detailed discussion of these results will be published soon (Gathier and Pottasch, 1987, Astron. Astrophys.)

REFERENCES

Atherton, P.D., Reay, N.K., Pottasch, S.R. 1986, Nature, 120, 423
Curtis, H.B. 1918, Lick Obs. Publ. XIII, 55
Gathier, R., Pottasch, S.R., Pel, J.W. 1986a, Astron. Astrophys. 157, 171
Gathier, R., Pottasch, S.R., Goss, W.M. 1986b, Astron. Astrophys. 157, 191
Kaler, J.B. 1976, Astrophys. J. 210, 113
Kaler, J.B. 1983, Astrophys. J. 271, 188
Paczynski, B. 1971, Acta Astron. 21, 417
Pottasch, S.R. 1984, Planetary Nebulae (Reidel, Dordrecht)
Preite-Martinez, A., Pottasch, S.R. 1983, Astron. Astrophys. 126, 31
Reay, N.K., Pottasch, S.R. Atherton, P.D. Taylor, K. 1984, Astron. Astrophys. 137, 113
Schonberner, D. 1981, Astron. Astrophys. 103, 119
Shaw, R.A., Kaler, J.B. 1985, Astrophys. J. 295, 537
Stasinska, G., Tylenda, R. 1986, Astron. Astrophys. 155, 137

DISCUSSION

<u>Giovanelli</u>: You have shown a log L - log T diagram in which you put about 20 objects, for which you know better the distance. It seems that the problem of the distance is crucial. This is a general question: can somebody tell me what people are doing in order to improve the determinations of the distances?

Pottasch: In the near future the Hipparcos satellite will be able to give trigonometric distances for between 20 and 30 nebulae. Secondly, expansion distances based on VLA measurements of strong nebulae measured at two epochs separated be several years, as Masson has done for NGC 7027, will become available for 5 to 10 nebulae. Thirdly, extinction distances are being measured for a few more nebulae and older values are being improved. Fourthly, 21 cm absorption distances are also still being made.

D'Antona: I will comment this afternoon on the discrepancy between expansion ages and theoretical ages, basing the comparison on Sabbadin (1986) sample of PNae of known distances, but which similar results to yours. The real problem for the theory are the very low luminosity objects below 10^3 $L_\odot$. I suggest, although I have reservations on this interpretation, that they should be checked for binarity (spectroscopic binaries) as they could be a result of common envelope evolution occurring during the H-shell burning stage.

Peimbert: Where would you place NGC 6302 in the HR diagram?

Pottasch: It is undoubtably a very hot star, with a temperature well in excess of 3×10^5 K. I would expect that it has a luminosity similar to NGC 2440 or 7027. This would put it at a distance of 1 to 2 kpc which I do not think is unreasonable.

Renzini: I am suspicious about the location in the HR diagram of He 2-131, which is the most discrepant case — I had a look at this case in the Gathier's thesis, and found that both the reddening, and the reddening-distance relation are most uncertain. My impression is that an underestimate of the distance by perhaps a factor of 3 cannot be ruled out. Let me also mention that small errors in temperature led to large variation in the inferred "age", particularly at low luminosities. I would then not attach much meaning to some wild discrepancies between kinematical, and "evolutionary" - inferred ages.

Pottasch: The distance to He 2-131 has also been determined by Maciel (Rev. Mex. Astron. Astrophys. 10, 199, 1985) who found a distance of 630 ± 250 pc from the extinction method. This is very similar to the distance found by Gathier et al. In spite of the fact that individual distances may always prove to be in error, I think these results should be treated seriously. As additional evidence there are other nebulae with low temperature central stars that fall under the lowest theoretical track in the HR diagram (fig. 1). These are BD +30 3639 and IC 418. These nebulae all have very bright central stars, are quite young and are very close (the distance are more uncertain than those given in the diagram, but much evidence places IC 418 at 500 pc and BD +30 3639 at about 600 pc).

Hunger: You have not referred to the effective temperatures derived from direct spectral analyses of the central stars. Don't you believe in these temperatures?

Pottasch: The work of Mendez, Kudritzki and co-workers has been very interesting in this respect. The results are still in an early stage since they compare the observational profiles with models containing only hydrogen and helium. The temperatures they have found are consistent with those determined from other methods, especially the Zanstra method. I think that they are essentially correct although their errors might be underestimated. For example, for NGC 7293 the Zanstra temperature is 115,000K while they derive 95,000K.

The reason I haven't mentioned their work is different. Their method limits them in principal to O type spectra with absorption lines, and in practice to only the very brightest of these stars because they need high dispersion spectra. Thus their sample has essentially no overlap with those nebulae for which independent values of distance are available. Therefore there is no independent check on the surface gravities they derive. If I use their values of surface gravity and assume the star has a mass of 0.55 $M_\odot$ they can be placed on an HR diagram. I have done this in Fig. 2, which may be compared to my selection in Fig. 1. There are three things to notice in such a comparison. Firstly, their selection covers only a very limited temperature range (between 6×10^4 K and 10^5 K). Secondly, within this temperature range, the distribution of their nebulae in the HR diagram is essentially the same as mine, and would lead to the same conclusion. Thirdly, the one nebula we have in common (NGC 3242) is in a very different position. This is because the surface gravity they have found leads to a distance of 2.2 kpc, while the angular expansion measured by Liller leads to a distance of 0.7 ± 0.2 kpc. Since the latter is the only direct evidence for the distance it must be checked.

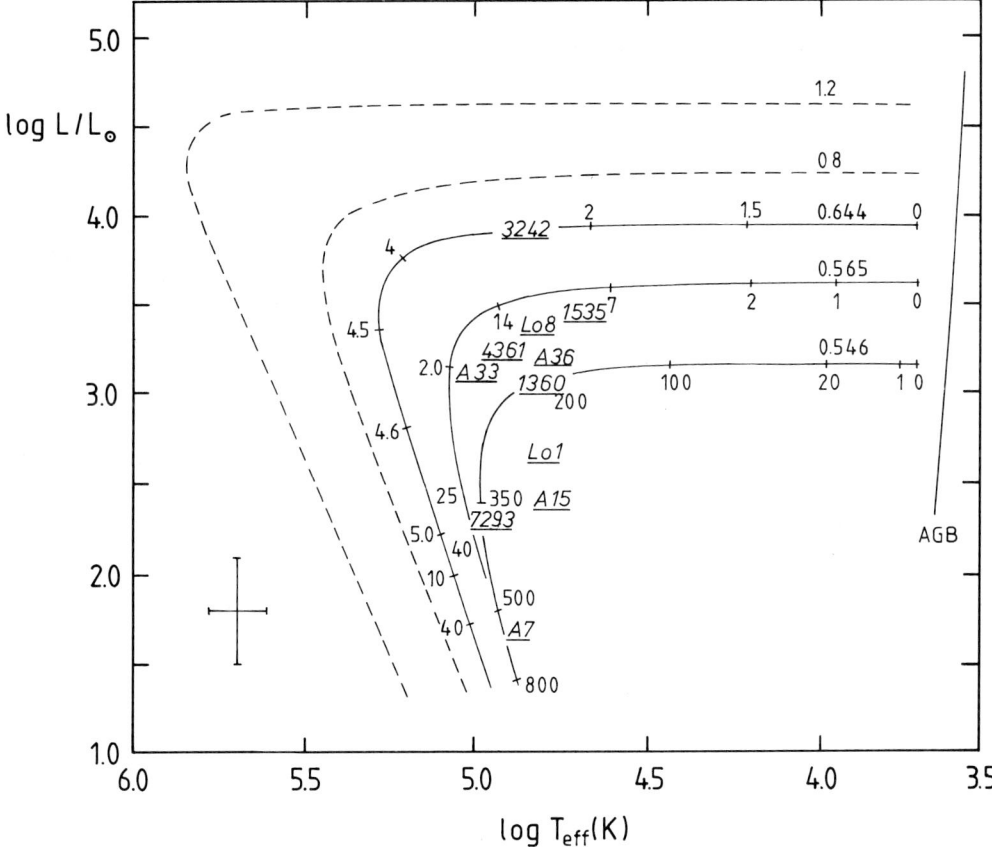

Fig. 2 Same as Fig. 1, but with the nebulae whose temperature and gravities were found by Mendez et al. plotted on the diagram.

ON THE NITROGEN AND HELIUM ENRICHMENT OF THE INTERSTELLAR MEDIUM

Manuel Peimbert
Instituto de Astronomía, UNAM
Apdo. Postal 70-264
04510 México, D.F., MEXICO

ABSTRACT. Based on the He/H and N/H ratios in the interstellar medium (ISM) of galaxies with different heavy element abundances, and in planetary nebulae (PN) of different types, some aspects of the N enrichment of the ISM are discussed. It is found that the predicted secondary N production by massive stars is roughly consistent with the observed Y versus X_N diagram. For oxygen poor objects there is an excess of N relative to He which could imply an additional source of N or that infall has been important during the history of these systems. The contribution to the He and N enrichment of the ISM by PN of Types II and III is negligible in comparison to that provided by PN of Type I.

1. INTRODUCTION

It has been known for a long time that PN are enriching the ISM with N and He (e.g. Kaler 1979, Peimbert and Serrano 1980 and references therein). The progenitors of PN are intermediate mass stars, IMS, with main sequence masses in the $1 \lesssim M/M_\odot \lesssim 8$ range. PN of Type I, those that are He and N rich, correspond to $M \gtrsim 2.4\ M_\odot$ while those of Types II and III to $M \lesssim 2.4\ M_\odot$. It has been argued that most of the N produced by PN of Types II and III is of secondary origin (e.g. Peimbert and Serrano 1980), while for those of Type I part of the ejected N might be of primary origin.

Massive stars, those with $M \gtrsim 10\ M_\odot$, are also enriching the ISM with N and He, typical examples are the WN central stars of NGC 2359 and NGC 6888 (e.g. Parker 1978, Willis and Wilson 1978, Peimbert 1979, Talent and Dufour 1979, Kwitter 1981). Also in these objects apparently most of the N is of secondary origin.

A relationship of the type

$$\log (N/O) = a \log (O/H) + b \qquad , \qquad (1)$$

with $a = 1$ is predicted by the simple model of galactic chemical evolution for the ISM under the assumptions that N is of secondary origin and O is of primary origin (e.g. Talbot and Arnett 1973). The simple model is based on the following assumptions: a) there are no gas flows, b) the initial mass function (IMF) is constant and c) the instant recycling approximation applies. Determinations of a yielded values considerably smaller than unity for several spiral galaxies and for a set of irregulars and blue compact galaxies. These results led several authors to suggest that a significant fraction of N is of primary origin (e.g. Smith 1975, Edmunds and Pagel 1978, Peimbert 1979, Alloin et al. 1979, Lequeux et al. 1979).

Edmunds and Pagel (1978) and Peimbert (1979) noticed that H II regions in irregular and dwarf blue galaxies have a smaller N/O ratio for a given O/H ratio than galactic H II regions. To explain this variation at least one of the assumptions of the simple model had to be dropped; possible explanations are that: a) the galactic initial mass function is different from those of irregular and dwarf blue galaxies, b) infall in the Galaxy of material with pregalactic abundances has reduced the O/H ratio of regions with substantial amounts of N produced by secondary mechanisms (Peimbert 1979), and c) most of the N is of primary origin produced by stars in the $1 - 2.5$ $M_\odot$ range and the differences in N/O for a given O/H are due to age effects (Edmunds and Pagel 1978).

Several authors have considered the possibility that in O poor galaxies, most of the N is of primary nature and that, as the O/H ratio increases, the ratio of secondary to primary N also increases; thus that in O rich galaxies most of the N is of secondary nature (Dufour et al. 1982, Dufour 1984a, Renzini 1984, Matteucci 1986).

Serrano and Peimbert (1983) found that the N/O versus O/H diagram can be explained under the following assumptions: a) most of the N is of secondary origin, b) most of the N is produced by stars in the $1.3 \lesssim M/M_\odot \lesssim 5$ mass range, c) accretion plays an important part in the chemical evolution of galaxies and, d) the yield increases with metallicity. White and Audouze (1983) found that the N/O versus O/H diagram can be explained with: a) secondary N production, b) inflow of fresh gas into a galaxy and c) stochastic mixing. Matteucci and Tosi (1985) have shown that the N/O versus O/H diagram can be reproduced by means of a model with: a) bursts of star formation, b) galactic winds powered by supernovae, and c) that N is both a primary and a secondary element.

Pagel (1987) based on a correlation between the N and He abundances in irregular and blue compact galaxies of low O abundance and on the presence of WN features in their spectra, suggested that most of the N in these objects was secondary but produced by WN stars, that is by massive stars.

In what follows, based on observations of galactic and extragalactic H II regions and PN, we will explore further the suggestion by Pagel (1987), that N is secondary and produced by massive stars, and we will try to see if it is possible to apply this idea to O/H rich systems.

2. THE HELIUM AND NITROGEN PRODUCTION

2.1. Massive Stars

WN stars are massive stars that have transformed, in the exposed atmospheric layers, all the C and O into N and all the H into He (*e.g.* Maeder 1984). Their stellar wind is characterized by freshly made secondary N.

The change in the N and He abundances in the ISM by a generation of WN stars will be given by

$$dX_N = (X_C + X_O) \, dY/(Y_f - Y_i) , \qquad (2)$$

where X_N, X_C, X_O are the N, C and O mass fractions, dX_N and dY are the freshly made N and He by mass, and Y_i and Y_f are the initial and final helium mass fractions of the outer layers of the massive star where H → He. For the solar neighborhood and for metal rich and metal poor H II regions $(X_C + X_O) \simeq 75\% \, Z$ (*e.g.* Peimbert *et al.* 1986); moreover $Y_f \simeq 98\%$ (*e.g.* Maeder 1984) and $Y_i \simeq 23\%$ (*e.g.* Peimbert 1986). From these considerations it follows that equation (2) can be expressed as

$$dX_N = Z \, dY . \qquad (3)$$

From observations of galactic and extragalactic H II regions it has been found that the total enrichment of freshly made helium during the history of a galaxy is given by

$$\Delta Y = (Y - Y_p) = \alpha Z , \qquad (4)$$

where Y is the observed He abundance by mass, Y_p is the pregalactic He abundance, and α is in the 3 to 4 range (Peimbert 1986, Pagel 1986). From equations (3) and (4) it follows that

$$\Delta X_N = \int dX_N = \int \frac{\Delta Y}{\alpha} dY = \frac{\Delta Y^2}{2\alpha} = \frac{Z}{2} \Delta Y , \qquad (5)$$

where ΔX_N is the total N enrichment during the history of the system under the assumptions that: a) N is of secondary origin, b) it has been produced by massive stars and c) that the instant recycling approximation applies.

In Fig. 1 we have plotted equation (5) for $\alpha = 3$ and $\alpha = 4$. Also in Fig. 1, under the assumption that $Y_p = 0.23$, we present the N and He abundances for: H II regions in the Large and Small Magellanic Clouds, Orion ($t^2 = 0.02$) and M17 (Peimbert and Torres-Peimbert 1974, 1976, 1977, 1987). The X_N and ΔY values for NGC 2363, an H II region in NGC 2366, are very similar to those of the H II regions in the SMC and are not presented in Fig. 1 (Peimbert *et al.* 1986).

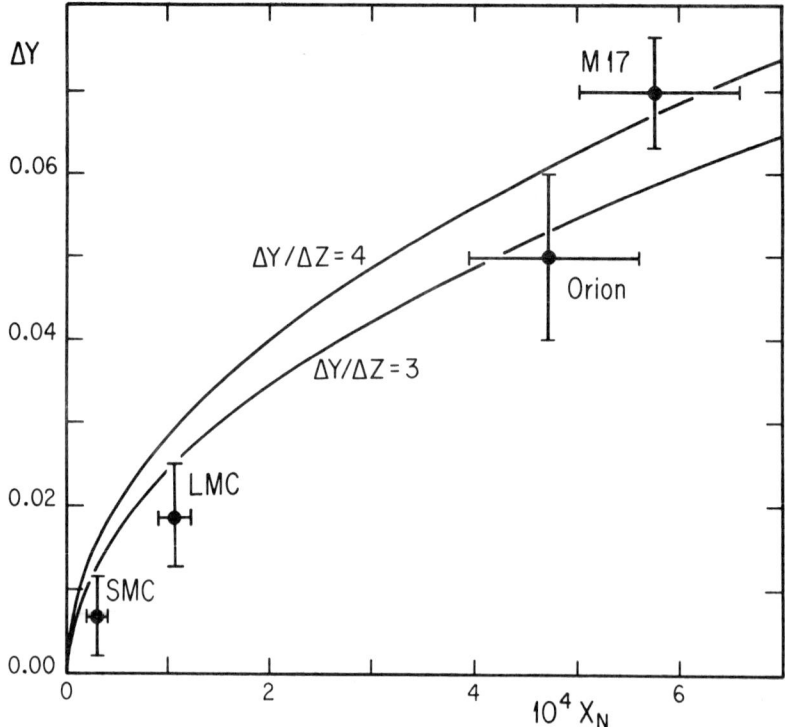

Figure 1. Helium *versus* Nitrogen diagram for galactic H II regions (the Orion nebula and M17), and extragalactic H II regions (in the Small Magellanic Cloud and in the Large Magellanic Cloud). The two curves are plotted under the assumptions that N and He have been produced only by massive stars and that N is secondary.

From Fig. 1 it can be seen that equation (5) provides a good fit to the data. That does not imply that all the N is produced by massive stars in a secondary fashion, but only that the observations are consistent with such an idea.

For very low values of X_N, like those of the SMC, the LMC and NGC 2363, the N abundance is higher than that predicted by equation (5), this could be due to one of the following causes: a) that probably some of the N is of primary origin coming from massive stars, Matteucci (1986) reached a similar conclusion based on a study of the N/Fe ratio in galactic metal poor stars, b) that PN of Type I are responsible for a substantial fraction of the N enrichment (see §2.3), and c) that infall, of material with $\Delta Y = 0$ and $\Delta X_N = 0$, plays an important role in the evolution of these systems, that is that the infall rate is comparable to the star formation rate.

Under the assumption that all the N is secondary and produced by massive stars Pagel (1987) finds that

$$\Delta X_N = 1.3 \: Z \: \Delta Y \tag{6}$$

From a sample of O/H poor galaxies with $\langle Z \rangle$ = 0.003 Pagel found that equation (6) was in very good agreement with the observed values, probably implying that most of the N is of secondary origin produced by massive stars. From our previous discussion we prefer equation (5) to equation (6), therefore the observed N from Pagel's sample is also larger than predicted by equation (5). This result would also indicate an extra source of N or that infall of unprocessed material plays an important role in the evolution of these objects. On the other hand most of the He could have been produced by massive stars.

2.2. PN of Type I in the Galaxy

PN of Type I presumably are very young objects that are He and N rich (Peimbert 1978, 1984). In Table I we present a list of PN of Type I together with their He and N abundances (Peimbert and Torres-Peimbert 1983, 1987). Under the assumption that at present the ISM abundances are Y = 0.29 and X_N = 8 x 10^{-4}, the values of δY and δX_N given in Table I were obtained; δY and δX_N are the differences between the observed PN values and those of the ISM. The average values for the thirteen objects are δY = 0.076 and δX_N = 0.00204; δX_N is just marginally higher than predicted by equation (3) (δX_N = 0.00153). This result implies that at present Type I PN could be responsible for a large fraction of the N enrichment of the ISM without substantially affecting Figure 1.

TABLE I. He and N abundances in Type I PN

Object	Y	12+log N/H	δY	δX_N (10^{-3})
NGC 650	0.345	8.30	0.055	0.98
NGC 2346	0.345	8.06	0.055	0.22
NGC 2371	0.345	8.58	0.055	2.58
NGC 2440	0.340	8.80	0.050	4.85
NGC 2818	0.368	8.43	0.078	1.51
NGC 5315	0.328	8.38	0.038	1.31
NGC 6302	0.416	8.70	0.126	3.16
NGC 6778	0.380	8.51	0.090	1.92
NGC 6853	0.345	8.52	0.055	2.14
Hu 1-2	0.382	8.43	0.092	1.45
M 3-3	0.337	8.66	0.047	3.31
Me 2-2	0.400	8.40	0.110	1.24
PB 6	0.421	8.53	0.131	1.85
Average	0.366	...	0.076	2.04

There are some objects where δX_N is considerably larger than predicted by equation (3), particularly NGC 2440, NGC 2371 and M 3-3; in these cases probably part of the N is of primary origin, *i.e.* produced by freshly made carbon, or O and C went into N in a region where not all of the H was transformed into He.

Galactic and extragalactic Type I PN seem to be O deficient (Peimbert and Torres-Peimbert 1983, Peimbert 1984); infrared evidence in favor of this result has been also found by Pottasch (1986) from low resolution IRAS spectra. This result is in agreement with the idea that part of the O has been transformed into N.

2.3. Type I PN in O Poor Galaxies

PN of Type I have been found in NGC 6822, the SMC and the LMC, and as mentioned above they seem to be O deficient which might imply that the missing O has been converted into N.

In Table II we present the Y and N/H values for the ISM derived from H II regions, in three O poor galaxies, as well as the same ratios derived from Type I PN. The derived δX_N values are a factor of 2 (LMC) a factor of 1 (SMC) and a factor of 6 (NGC 6822) higher than those predicted by equation (3). Therefore PN of Type I could be responsible for the excess N in metal poor galaxies.

TABLE II. He and N abundances in extragalactic H II regions and Type I PN.

Object	Y	12+log N/H	Z (10^{-3})	δX_N (10^{-3})	δY	Source
⟨H II⟩ LMC	0.249	7.03	5.3	...	...	1,2,3
⟨Type I⟩ LMC	0.404	8.4	...	1.99	0.155	4,5,6
⟨H II⟩ SMC	0.237	6.41	1.6	...	...	2,3,7
N 67 SMC	0.427	7.6	...	0.29	0.190	5
⟨H II⟩ 6822	0.249	6.6	4.6	...	...	2
PN 6822	0.427	8.8	...	5.00	0.178	8

1. Peimbert and Torres-Peimbert 1974; 2. Lequeux *et al.* 1979; 3. Dufour *et al.* 1982; 4. Aller 1983; 5. Dufour and Killen 1977; 6. Barlow *et al.* 1983; 7. Peimbert and Torres-Peimbert 1976; 8. Dufour and Talent 1980.

2.4. PN of Types II and III in the Solar Neighborhood

From the 15 best observed PN of Types II and III of the list by Peimbert and Serrano (1980), for $t^2 = 0.00$, it is found that ⟨Y⟩ = 0.292 and that ⟨X_N⟩ = 1.3 x 10^{-3}. By adopting Y = 0.29 and X_N = 8 x 10^{-4} for the present values of the ISM we obtain ⟨δY⟩ = 0.002, a negligible value compared to that derived from Type I PN, and ⟨δX_N⟩ = 5 x 10^{-4}, a value a factor of four smaller than that derived from

Type I PN (see Table I).

We will estimate the fraction of Type I PN relative to those of Types I, II and III by two different methods. By adopting the IMF due to Serrano (e.g. Lequeux et al. 1979), it is found that 22% of the IMS are located in the 2.4 to 8 $M_\odot$ range, while from the Salpeter IMF the fraction of the IMS in that range is of 26%. It is also estimated that about 10-15% of the PN in the solar neighborhood and the Magellanic Clouds are of Type I; considering that the central stars of Type I PN evolve faster than those of other PN, the formation rate of Type I PN, relative to those of Types II and III PN, should be higher than the observed ratio. Therefore from the IMF predictions and from the observed fractions it follows that at present the rate of production of PN of type I is about 20%, or even larger. In what follows we will assume that the production rate of PN of Type I relative to those of PN of Types I, II and III is of 20%.

The envelope masses of PN of Type I are expected to be higher than those of Types II and III and in what follows we will assume that they are a factor of three higher (e.g. Renzini and Voli 1981, Dufour 1984b).

From these considerations it follows that the N enrichment per unit time due to PN of Type I (given by $\dot{M}\delta X_N$) would be about a factor of 3 larger than that due to PN of Types II and III.

If instead of $Y = 0.29$ we adopt $Y = 0.28$ for the present value in the ISM, the He enrichment per unit time due to Type I PN, $\dot{M}\delta Y$, would be a factor of 5 to 6 times larger than that due to PN of Types II and III.

3. CONCLUSIONS

i) The ΔY versus X_N diagram is consistent with the idea that most of the N is of secondary origin produced by massive stars, or produced in regions where all the H has been converted into He and all the C and O into N.

ii) In O poor galaxies there is an X_N excess for a given ΔY value. The excess could be due to: a) PN of Type I, b) to infall of material with $\Delta Y = 0$ and $X_N = 0$, or c) to primary production of N in massive stars.

iii) In the solar neighborhood the enrichment of He and N, given by δY and δX_N, due to Type I PN is similar to that produced by massive stars; therefore the ΔY, X_N diagram is consistent with a substantial contribution to the He and N enrichment of the ISM by PN of Type I.

iv) The ΔY and ΔX_N values for the solar neighborhood agree with the theoretical predictions for secondary production of N by massive stars. If there is infall with $\Delta Y = 0$ and $X_N = 0$ then there must be an additional source of He-rich material otherwise we would have detected an excess in X_N.

v) The He enrichment produced by PN of Types II and III of the solar neighborhood is negligible in comparison to that produced by Type I PN, i.e. $(\dot{M}\delta Y)_{II + III} \ll (\dot{M}\delta Y)_{I}$.

vi) The N enrichment produced by PN of Types II and III of the solar neighborhood is small in comparison to that produced by Type I PN, i.e. $(\dot{M}\delta X_N)_{II + III} \approx (\dot{M}\delta X_N)_{I}/3$.

vii) Type I PN apparently show a defficiency in O/Ne which would imply that a fraction of the O has been converted into N. Consequently that N is of secondary origin but coming from O and C instead of only from C.

REFERENCES

Aller, L.H.: 1983, Astrophys. J. **273**, 590.
Alloin, D., Collin-Souffrin, S., Joly, M. and Vigroux, L.: 1979, Astron. Astrophys. **78**, 200.
Barlow, M.J., Adams, S., Seaton, M.J., Willis, A.J., and Walker, A.R.: 1983, in D.R. Flower (ed.), "Planetary Nebulae", IAU Symp. 103, Dordrecht: Reidel, p. 538.
Dufour, R.J.: 1984a, in S. van den Bergh, K.S. de Boer (eds.), "Structure and Evolution of the Magellanic Clouds", IAU Symp. No. 108, Dordrecht: Reidel, p. 353.
Dufour, R.J.: 1984b, Astrophys. J., **287**, 341.
Dufour, R.J. and Killen, R.M.: 1977, Astrophys. J. **211**, 68.
Dufour, R.J., Shields, G.A. and Talbot, R.J. Jr.: 1982, Astrophys. J. **252**, 461.
Dufour, R.J. and Talent, D.L.: 1980, Astrophys. J. **235**, 22.
Edmunds, M.G. and Pagel, B.E.J.: 1978, Monthly Notices Roy. Astron. Soc. **185**, 77.
Kaler, J.B.: 1979, Astrophys. J. **228**, 163.
Kwitter, K.B.: 1981, Astrophys. J. **245**, 154.
Lequeux, J., Peimbert, M., Rayo, J.F., Serrano, A. and Torres-Peimbert, S.: 1979, Astron. Astrophys. **80**, 155.
Maeder, M.: 1984, in C. Chiosi and A. Renzini (eds.), "Stellar Nucleo-synthesis", Dordrecht: Reidel, p. 115.
Matteucci, F.: 1986, Monthly Notices Roy. Astron. Soc **221**, 911.
Matteucci, F. and Tosi, M.: 1985, Monthly Notices Roy. Astron. Soc. **217**, 391.
Pagel, B.E.J.: 1986, in J.P. Swings (ed.), "Highlights of Astronomy", Dordrecht: Reidel, p. 377.
Pagel, B.E.J.: 1987, paper delivered to the Paris (IAP) Conference on *Nucleosynthesis*.
Parker, R.A.R.: 1978, Astrophys. J. **224**, 873.
Peimbert, M.: 1978, in Y. Terzian (ed.), "Planetary Nebulae: Observations and Theory", IAU Symp. No. 76, Dordrecht: Reidel, p. 215.
Peimbert, M.: 1979, in W.R. Burton (ed.), "The Large-Scale Characteristics of the Galaxy", IAU Symp. No. 84, Dordrecht: Reidel, p. 307.
Peimbert, M.: 1984, in S. van den Bergh, K.S. de Boer (eds.), "Structure and Evolution of the Magellanic Clouds", IAU Symp. No. 108, Dordrecht: Reidel, p. 363.
Peimbert, M.: 1986, Publ. Astron. Soc. Pac., in press.

Peimbert, M., Peña, M. and Torres-Peimbert, S.: 1986, Astron. Astrophys.
    158, 266.
Peimbert, M. and Serrano, A.: 1980, Rev. Mexicana Astron. Astrof. **5**, 9.
Peimbert, M. and Torres-Peimbert, S.: 1974, Astrophys. J. **193**, 327.
Peimbert, M. and Torres-Peimbert, S.: 1976, Astrophys. J. **203**, 581.
Peimbert, M. and Torres-Peimbert, S.: 1977, Monthly Notices Roy. Astron.
    Soc. **179**, 217.
Peimbert, M. and Torres-Peimbert, S.: 1983, in D.R. Flower (ed.),
    "Planetary Nebulae", IAU Symp. No. 103, Dordrecht: Reidel,
    p. 233.
Peimbert, M. and Torres-Peimbert, S.: 1987, in preparation.
Pottasch, S.R.: 1986, in D. Pequignot (ed.), "Model Nebulae",
    Observatoire de Paris-Meudon, p. 102.
Renzini, A. 1984, in C. Chiosi and A. Renzini (eds.), "Stellar Nucleo-
    synthesis", Dordrecht: Reidel, p. 99.
Renzini, A. and Voli, M. 1981, Astron. Astrophys. **94**, 175.
Smith, H.E.: 1975, Astrophys. J. **199**, 591.
Talbot, R.J. and Arnett, W.D. 1973, Astrophys. J. **186**, 51.
Talent, D.L. and Dufour, R.J.: 1979, Astrophys. J. **233**, 888.
White, S.D. and Audouze, J.: 1983, Monthly Notices Roy. Astron. Soc.,
    203, 603.
Willis, A.J. and Wilson, R.: 1978, Monthly Notices Roy. Astron. Soc.,
    182, 559.

DISCUSSION

PANAGIA: About the determination of $\Delta Y/\Delta Z$ it is clear that no reliable value can be derived from data on blue dwarf galaxies (BDG) because their enrichment in He is generally lower than the observational uncertainties (generally $\geq 10\%$). On the other hand one can combine data on BDG with those of H II regions in spiral galaxies and derive both the primordial He and the slope $\Delta Y/\Delta Z$. Paolo Lenzuni and I have just started the analysis of data collected for 173 H II regions in 25 spiral galaxies plus 43 regions in BDGs. The preliminary results are that $Y_p \simeq 0.235$ and that $\Delta Y/\Delta Z \sim 3$ can explain the data for all the galaxies except NGC 5236 (M83), NGC 1566 and the Magellanic Clouds. For these a value of $\Delta Y/\Delta Z \sim 6$ seems more appropiate.

ROCHE: When you determine the abundances in PN, only the ionized material (of the order of 0.2 $M_\odot$) is observed, but before reaching that stage, the precursor star may have ejected up to 2 or 3 $M_\odot$. What do we know of the enrichment of the ISM by the precursor object?

PEIMBERT: Those objects that have lost up to 2 or 3 $M_\odot$ are Type I PN and their envelope masses might be of the order of 0.6 $M_\odot$ (e.g. Dufour 1984b). On the other hand I agree that we do not have observational information on the chemical composition of the matter lost before the AGB wind phase.

KOPPEN: From hydrodynamical models of PN we recognize that accretion from AGB wind material into the nebula proper is most important. So by analyzing the nebula we analyze material of the same composition as the dilute, invisible remnant of the AGB wind.

TEMPERATURES, LUMINOSITIES AND MASS LOSS RATES FOR PN NUCLEI

M. Grewing(1), L. Bianchi(2), M. Gutekunst(1)
(1)Astronomisches Institut der Universitaet
D-7400 Tuebingen, W.Germany
(2)Osservatorio Astronomico di Torino
I-10025 Pino Torinese, Italy

ABSTRACT. Low and high resolution ultraviolet spectra obtained with the IUE satellite offer a unique opportunity to study the properties of the nuclei of planetary nebulae (PNN) in a more direct way than has been possible in the past. While the low resolution data give the temperatures and luminosities -after proper correction of the observed fluxes for interstellar extinction-, the high resolution spectra allow to study line profiles from which one can derive e.g. wind velocities and estimates of the mass loss rates. Improved knowledge of the properties of the central stars is essential to constrain the ionization models for the surrounding nebular shells.
In this note we report results obtained from IUE and optical data for the two objects NGC 40 and NGC 6543.

1. INTRODUCTION

While for many decades optical data were the only source of information about the properties of the nuclei of planetary nebulae (PNN), satellite experiments at ultraviolet wavelengths, in particular the ANS, and more recently the IUE have significantly widened the possibilities of studying these stars, their temperatures, luminosities and detailed atmospheric properties. At the same time, greatly improved optical detectors have become available, allowing high resolution spectroscopy and deep imaging with large dynamic range of both the central stars and the surrounding nebulae, including their outermost (halo) shells.

Here we shall confine the discussion largely to results obtained from low and high resolution ultraviolet spectra taken with the IUE satellite. From the absolutely calibrated low resolution data, after their proper correction for interstellar extinction, one can derive temperatures and luminosities for the central stars, i.e. parameters which can directly be compared to evolutionary tracks, e.g. those of Schoenberner (1983 and earlier references cited therein). High resolution spectra, on the other hand, give detailed information on emission and absorption line profiles, of which those indicative of stellar winds (P Cygni lines) are particularly important. While these lines can also be recognised in low-resolution spectra, the high resolution is needed to determine quantitatively the terminal wind velocities and the mass loss rates.

2. NGC 40

In Fig. 1a we show the de-reddened, absolutely calibrated low-resolution spectrum of the nucleus of NGC 40 (m_v=11.8). The observations have been corrected for an E(B-V)=0.5 (Bianchi and Grewing 1986), assuming a standard extinction law as determined by Mathis and Savage (1979). The curve drawn through the observations corresponds to a pure-Helium atmosphere model of T=90.000 K (Wesemael 1980). To fit the absolute flux observed, the star's radius must then be R=0.66R$_\odot$ for an assumed distance of NGC 40 of 980pc. The corresponding luminosity would be log L/L$_\odot$=4.4. This places the object in the upper left part of the region in the H-R-diagram that is occupied by the central stars of PNe. As argued by Bianchi and Grewing (1986), the current mass of the nucleus must be about 1.1±0.1 M$_\odot$. Such a heavy nucleus can only have arisen from a progenitor with an original mass M>6 M$_\odot$ (cf. e.g. Bertelli et al. 1986).

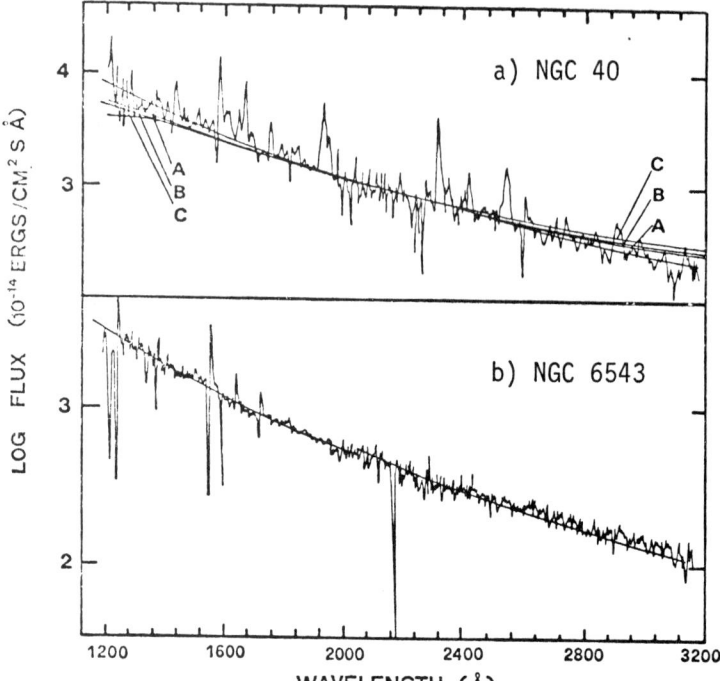

Fig.1: IUE low resolution spectra of the nuclei of the low and medium excitation nebulae NGC 40 and NGC 6543, respectively. The unlabbeled curves refer to pure-Helium model atmosphere energy distributions of 90.000 and 80.000 K. The lettering A,B,C refers to further models discussed by Bianchi and Grewing (1986).

The temperature of 90.000 K contrasts sharply with the observation that NGC 40 is a low excitation nebula. This well known fact is illustrated in Fig.2 which shows the [O III]500.7/H-beta-line ratio which with the exception of a very small region nowhere exceeds a value of 1.0 whereas for higher excitation nebulae it is of order 10. This isophote-ratio map was derived by Gutekunst (1986) from direct images obtained through narrow band interference filters in front of the photon-counting imaging detector developed at the AIT (e.g. Barnstedt 1985). The detector was placed at the Cassegrain focus of the 1.8m telescope of the Asiago Observatory. The images were absolutely calibrated with the help of the absolutely calibrated spectra published by Clegg et al. (1983).

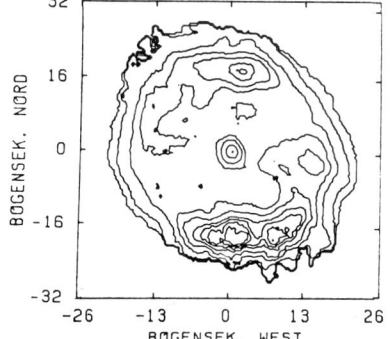
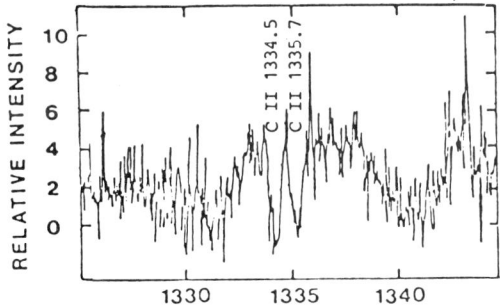

Fig.2: Flux ratio [O III]500.7 to H-beta. Contours are at 0.01, 0.1, 0.25, 0.38, 0.5, 0.65, 0.75, 0.85, and 0.95.

Fig.3: Section from the high resolution IUE spectrum showing the narrow C II absorption superimposed on the much wider stellar features.

A clue as to why the nebula shows such low exciation characteristics might come from the high resolution IUE spectrum. In Fig.3 we show a section from the short-wavelength spectrum in the neighborhood of the C II 133.5nm resonance line. There is clear evidence for fairly narrow absorption components superimposed onto the broader emission and absorption structures. These absorptions arise from the $^2P_{1/2}$ and $^2P_{3/2}$ level, respectively. The equivalent widths of the two lines are 0.676 and 0.469 Å. These numbers translate for a wide range of b-values to a C II column density of $(7\pm3)10^{18}$ cm^{-2}. If located between the central object and the nebula, this column of C II would provide enough shielding at intermediate photon energies to account for the observed nebular excitation as shown in a detailed ionization model by Gutekunst (1986). Supporting evidence for such model comes from the fact that the radial velocity of the absorption components corresponds to the expansion velocity of the nebular shell, and that the electron density in the region where the lines arise must be in the order of 5000±3000 cm^{-3} from the ratio of the C II ground-state fine-structure level population. Such a density which is definitely higher than the density within the nebula could arise at the shock interface between the fast wind from the central star and the nebular shells. Evidence

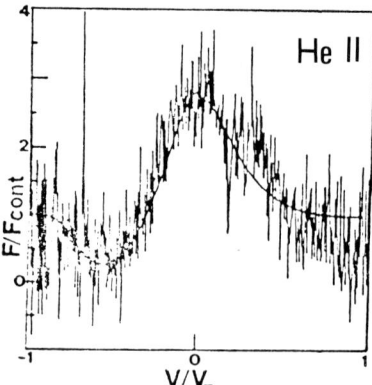

for such a wind is shown in Fig.4 where we have plotted the observed He II line profile from the high resolution spectrum of NGC 40 together with a theoretical line fit. The wind from the current nucleus has a terminal velocity of $v_\infty = -1800$ km/s.

Fig.4: Section from the high resolution IUE spectrum of NGC 40 showing a P Cygni type line profile of He II. Superimposed is a theoretical fit to the observation.

In Table 1 we show results from model calculations for NGC 40 which are based on two different sets of parameters for the central star and nebular abundances as specified in the Table.

Table 1
Comparison of observed and calculated line intensities

Transition		Obs. Intens.[1]	T=38000 K [2] L=250 L☉ OSA	FRT	T=90000K L=25000 L☉ +carbon curtain
H-β	4861	100.0 !	100	100	100
He I	4471	2.29	2.2	2.2	0.4
	5876	6.17	6.1	6.0	1.3
C II	2326	60.3	1.2	10.1	44.2
III	1908	55.0	0.1	0.6	58.1
IV	1549	25.7	0.0..	0.0..	0.1
N I	5198/200	0.62	0.7	0.7	0.0..
II	6548	81.3	71.0	67.1	103.1
	6584	257.1	209.2	197.4	303.5
III	1749	9.71	0.0..	0.0..	0.0..
O I	6300	2.63	0.8	1.1	0.0..
	6363	0.87	0.3	0.4	0.0..
II	3726/29	436.5	284.6	330.0	481.4
III	4959	5.89	7.6	8.5	10.7
	5007	18.2	21.9	24.5	30.8
S II	6718	8.31	4.9	4.8	3.3
	6731	12.6	6.2	6.0	4.1

----

1) from Clegg et al. (1983)

2) the following abundances had to be assumed:

	H	He	C	N	O	S
OSA model:	5(-2)	5(-2)	1(-4)	2(-4)	1.5(-3)	2(-6)
FRT[+] model:	"	"	1(-3)	"	2.0(-3)	"
mean PNe:	1.1(-1)	1.1(-1)	7.1(-4)	1.2(-4)	3.8(-4)	1.0(-5)

+)FRT=full radiation transfer model for the nebular shells

Table 1 shows that the observed emission line strengths can roughly be accounted for if one assumes a low temperature ionizing star and at the same time chemical abundances which differ significantly from those found normally in planetary nebulae. Assuming a very hot central star of high luminosity and at the same time a 'carbon curtain' with the properties as discussed above, i.e. a C II column density of about 10^{19} cm^{-2}, one can reproduce the observations equally well without deviating from the mean abundances. Obviously, further studies are needed to arrive at a fully self-consistent model for NGC 40.

3. NGC 6543

A wind of similar magnitude ($v_\infty = -1900$ km/s) is also found in NGC 6543 as evidenced by Fig.5. The nucleus of this medium excitation nebula shows an UV-continuum very similar to that of NGC 40 as can be seen from Fig.1b where we have corrected the observed IUE spectrum for an

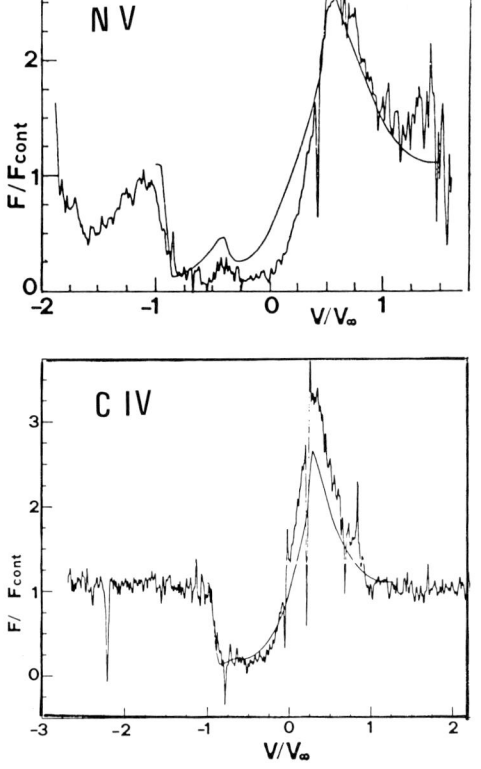

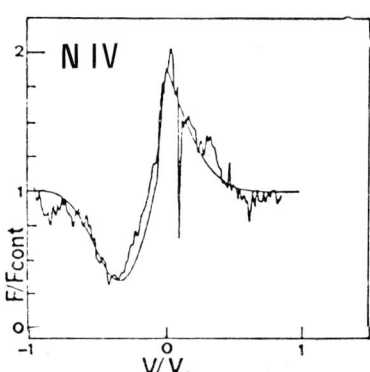

Fig.5: Sections from the high resolution IUE spectrum of NGC 6543 showing the P Cygni profiles observed in N V, N IV, and C IV. Superimposed are theoretical fits (for details see Bianchi et al. 1986).

extinction of E(B-V)=0.08. The best fit to the de-reddened spectrum is obtained for a pure-Helium atmosphere model with T_{eff}=80.000±10.000 K (Bianchi et al. 1986).

Fitting the P Cygni profiles of the resonance doublets of N V, Si IV, and C IV and from other transitions of He II, O IV, and O V, by applying the method of Olson (1982), and/or the method of Surdej (1982), we can determine quantitatively the mass loss rate for the current nucleus of NGC 6543. A consistent solution is found from all lines considered for $M=3.2 \cdot 10^{-7}$ M$_\odot$/yr. A detailed discussion of this object has been given by Bianchi et al.(1986).

REFERENCES

Barnstedt,J. 1985,Thesis University of Tuebingen
Bertelli, G.,Bressan,A.,Chiosi,C.,Angerer,K. 1986,Astron.Astrophys.(in press)
Bianchi,L.,Cerrato,S.,Grewing,M. 1986,Astron.Astrophys.(in press)
Bianchi,L., Grewing,M. 1986,Astron.Astrophys.(submitted)
Clegg,R.,Seaton,M.,Peimbert,M.,Torres-Peimbert,S. 1983, M.N.R.A.S. 205,417
Gutekunst,M. 1986,Thesis University of Tuebingen
Mathis,J., Savage,B.D. 1979,Ann.Rev.Astron.Astrophys. 17,173
Schoenberner, D. 1983, Astrophys.J. 272, 708.
Wesemael,F. 1981,Astrophys.J.Suppl.Ser. 45,2

DISCUSSION

<u>Perinotto:</u> Apparently you used a T_e of about 20.000 K which seems a bit low relative to the high T_{rad} that you adopt. Which may be the results if you use a T_e about 4 times higher ?

<u>Bianchi:</u> The graph which I showed was computed for T_e=30.000 K. In the approximation which is used to calculate the ionization equilibrium, T_e enters only weakly (via the recombination coefficient).

<u>Clegg:</u> At the London PN meeting, Bignell reported a search for radio emission from the NGC 40 wind, using the VLA. Do you know if a detection or an interesting upper limit was obtained ?

<u>Bianchi:</u> No, we did not yet see anything appear in the literature.

AN OUTFLOW MODEL FOR BIPOLAR PLANETARY NEBULAE AND THE CASE OF NGC 6302

G. Silvestro and M. Robberto
Istituto di Fisica Generale, Università di Torino
Torino, Italy

1. INTRODUCTION

NGC 6302 is a bright, bipolar planetary nebula (PN) that exhibits a wide range of excitation and ionization conditions, together with complex structure and kinematics. High-velocity (>300 Km s^{-1}) flows were detected in the nebula, which are believed to originate as a wind from the central star (Elliot and Meaburn, 1977). A very high effective temperature $\sim 2 \times 10^5$K is estimated (Rodriguez et al., 1985) for the central star, which is heavily obscured from view, being located within a neutral filament in the nebular core.

Considerable uncertainty exists on the source distance. D. Meaburn and Walsh (1980a) obtained D<150 pc by interpreting the Ne V 3426 Å line as emission from a radiatively ionized stellar wind. Rodriguez et al. (1985), by combining radio and optical data, estimate D$\sim$2 kpc. Altschuler et al. (1986) derive a radio distance D$\sim$0.5 kpc, following a method suggested by Milne (1982) for young PNs that are optically thick to the Lyα radiation.

NGC 6302 is a strong continuum radio source ($S_\nu \sim 3.5$ Jy at $\lambda=6$ cm). Rodriguez et al. (1985) found that most of the radio emission is concentrated into the central 10 arcsec, while the optical nebula has an extent of almost 5'. Lester and Dinerstein (1984) detected an infrared disk with its long axis perpendicular to the long axis of the bipolar structure. Near-infrared spectroscopy (Phillips et al., 1983) shows evidence for emission in the S(1) v=1→0 vibrational line of the H$_2$ molecule. The emission turns out to be very strongly concentrated towards the inner region of the nebula ($\lesssim 10"$ angular diameter). An image (Meaburn, 1978) of the central region in [SII] 6716,6731 Å lines shows the presence of two hyperbolic arcs, extending 30" to 1' from the centre (Fig. 1).

Meaburn and Walsh (1980 b) obtained evidence for complex flows in the outer parts of the nebula, which they interpreted within the model (Cantò, 1980) of a stellar wind from a star embedded in a dense disk, forming a cavity with flowing walls, which are defined by standing shocks (see Section 2). In this interpretation, the hyperbolic [SII] arcs would delineate the boundary of the flow. We notice these [SII] lines are typical of emission seen in the Herbig-Haro objects, and have been interpreted as originated in the cooling regions of high-velocity shock waves (Schwartz, 1981; Brugel et al., 1984).

We shall present (Section 3) a numerical analysis of the morphology of the inner outflow region, based on a generalization of the equations given by Cantò (1980) and Barral and Cantò (1981). Section 4 gives a discussion of the interpretation of NGC 6302 within a model of disk-confined stellar wind. Finally, Section 5 summarizes the results.

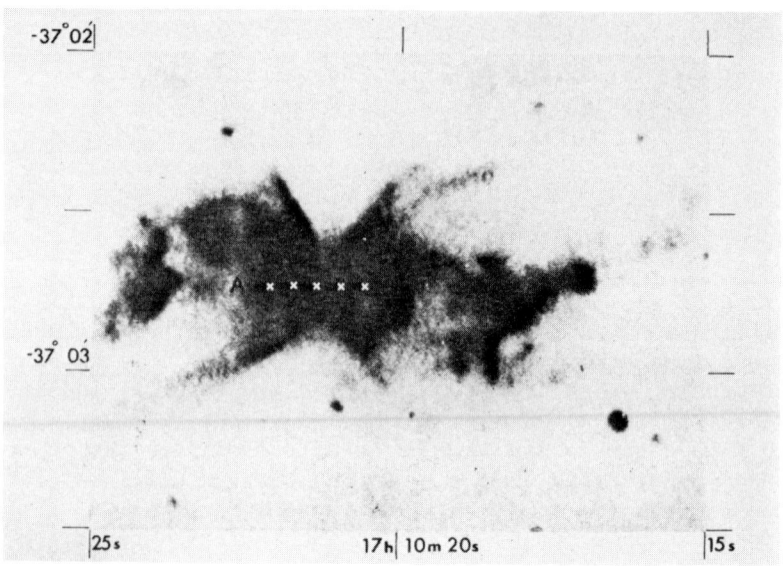

Fig. 1. An [S II] SRC Schmidt photograph of the core of NGC 6302. (Reproduced from Meaburn and Walsh, 1980a)

2. THE BARRAL AND CANTO' MODEL

The isotropic stellar wind from a star embedded in an anisotropic medium creates a bipolar cavity in the surrounding cloud. The stellar wind suffers an oblique shock and is refracted, sliding along the walls of the cavity. In a steady-state configuration, the locus of the shock is given by the equation

$$\rho_s v_{*n}^2 = P - P_c \qquad (1)$$

where: ρ_s is the wind density immediately before the shock, $v_{*n} = v_* \sin\alpha$ is the normal component of the wind velocity, P the ambient pressure outside, and the centrifugal correction P_c (= momentum flux per unit depth times the shock front curvature) accounts for the decrease in external pressure due to the curvature of the shock front (Hayes and Probstein, 1966).

Barral and Cantò (1981) modeled the source NGC 6302 assuming an isotropic wind which expands in an infinite and isotropic disk of material in gravitational equilibrium, with a pressure distribution (Spitzer, 1978)

$$P(z) = P_o \text{sech}^2 (z/\Lambda),$$

the disk being surrounded by a more tenuous medium with uniform pressure. An adimensional solution is given of Eq.(1), in which the geometry of the cavity is specified by one parameter, the ratio $\lambda=\Lambda/R_0$ (R_0: radius of the shock at $\Theta = 0$). Characteristic feature of the model: each lobe has an "ovoid" configuration ending in an acute tip, since the annular flow of shocked wind is always forced by outside pressure to turn towards the nebular axis.

3. NUMERICAL ANALYSIS OF THE OUTFLOWING LOBE

We have been developing a two-dimensional numerical code, which computes the locus of the time-stationary shock surface for arbitrary orientations of the axis of the outflow, estimates for each transition the line emissivity and computes the line structure at each point of a (90x180) array, with steps of 1° in latitude and 2° in azimuth. We evaluated with such code position-velocity maps for high-velocity molecular outflows, which appear to be in good agreement with observational data for the source Mon R2 (Silvestro and Robberto, 1987).

The results of our computations for the source NGC 6302 are presented in Figs. 2 and 3.

4. RESULTS AND DISCUSSION

Our main purpose is to see whether the morphological features of the inner flow region, namely a hyperbolic lobe profile, as delineated by arcs seen in the [SII] emission lines (Fig. 1), can allow to put some constraints on the hydrodynamic flux and the confinement mechanism. The Barral and Cantò (1981) model cannot account for the observed profile, due to the restrictive conditions imposed on the analytic law for density distribution within the disk and the ambient medium. Our analysis (cfr. Figs. 2 and 3) allows to see that a hyperbolic profile can be obtained as a result of two distinct confinement processes:

(a) a 'local' one by a dense circumstellar disk, which influences the cone opening angle simply through its depth λ. Fig. 2 shows the profiles obtained for $\lambda=0.1$ and 0.5, using density distributions $n(z) \sim \text{sech}^2(z/\Lambda)$ (self-gravitating disk), and $n(z) \sim \exp(-z^2/\Lambda^2)$, with ambient medium density equal to zero. One can see the opening angle does not, for a given λ, depend critically on the density distribution law. For modeling our source we choose $\lambda=0.1$, a value (1/5) that previously assumed (Silvestro and Robberto, 1987) for outflows in molecular clouds: it is of course reasonable that a PN disk be less conspicuous than the accretion disk in a protostellar system.

(b) a second confinement process that tilts the profile towards the axis of the outflow must be expected at large distances from the star, due to the influence of the ambient medium. Barral and Cantò (1981) pointed out that any homogeneous distribution of material will cause convergence of the beam. The curves in Fig. 3, which were obtained assuming $n = 0$ when $R<R_1 (=4 R_0)$, and at distances $R \geq R_1$ a tenuous ambient medium with constant density $n = 0.0010\ n_0$, $0.0025\ n_0$, and $0.0040\ n_0$, where n_0

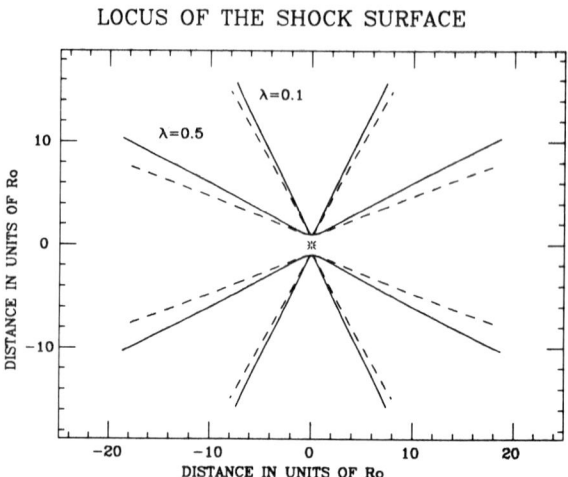

Fig. 2. Locus of the shock surface for two different values of λ (see text), and two density distributions in the disk: (a) gaussian (solid line); (b) self-gravitating disk (dashed line). The ambient density is zero for all cases.

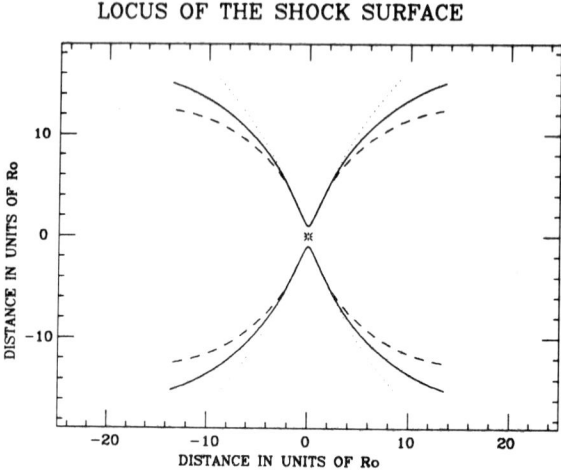

Fig. 3. Locus of the shock surface for a gaussian density distribution ($\lambda=0.1$) within the disk, and different constant ambient densities at $R \geq R_1 (=4\ R_0)$: (a) $n=0.0010\ n_0$ (dotted line); (b) $n=0.0025\ n_0$ (solid line); (c) $n=0.0040\ n_0$ (dashed line). The ambient density is $n=0$ when $R<R_1$.

is the maximum density in the disk (such matter distribution could be due to an earlier phase of envelope ejection), display a suggestive similarity with the observed morphology.

5. CONCLUSIONS

Our numerical analysis allowed us to see that an outflow model, based on less restrictive analytic conditions than those of Barral and Cantò, gives results in satisfactory agreement with the morphology of the inner lobe of the bipolar planetary nebula NGC 6302. On the other hand, our results do not give a good description of the outer part of the nebula, which, as already pointed out by Barral et al. (1982), is exceptionally complex. Further study is therefore needed in order to obtain the physical conditions in the region and to determine whether the flow is confined along the whole extent of the nebula. Our results were limited to a time-stationary situation, we are presently developing a time-dependent code which could allow more detailed comparison with observations.

REFERENCES

Altschuler, D.R., Schneider, S.E., Giovanardi, C. and Silvergate, P.R., 1986. Astrophys. J., 305, L85.
Barral, J.F. and Cantò, J., 1981. Rev. Mex. Astr. Astrof., 5, 101.
Barral, J.F., Cantò, J., Meaburn, J. and Walsh, J.R., 1982. Mon. Not. R. astr. Soc., 199, 817.
Brugel, E.W., Mundt, R. and Buhrke, T., 1984. Astrophys. J., 287, L73.
Cantò, J., 1980. Astr. Astrophys., 86, 327.
Elliot, K.H. and Meaburn, J., 1977. Mon. Not. R. astr. Soc., 181, 499.
Hayes, W. and Probstein, R., 1966. Hypersonic Flow Theory (Academic Press, New York), p. 133.
Lester, D.F. and Dinerstein, H.L., 1984. Astrophys. J., 281, L67.
Meaburn, J., 1978. Appl. Opt., 17, 1271.
Meaburn, J. and Walsh, J.R., 1980a. Mon. Not. R. astr. Soc., 191, 5P.
Meaburn, J. and Walsh, J.R., 1980b. Mon. Not. R. astr. Soc., 193, 631.
Milne, D.K., 1982. Mon. Not. R. astr. Soc., 200, 51P.
Phillips, J.P., Reay, N.K. and White, G.J., 1983. Mon. Not. R. astr. Soc., 203, 977.
Rodriguez, L.F., Garcia-Barreto, J.A., Cantò, J., Moreno, M.A., Torres-Peimbert, S., Costero, R., Serrano, A., Moran, J.M. and Garay, G., 1985. Mon. Not. R. astr. Soc., 215, 353.
Schwartz, R.D., 1981. Astrophys. J., 243, 197.
Silvestro, G. and Robberto, M., 1987. In IAU Symposium N. 122 "Circumstellar Matter", Heidelberg.
Spitzer, L., 1978. Physical Processes in the Interstellar Medium (Wiley, New York).

DISCUSSION

Danziger: D. Baade and I have made velocity maps covering a major portion of NGC 6302. In the inner region (<1 arcminute) that you discuss, we find evidence for what could be interpreted as flow outwards along a conical surface aligned with the major axis. Further, what is the evidence from H_2 observations that the gas is shocked? We do not believe that the [S II]/Hα ratio observed over the surface of NGC 6302 gives convincing evidence of shock excitation.

Silvestro: Measurements of relative intensities of $S(1)v=1 \rightarrow 0$ and Q-branch vibrational lines of molecular hydrogen might allow to discriminate between shock and UV radiative excitation. Such data are at present not available for NGC 6302.

Anon.: There is a simple way to distinguish as whether [S II] lines are excited by shocks, or to exclude that?

Danziger: Apart from [S II] there are not any very clear cut criteria in the optical region (if O III 5007/ O III 4363 gave a very high temperature $\gtrsim$ 50 000 K it might indicate shock heating). However, in the IR region relative strength of key molecular hydrogen lines can in principle provide clear cut discrimination between shock and UV radiation excitation.

Viotti: There is any evidence of binarity?

Silvestro: There is no such evidence for NGC 6302.

MASS LOSS AND THE TRANSITION OF AN AGB STAR
TO A CENTRAL STAR OF A PLANETARY NEBULA

Detlef Schönberner
Institut für Theoretische Physik
und Sternwarte der Universität
Olshausenstraße 40
2300 Kiel, F.R.G.

SUMMARY. The evolution of hydrogen-burning post-AGB stars in thermal equilibrium is controlled by nuclear burning and mass loss. It is independent of earlier mass-loss episodes since, for a given core mass M_H, the effective temperature is only a function of the actual envelope mass M_e. Helium-burning post-AGB stars behave differently. Their evolution depends on earlier mass-loss episodes, and no unique relation between T_{eff} and M_e exists.
A strong stellar wind ("superwind") at the tip of the AGB is necessary to create a nebular shell and to force the remnant to contract to a post-AGB star within, say, about 10^3 yr. If, however, this superwind is too strong, only helium-burning remnants will be created during the luminosity peaks of thermal pulses.

1. INTRODUCTION

It is now well established that post-asymptotic giant branch (PAGB) stars can account for the observed temporal evolution of central stars of planetary nebulae (CPN). Two major modes of PAGB evolution to the white dwarf stage are possible, according to the two main phases of an thermally pulsing AGB star: the hydrogen-burning or helium-burning mode. If PN formation occurs during the quiescent hydrogen-burning phase on the AGB, the remnant continues to burn mainly hydrogen on its way to becoming a white dwarf. The computations now show that a hydrogen-burning PAGB model evolves about 3 times faster through the region occupied by the central stars than a helium-burning model of the same mass. Using hydrogen-burning models, Schönberner (1981), Schönberner and Weidemann (1981), and Schönberner (1984) demonstrated that the temporal evolution of central stars can be very well explained by models with masses between 0.55 and 0.64 $M_\odot$. The conclusion then follows that obviously the PN ejection is generally not initiated by a thermal pulse, but occurs during the quiescent hydrogen-burning phase on the AGB.
   Assuming a typical CPN mass of 0.6 $M_\odot$ and a typical PN lifetime of 3.10^4 yr, it can be predicted in the case of hydrogen burning that most

CPN are intrinsically faint with $L \lesssim 10^2$ $L_\odot$ or $M_v \gtrsim 5$. If, on the other hand, the PN formation, i.e. the removal of the stellar envelope by mass loss, happens during a luminosity peak that follows a thermal pulse of the helium-burning shell, the remnant leaves the AGB while still burning helium as the main energy supplier (Härm and Schwarzschild 1975, Iben 1984). In this case of helium burning, we expect, for a typical PN of 0.6 $M_\odot$ and a typical PN lifetime of 3.10^4 yr, no CPN with $L \lesssim 10^3$ $L_\odot$, or $M_v \gtrsim 4$. Since the observations indicate that most CPN are in the hydrogen-burning mode (cf. also Schönberner, this volume), we concentrate in the following mainly on this evolutionary state.

2. TERMINATION OF THE AGB EVOLUTION

The evolution along the AGB is terminated if the envelope mass M_e becomes very small by the combined effect of nuclear burning in the hydrogen-burning shell and mass loss from the surface. The transition from an AGB star to a white dwarf can be split into two steps:

i) If M_e is only of the order of several percent of the total mass, the envelope starts to shrink, but is still able to release enough gravitational energy as to maintain the burning temperatures at its base. Consequently, the luminosity stays about constant, and the star evolves horizontally across the HR diagram. The core evolution is still independent from that of the envelope.

ii) If $M_e/M_\odot$ becomes about 10^{-4}, the hydrogen-burning shell starts to cool, the luminosity drops rapidly and the star enters the WD regime.

Fig. 1 shows the variation of the envelope mass M_e with effective temperature for different PAGB models as given by Schönberner (1983). Note that in these evolutionary phases the stellar core M_H practically equals the total stellar mass M because of the smallness of M_e ($M = M_e + M_H$). For a given M_H, a unique relation $T_{eff}(M_e)$ for the horizontal evolution from the AGB till the turn-around point at $T_{eff} \approx 10^5$ K exists. The shapes of the relations $T_{eff}(M_e)$ are similar, but M_e increases with decreasing M_H. Similar relations for a larger range of M_H are given in Paczynski (1971).

The timescale for the crossing of the HR diagram is determined by the total amount of the available fuel M_e and the total fuel consumption $\dot{M}_e = -(\dot{M}_H + \dot{M}_W)$, with $\dot{M}_W$ being the mass loss from the surface and $\dot{M}_H$ the nuclear fuel consumption. The function $T_{eff}(M_e)$ determines the variation of the speed $\dot{T}_{eff}$ during the horizontal evolution:

$$\dot{T}_{eff} = (dT_{eff}/dM_e)\,\dot{M}_e = -(\dot{M}_H + \dot{M}_W)\,dT_{eff}/dM_e.$$

A full discussion of the crossing times for different masses M_H is given in Schönberner (1987).

THE TRANSITION OF AN AGB STAR TO A CENTRAL STAR

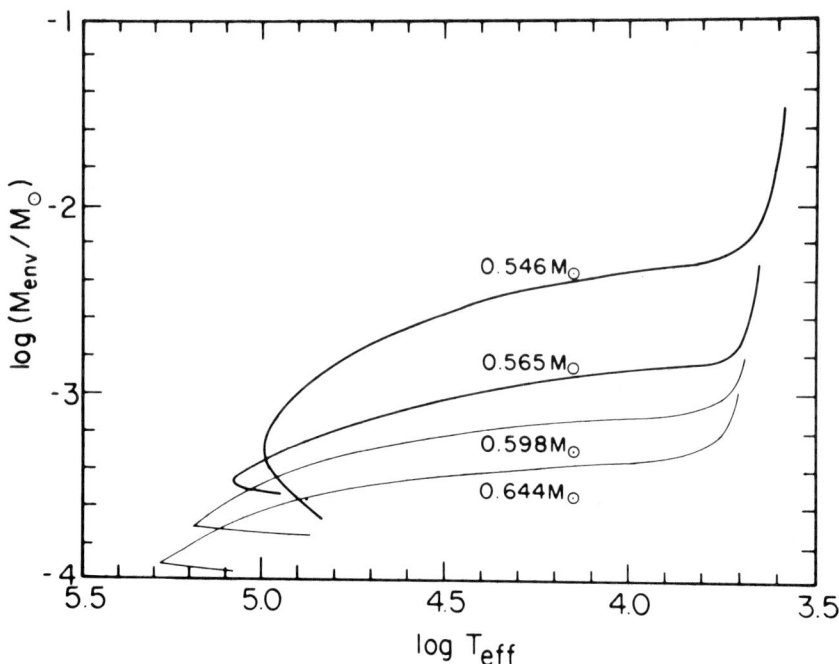

Fig. 1: Envelope mass M_e vs. T_{eff} for PAGB models of different core masses M_H according to Schönberner (1983).

If we describe the mass loss on the AGB by a Reimers' type law, we have $\dot{M}_W \sim L^{1.7} M^{-1}$ (Reimers 1975, Schönberner 1979), whereas $\dot{M}_H \sim L$. Consequently, mass loss will dominate at large luminosities. Numerical estimates show that $\dot{M}_H \approx \dot{M}_W$ is already reached for $L \approx 10^3 L_\odot$ ($\eta = 1$). We conclude that mass loss reduces the envelope on the luminous AGB ($L > 10^3 L_\odot$) more effectivly than nuclear burning. With other words, mass loss terminates the AGB evolution and reduces the maximum possible luminosity for a given initial mass.

The effect on the evolution is as follows: as long as the envelope is massive enough, the evolution along the AGB is not affected since it is determined by the core mass M_H. Of course, the envelope starts to shrink if M_e falls below a certain small value, as discussed above. Mass loss, however, speeds up the horizontal evolution, compared to the case of $\dot{M}_W = 0$, and even determines this speed if $\dot{M}_W \gtrsim \dot{M}_H$. Please note that these considerations are only valid for hydrogen-burning PAGB models in thermal equilibrium. In this case the evolutionary speed depends only on the actual value of $\dot{M}_W$ and is independent of the earlier mass loss history, as long as the thermal equilibrium has not been destroyed. This fact allows the modelling of CPN evolution by means of hydrogen-burning PAGB stars without knowing the details of the mass loss phases which lead to the PN formation at the tip of the AGB, provided that the PN-formation time is small compared to the PN lifetime. Another fact also deserves attention: heavy mass loss cannot completely remove the hydrogen-rich envelope as long as hydrogen is still burning. Instead, the star shrinks more quickly, and the time Δt elapsed between two positions along the horizontal track is given by

$$\Delta t = \Delta M_e / (\dot{M}_H + \dot{M}_W) \approx \Delta M_e / \dot{M}_W \quad \text{for} \quad \dot{M}_W \gg \dot{M}_H.$$

It can be estimated from the the observations of young PN that mass loss rates not much less than 10^{-4} $M_\odot \, yr^{-1}$, lasting for some 10^3 yr, are necessary to create a typical PN of about 0.2 $M_\odot$ (Renzini 1981). Such rates are about two orders of magnitudes larger than what one would expect from the application of Reimers' formula (for $\eta = 1$). and it is what we now call the "superwind" (Renzini 1981). A mass loss of about 10^{-4} $M_\odot \, yr^{-1}$ leads to a somewhat 3 orders of magnitudes larger envelope depletion than nuclear burning, and the horizontal speed is increased accordingly. The transition from the tip of the AGB to 5000 K (log T_{eff} = 3.7) is shortened by the same amount from about $5 \cdot 10^5$ yr to about 10^3 yr. The superwind forces the envelope to shrink within the same time until the physical mechanisms that lead to the enhanced mass loss are shut off.

We can, however, safely assume that some sort of mass loss will persist, and its order of magnitude can certainly be estimated by Reimers' law. Such an estimation shows that $\dot{M}_W$ dominates over $\dot{M}_H$ as long as $T_{eff} \lesssim 10^4$ K, and the contraction is speeded up accordingly (cf. Fig. 1 in Schönberner 1983). In the CPN region, i.e. for $T_{eff} \gtrsim 30000$ K (or log $T_{eff} \gtrsim 4.5$), we have $\dot{M}_W \lesssim 0.1 \dot{M}_H$, and the evolution is practically only controlled by the nuclear term.

Of importance for the PAGB evolution is the question of when the "superwind" stops. In Schönberner (1983) it has been assumed this happens between $10^{3.7}$ K and $10^{3.75}$ K, and this instant can then be considered as the beginning of the PAGB phase of evolution. We conclude from Fig. 1 that this is an essential assumption, otherwise too much envelope mass would remain and consequently the model would spend too much time near the AGB ("lazy" remnants according to Renzini 1981; Iben and Renzini 1983). Now the hydrodynamical calculations of Tuchman et al. (1979) show that the pulsational instability ceases only when all but 0.001 $M_\odot$ of the envelope has been ejected. A similar result was obtained earlier by Härm and Schwarzschild (1975). For the core masses involved (≈ 0.6 $M_\odot$), this residual envelope mass corresponds roughly to $10^{3.7}$ K (Fig. 1). But, as can be also seen from Figure 1, at this temperature a change in the behavior of PAGB models occurs: below, their effective temperature are only weakly dependent on envelope masses, whereas above this temperature the opposite is true. The location of this transition region does not change for somewhat higher (core) masses (Fig. 1). Therefore, we suggest that any mechanism which leads to enhanced mass loss remains in effect as long as no drastic changes in the envelope structure occur. Because this mass loss "pushes" the star toward higher effective temperatures by reducing its envelope mass, it will then continue until the star reaches a configuration which leads to a rapid shrinking of the envelope and correspondingly to a disruption of the physical mechanism responsible for the "superwind". In this scenario there is no room for "lazy" remnants.

In conclusion we argue that, although we do not know exactly the moment when a "superwind" stops, it will most likely not be before the envelope is reduced to 10^{-3} $M_\odot$. Then any residual "normal" wind, together with hydrogen burning, will be efficient enough to provide a sufficiently fast transition such as to convert an AGB remnant into a CPN with an observable PN.

3. THERMAL PULSES AND MASS LOSS

Finally, we would like to discuss the importance of thermal pulses. It was originally thought that the short-lived luminosity peak during a thermal pulse triggers an (pulsational) instability which leads to the loss of the whole envelope (cf. Härm and Schwarzschild 1975). The computations show that the remaining envelope contracts very rapidly on a thermal timescale because the star is far out of thermal equilibrium. Later on, the evolution slows down when thermal equilibrium is gradually restored, but the whole evolution through the CPN region is about 3 times slower when compared to that of a hydrogen-burning model of the same mass. Several important facts should be noted:

i) The remnant envelope mass depends on the details how the mass is removed (i.e. on the mass loss rate and on its duration), and this envelope mass M_e remains constant during the following helium-burning phase, so that, contrary to the case of hydrogen burning, the location in the HR diagram is not uniquely related to M_e.

ii) Despite the fact that M_e can be smaller than for a corresponding hydrogen-burning remnant, it cannot be removed completely, as one might think at a first glance. It is, however, possible that a hot stellar wind removes this small hydrogen-rich envelope during the helium-burning phase and before hydrogen is re-ignited. A thorough discussion of the properties of helium-burning PAGB models can be found in Iben (1984).

The problem that we encounter with the occurrence of thermal pulses is as follows: It is normally assumed that the "superwind" sets in if the star crosses a certain threshold luminosity L_{PN} (cf. Tuchman, Sack and Barkat 1979). This will certainly happen first during the luminosity peak of a thermal pulse, and with mass-loss rates of 10^{-4} $M_\odot yr^{-1}$ or more, the outcome will be a helium-burning CPN. Since, however, most CPN appear to be hydrogen burners (cf. chapter 1), we tried to investigate this problem more closely by a numerical experiment.

We assumed on the AGB a PN-ejection luminosity L_{PN} as proposed by Tuchman, Sack and Barkat (1979), above which a "superwind" sets in, with a rate yet to be specified, and below which only an "ordinary" wind according to Reimers' (1975) formula with $\eta = 1$ is in effect. We evolved a 1 $M_\odot$ model along the AGB (for the moment without any mass loss) until it crossed L_{PN} during the luminosity peak of the 3rd thermal pulse. Then a "superwind" rate of $2\ 10^4$ $M_\odot yr^{-1}$ was switched on. The situation is illustrated in the (L, M)-diagram of Fig. 2. As

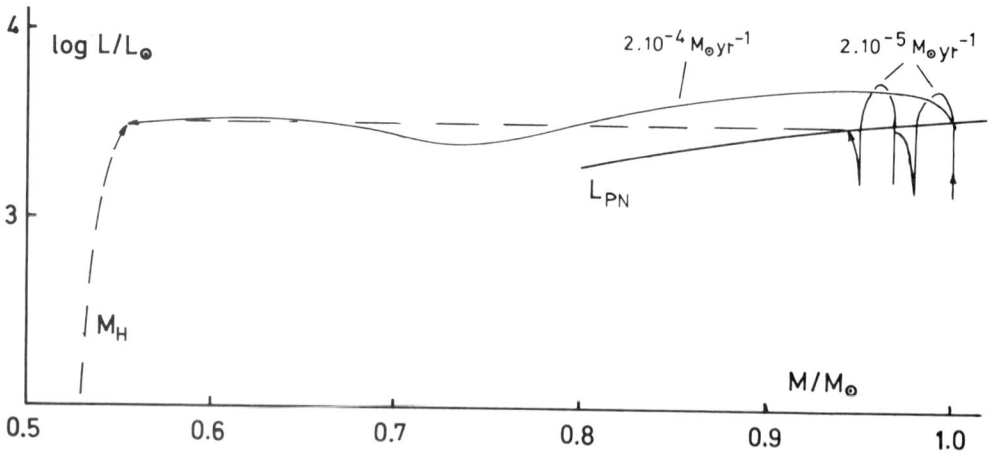

Fig. 2: Luminosity vs. mass for two mass loosing AGB models. For $L \geq L_{PN}$, the mass loss rates are $\dot{M}_W = 2\,10^{-4}$ and $2\,10^{-5}$ $M_\odot\,yr^{-1}$, for $L < L_{PN}$, Reimers formula is used ($\eta = 1$). The evolution of the core mass M_H is also shown (dashed line).

expected from earlier experiments, almost the whole envelope was removed within $2.2\,10^3$ yr, and a helium-burning PAGB star of 0.555 $M_\odot$ with an hydrogen-rich envelope of $M_e = 1.2\,10^{-3}\,M_\odot$ was created. Lowering the "superwind" mass loss rate to $6.7\,10^{-5}\,M_\odot\,yr^{-1}$ did not change the situation qualitatively, and the result was still a helium-burning CPN. (not shown in Fig. 2). At a rate of only $2\,10^{-5}\,M_\odot\,yr^{-1}$, however, the model stayed only 10^3 yr above L_{PN}, lost 0.02 $M_\odot$ and then returned into the quiescent hydrogen-burning state. The same business started again during the 4th thermal pulse. But later, after the model's return into the quiescent hydrogen-burning state, it reached the threshold luminosity L_{PN} while still burning hydrogen, thereby giving birth to a hydrogen-burning remnant.

This numerical experiment shows clearly that an AGB star, which is far out of thermal equilibrium as a consequence of a thermal pulse, reacts quite sensitively on the mass-loss rate. In the present case of a 1 $M_\odot$ AGB model with an envelope mass $M_e = 0.45\,M_\odot$, mass loss rates in excess of about $5\,10^{-5}\,M_\odot\,yr^{-1}$ lead to a nearly complete loss of the envelope and force the model to leave the AGB while still burning helium. With smaller mass-loss rates, however, the model remains on the AGB and returns to the thermal equilibrium phase of hydrogen burning.

In the particular case shown in Fig. 2, $\dot{M}_W = 2 \cdot 10^{-5}$ $M_\odot yr^{-1}$, and only $\Delta M_e \approx 0.02$ $M_\odot yr^{-1}$, or $\Delta M_e/20$, is lost during each of the two thermal pulses.

It is clear that the results of this numerical experiment depend on both the shape of $L_{PN}(M)$ and the total stellar or envelope mass. They are only preliminary, but we feel, however, that "superwind" mass loss rates larger than 10^{-4} $M_\odot yr^{-1}$ can safely be excluded under the assumptions made above because the majority of the CPN population is obviously not in the helium-burning mode (cf. also Schönberner, this volume).

REFERENCES

Härm, R., Schwarzschild, M.: 1975, Astrophys. J. **200**, 324
Iben, I.Jr.: 1984, Astrophys. J. **277**, 333
Iben, I.Jr., Renzini, A.: 1983, Ann. Rev. Astron. Astrophys. **21**, 271
Paczynski, B.: 1971, Acta Astron. **21**, 417
Reimers, D.: 1975, "Problems in Stellar Atmospheres and Envelopes", B. Baschek, W.H. Kegel, G. Traving Eds. Springer, Berlin, p. 229
Renzini, A.: 1981, "Physical Processes in Red Giants", I. Iben Jr. and A. Renzini Eds. Reidel, Dordrecht, p. 431
Schönberner, D.: 1979, Astron. Astrophys. **79**, 108
Schönberner, D.: 1981, Astron. Astrophys. **103**, 119
Schönberner, D.: 1983, Astrophys. J. **272**, 708
Schönberner, D.: 1984, IAU Symp. No. 105 "Observational Tests of the Stellar Evolution Theory", A. Maeder and A. Renzini Eds., Reidel, Dordrecht, p. 209
Schönberner, D.: 1987, "Late Stages of Stellar Evolution", S.R. Pottasch and S. Kwok Eds., Reidel, Dordrecht.
Schönberner, D., Weidemann, V.: 1981, "Physical Process in Red Giants", I. Iben Jr. and A. Renzini Eds., Dordrecht, p. 463
Tuchman, Y., Sack, N., Barkat, Z.: 1979: Astrophys. J. **234**, 217

OBSERVATIONAL CONSTRAINTS TO THE THEORY
OF PLANETARY NEBULAE EVOLUTION

F.D'Antona (1)(2), I.Mazzitelli (2) and F.Sabbadin (3)
(1) Osservatorio Astronomico di Roma, Monte Porzio
(2) Istituto di Astrofisica Spaziale del CNR, Frascati
(3) Osservatorio Astronomico di Padova - Asiago

ABSTRACT. Two main observational points regarding Planetary Nebulae
(PNe) are discussed in the framework of our theoretical knowledge of
stellar evolution, namely:
i) the expansion age of young PNe is much shorter than the age which
the PN nuclei (PNN) would have if they were in the stage of shell
hydrogen burning, and compare much better with the thermal timescales of
the PNN. This indicates that the expulsion of the PN leaves on the PNN a
very small or even negligible hydrogen envelope, as it is expected on
the basis of the few existing hydrodynamic models.
ii) the morphological subdivision of PNe into two main classes (B and
C) appears to correspond also to a subdivision in terms of physical
parameters both of the nebula and of the central star, indicating that
asymmetric "bipolar" nebulae, at least statistically descend from more
massive progenitors with respect to symmetric PNe. It is discussed
whether the asymmetries can be imputed to common envelope evolution in
binaries, which would give origin to the B nebulae, or to the different
core rotation of more massive progenitors. In any case, the two classes
B and C seem to correspond to a subdivision between progenitors which
suffer or do not suffer the helium flash.

1. THE FIRST OBSERVATIONAL PROBLEM: KINEMATIC AGES OF YOUNG
   PLANETARIES

After two classic papers by Paczynski (1971) and Härm and Schwarzschild
(1975), the theoretical evolution through the PNe region has been
recently studied in a series of works by Schönberner (1979, 1981, 1983),
Iben (1984), Wood and Faulkner (1986). Also Mazzitelli and D'Antona
(1986a) studied the problem of the fast mass-loss phases from a 3M$_\odot$ star
on the Asymptotic Giant Branch (AGB). All these computations are made,
in principle, in the assumption of hydrostatic equilibrium, but the
substantial improvement of the recent works with respect to Paczynski
(1971) or Harm and Schwarzschild (1975) is that the starting models take
into account the effect of previous evolution, including the helium
thermal pulses, and sometimes of mass-loss, on the structure of the

hydrogen and helium layers which are on top of the carbon-oxygen core at the time the PN is ejected.
Although the "philosophy" implicit in these computations has been often reviewed (e.g. Renzini 1983, Schonberner 1987a), we recall it here shortly:

i) it is assumed that Asymptotic Giant Branch (AGB) stars suffer a stage of "rapid" mass loss beginning at a still uncertain point of their evolution;

ii) during this phase the stellar Teff decreases, following the accomodation of the stellar model on the "Hayashi track" corresponding to the value of smaller masses, at least as long as the hydrogen rich layer is large enough ($\gtrsim 0.1 M_\odot$) to have a deep convective envelope;

iii) when such a small envelope mass is left that the star can not be any longer on the Hayashi track ("departure" hydrogen mass Mhd), the Teff of the models begins to increase;

iv) as the envelope mass decreases to a few hundreths of $M_\odot$, the details of the relation between remnant hydrogen mass and Teff begin to depend on the efficiency of the Helium shell, that is on the phase during the thermal pulse (e.g. Mazzitelli and D'Antona 1986a, Wood and Faulkner 1986);

v) when the model reaches a relatively high Teff (generally logTeff= 3.70 - 3.80), the shut-down of the fast mass loss mechanisms is assumed.

This procedure of computation is followed simply because the mechanisms of expulsion of the PNe are still not well known. In particular it is not yet clear whether it is necessary to assume that the PN ejection is a hydrodynamic event or not. For instance, taking the mass loss rates inferred from observation of OH-IR masers ($10^{-4} - 10^{-5}$ $M_\odot$/yr, e.g. Herman and Habing 1985) the "superwind" phase is not necessarily hydrodynamic (Mazzitelli and D'Antona 1986). In this case, the hypothesis of shut down of the superwind at a given Teff leaves the nucleus with a well defined amount of hydrogen on its surface, and the further evolution follows some of the schemes of the model computations by Schonberner, or Iben, as it is not possible to strip off as much mass as one wants from the stellar surface without raising the Teff.

Some other model computations, however, follow also another procedure, namely the "scaling" of the stellar mass to get rid in a few steps of most of the stellar envelope. The assumption underlying this procedure is that a hydrodynamic event has been responsible for stripping off a large fraction of the stellar envelope within a short timescale, since the "equivalent" mass loss rate is largely in excess of 10^{-4} $M_\odot$/yr, for which the dynamical term is dominant (Mazzitelli and D'Antona 1986). The consequences of such a scaling can be very different:

i) either the hydrogen envelope mass left (Mhr) is still larger than Mhd: in this case the star stops in the red, and superwind mass loss is resumed until the model departs to the blue (e.g. Wood and Faulkner 1986): this case thus follows the behaviour previously described;

ii) or Mhr is much smaller than the mass which allows nuclear burning in the "plateau" phase of the PN evolution, and the model suffers a very rapid evolution to the blue (e.g. Iben 1984). As, on the contrary, the "plateau" region of PNN is well populated, the brightest PNN in that phase are often assumed to be in an extended hydrogen burning stage.

On the other hand, these "scaled" models implicitly assume hydrostatic and, within some extent, thermal equilibrium, therefore the resulting evolutionary times are not realistic if the PN ejection is hydrodynamic. In fact, even if the hydrostatic equilibrium is soon restored in the star after the ejection, the corresponding thermal timescale critically depends on the (unknown) point of the plateau where the nucleus emerges after the ejection.

An approximate evaluation of the thermal timescale is given by:

$$t_{th} = 2 \times 10^7 \, M \, M_e / LR \, yr \quad (M, M_e, L \text{ and } R \text{ in solar units})$$

Here $M \cong 0.6 M_\odot$, $L \cong 3 \times 10^3 L_\odot$, $M_e \cong 0.02 M_\odot$ (taking into account that gravitational energy is available from the whole non degenerate region, and thus also from the helium intershell). Thus $t_{th} \cong 160 yr$ for $R \cong 0.5 R_\odot$, but $t_{th} \cong 800 yr$ for $R \cong 0.1 R_\odot$.

Actually, the kinematic ages of young compact PNe having low mass $(0.6 M_\odot)$ central stars are much closer to the thermal timescale than to the ages predicted by nuclearly powered evolution, which result to be an order of magnitude longer.

The kinematic ages are computed from the radius and expansion velocity of the nebulae. The systematic difference between "evolutionary" ages and kinematic ages of young nebulae was pointed out by Pottasch (1983), by the same Pottasch (1987) in the present workshop, and by Sabbadin (1986a). It rests mainly on the revision of the distance scale of compact PNe, which are optically thick, and for which the distance scale by Cahn and Kaler (1971) was thus incorrect. If we do not trust "statistical" distance scales, we find out that this problem is still present when we limit our consideration to the PNe for which several independent distance estimators exist (Sabbadin 1986a). It is also generally assumed that the kinematic age is an upper limit to the nebula age, as the average expansion velocity of the nebula increases with time. However, in the cases studied by Sabbadin (1986a) this age can be considered actually as a <u>lower</u> limit, as the expansion velocities adopted were quite low (10 - 20 $\overline{Km/s}$), of the order of the escape velocity from the red giant progenitor, so it is hard to expect that smaller velocities should have been assumed.

In several cases kinematic ages of 500 - 1000yr are found, in substantial agreement with the thermal timescale, but with a discrepancy of a factor at least 5 with respect to the nuclear evolution timescales, even if one applies the timescales relative to more massive PNN, in the conservative assumption that the distances are underestimated. This comparison leads us to a number of conclusions:

i) the PN ejection is actually hydrodynamic;

ii) it leaves the PNN with much a smaller amount of hydrogen than it would be left by the superwind quenching assumptions which are at the basis of the hydrostatic models discussed earlier;

iii) we cannot be much confident about the actual applicability of hydrostatic "scaled" model computations, as the thermal readjustment timescale depends on the unknown point where the nucleus emerges after the dynamic ejection.

At this stage it is clear that hydrodynamic computations of the PN ejection and evolution are badly needed. Do existing computations tell us anything about the remnant hydrogen mass left, if the ejection follows an hydrodynamic scheme?

As far as we know, the work by Kutter and Sparks (1974) is among the few pieces of information we have. The model described consists of a giant of 1.1M☉, in which the luminosity is provided by He-shell burning -thus coming from below the hydrogen envelope (it is irrelevant whether the burning is stationary, or it is due to the onset of a helium shell flash). A hydrogen envelope of 0.101M☉ is superimposed on a helium shell of 0.075M☉. When the matter acceleration at the bottom of the hydrogen layer overcomes the acceleration at the top of the helium layer, a "vacuum" is included in the computations to simulate the separation between envelope and shell.

The <u>initial</u> acceleration in the H-rich envelope is due to the discontinuity in opacity between H and He: the transfer of energy and momentum from the radiation to the gas is more efficient in the H-rich material. After about 1 yr, the radiation filling the vacuum further pushes the base of the hydrogen envelope, continuing the acceleration process.

Since the triggering parameter is the hydrogen opacity, <u>the whole</u> hydrogen envelope will be lost, or at least the envelope down to the point at which the hydrogen content begins to decrease, at the interface with the helium layer. Also hydrostatic computations indicate that the hydrogen opacity is a key factor for discriminating whether an envelope is stable or not: Wood and Faulkner (1986) for instance notice that when the luminosity corresponding to core mass $M_c=0.86$M☉ is reached, the hydrogen envelope of the AGB stars becomes unstable, as, at the basis of the hydrogen envelope the radiation pressure equals or is larger than the gas pressure: in other words, the Eddington luminosity <u>for the hydrogen envelope</u> is reached at that stage. Also Mazzitelli (1987a and b) suggests that the maximum PNN mass which may come out of the evolution of an intermediate mass star is 0.84M☉, leaving almost no hydrogen on the star. At least in these cases, therefore, the envelope should be ejected down to the layers for which the main source of opacity is still the hydrogen.

Although the eiection of the PN driven by the reaching of a critical luminosity (and thus a little after the peak of the thermal pulse) is appealing from a theoretical point of view (Wood and Faulkner 1986, Mazzitelli 1987b) and provides a direct interpretation of the very young ages of compact planetaries, it may be in contrast with the later stages of PNe evolution. Schönberner (1981, 1987b, and in this volume) presented evidence that the luminosity function of PNN shows a gap at $M_v=4-5$. This gap can be explained by the sudden drop of stellar luminosity occurring when hydrogen burning finishes (for a restrict range of stellar masses around 0.6M☉). In fact helium burning models do not experience any luminosity drop. This evidence leads Schönberner to conclude that most of PNN must be hydrogen burners. Notice that this observational test is largely independent from the one we have been discussing before: it refers to kinematic ages larger than ∼8000yr (Schönberner 1987b), for which the hydrostatic ejection and the

following thermal readjustment should in any case have been reset.
This test is crucial to our understanding of the PN phase, but it needs more attention, and it has been questioned e.g. by Wood and Faulkner (1986), as it is drastically dependent on the assumed distance scale. Both Wood and Faulkner (1986) and Sabbadin (1987) find that the situation is far from being settled. Thus, for the present, we favour the idea that the PN ejection takes place preferentially just after the peak of the helium flash, when the maximum total luminosity is reached, leaving most of PNN with a tiny hydrogen envelope. The details of the following evolution will further decide whether the remnant skin will be completely removed by wind, giving origin to a hydrogen-free white dwarf, or it will remain on the nucleus. In the latter case, the resulting DA white dwarf will cool down until very low Teff are reached (Teff<7000K) and only then the whole hydrogen envelope, becoming convective, will be mixed with the underlying helium, giving again origin to a helium dominated atmosphere (see Greenstein 1986, D'Antona 1987).

2. A SECOND OBSERVATIONAL PROBLEM: WHY MORPHOLOGY IMPLIES A SUBDIVISION IN TERMS OF PROGENITOR MASSES?

In 1971, 1972 Greig proposed a morphological subdivision of PNe into two main classes:
  B ("binebulae"), with large ansae and tubular or filamentary structure (or asymmetric nebulae); this class also includes Peimbert's (1978) type I nebulae;
  C ("centric increase in surface brightness"), which are highly symmetric;
  Evidence has been accumulated in recent years, and it has been recently discussed by Sabbadin (1986b), that this classification also corresponds to a subdivision in terms of physical and chemical parameters of the nebulae and of their nuclei. All parameters indicate that B nebulae come from more massive progenitors than C type nebulae.
  In regard to the nebular ionization, Greig (1971) noticed that the strength of forbidden low ionization lines relative to the Balmer lines (e.g. [OI], [OII], [NII], [SII]) is larger in B than in C nebulae. Greig (1972) noticed also that B nebulae tend to follow the double sine curve of radial velocity versus galactic longitude, while non-B have a large velocity lag behind the circular velocity. The average height on the galactic plane, z- component of the space velocity, and radial velocity dispersion are much lower for B than for C nebulae. These indications of younger population for the B type were later confirmed by Cudworth (1974) through the study of proper motions, and by Acker (1980) doing the ananlysis by Greig (1972) on a wider sample of PNe.
  Also the chemical composition differs: Almost all type I nebulae (Peimbert 1978, Peimbert and Torres Peimbert 1982), having enhanced helium and nitrogen (He/H$\gtrsim$0.125, logN/O>-0.3) belong to class B. This is confirmed by Sabbadin (1986b) through a statistical analysis.
  The nebular evolution follows different paths for B or C type: the evolutionary trends of the line intensity ratios [OII]/H$_\beta$, [NII]/H$_\alpha$

and [SII]/H$_\alpha$ (Sabbadin and Minello 1978), and of the ratio of the Zanstra temperature of HeII and HI (Sabbadin 1986b) with respect to the nebular radius indicate that B and C subclasses have different nebular evolution: both types start with opticaly thick nebulae, but, whereas the C type soon become and remain optically thin, the B type appears optically thin at intermediate stage, and becomes again optically thick for the whole following evolution. Further, the value of nebular mass (multiplied by the filling factor $\epsilon^{-1/2}$) is larger for B nebulae than in C nebulae (Sabbadin 1986a).

Finally, the position of B type nuclei in the HR diagram indicates that they are on average more massive than C type nuclei (Sabbadin 1986a).

Further studies on the statistical validity of this subdivision are undoubtedly necessary, but, if confirmed, it poses us a <u>second observational problem</u>: why nebulae coming from more massive stars are largely asymmetric? It has been recently suggested by Renzini (1983) that asymmetric nebulae could be the product of common envelope evolution of binaries. Assuming that this hypothesis is correct, the problem shifts to the following one: are there any reasons why the binary evolution should be more common among larger mass progenitors of PNe?

Common envelope evolution (Paczynski 1976) occurs when the radius of the evolving primary reaches the binary Roche lobe radius during the evolution of the star as giant: in this case the mass transfer becomes unstable and losses of mass and angular momentum from the system cause both the reduction of the orbital periods to the values (few days, or even hours) typical of cataclysmic variable stars, and the formation of a "common envelope" which, if the giant soon exposes its hot nucleus, may appear as a PN.

It is easy to understand that there is at least one good theoretical reason why larger mass stars could be preferred for this to happen: low mass stars develop degenerate helium cores durning the hydrogen shell burning phase, and are subject to the helium ignition by a flash. The limiting initial mass for the occurrence of a strong helium flash is about 2.2M$_\odot$ (e.g. Sweigart and Gross 1978; Mazzitelli and D'Antona 1986b find that a 2.5M$_\odot$ initial mass suffers a "mild" helium flash), but it can be even considerably smaller (down to 1.5M$_\odot$ in the most extreme case) if overshooting is acting in main sequence, as suggested by Barbaro and Pigatto (1984) and computed by Bertelli et al. (1986).

Contemporary to the growth of the degenerate helium core the low mass stars get very extended radii. As an example, the models by Mazzitelli and D'Antona (1986b) show that the radius of 1M$_\odot$ of population I at helium flash is about 150R$_\odot$, while helium ignition in a 3M$_\odot$ star takes place when the radius is smaller than 30R$_\odot$. The PN ejection in low mass stars occurs at luminosities and radii not much larger than those achieved during this hydrogen shell burning phase: thus, assuming a simple primordial distribution of binary periods, (e.g. dN = dlog $\underline{a}$, where $\underline{a}$ is the binary separation -Iben and Tutukov 1984), mass transfer from low mass primaries is probable to occur during the hydrogen shell burning phase, ending up with a helium core of M$\lesssim$0.5M$_\odot$, which still has probably a large hydrogen envelope mass on top: as we have seen, in this case hydrogen burning continues in the red for such a long time that the common envelope is dispersed before the central star may excite it as PN. We can also suggest that some low luminosity (L $\lesssim$ 1000L☉) OH - IR masers may be binaries in common envelope evolution.

On the contrary, stars having initial masses which do not develop a degenerate core may be subject, for a much wider range of binary periods, to common envelope evolution during the double shell burning phase, and one might speculate that most of these events give origin to asymmetric PNe.

Thus it is important to check whether the PNe which are the product of common envelope evolution, as testified by their having central binary nuclei of short period (P$\lesssim$ several days) belong preferentially to B-type. On the contrary, among the few PNe for which orbital period is known (Ritter 1987), only two belong clearly to class B, and at least two are surely of class C (see table 1). Many more data would be needed for a conclusive reject of the hypothesis.

It is worth however to examine shortly the hypothesis that the different shape of PNe is due to the magnetic field connected to the rotation. Also this hypothesis would lead to creation of asymmetric nebulae from more massive stars progenitors, which are fast rotators in main sequence. Surface layers of AGB stars rotate several hundred times slower than the initial main sequence rotation; thus Renzini (1983) asks why very small velocity asymmetries (as v surface << v expansion of the PN) should give origin to large asymmetries such as those of B type PNe. But we do not have information about core rotation. The main sequence phase is the longest in the stellar lifetime: it is thus possible that the slow rotation of the convective envelopes of low mass stars (M<2M☉) slows down also the core rotation. Viceversa, although larger masses have convective cores, they have also fastly rotating external envelopes, and the momentum of inertia of the external layers is dominant in the total stellar mass momentum. Since there is a period of the order of million years (following the core hydrogen exhaustion, and before the star arrives in the giant region) during which the stellar core is radiative, the external large rotational momentum may again be transferred to the core. If this is the case, the core rotation of AGB stars preserves memory of the main sequence rotation, and is faster for cores descendant of more massive progenitors.

Let us consider that, even if core rotation is the discriminant between the occurrence of B or C type nebulae, the transition between the two types would occurr around initial mass M $\simeq$ 2M☉: for smaller masses the main sequence rotation is much slower (e.g. Bernacca and Perinotto 1974).

So, in both scenarios, the dividing line between progenitors of type B and C nebulae is approximately the upper mass which is subject to a strong helium flash. Let us remark also that the low mass stars get such large radii and luminosities during the hydrogen shell burning stage, that they lose a large amount of mass by stellar wind, such that the total mass finally ejected as planetary nebula can result to be quite small, compared to the mass ejected by B type progenitors which, in the corresponding evolutionary stage, do not lose mass. Although it is obvious that more massive stars must have also more massive nebulae, the theory could be forced to predict even a <u>dicothomy</u> in the nebular mass passing from C type (M<2M☉) nebulae to B type (M>2M☉) nebulae, and helping to understand the difference between B type and C type nebular evolution (Sabbadin 1986b).

TABLE 1
CLOSE BINARY CENTRAL PLANETARY NEBULAE NUCLEI

Star	PN	Period	Type	Notes
AGK3-0965	NGC 2346	15.95d	B	
BD+26 2405	LT - 5	1.2:	C	
V477 Lyr	A46	0.4717	C	(1)
UU Sge	A63	0.4651	?	
BD+50 2869	NGC 6826	0.23768	C	
BD-12 1172	I 418	0.2:	B	
Abell 41	Abell 41	0.11323	?	
BD+66 1066	NGC 6543	0.06??	C	
CPD-26 389	NGC 1360	≈8.1	C	(2)
BD+30 623	NGC 1514	0.41	C	(3)

(1) It belongs to C type as NII is very faint (Kaler 1983)
(2) The period is from Mendez and Niemela (1977). It belongs to C type morphologically, and because of faint NII, strong HeII.
(3) Morphologically, and because of faint NII, it belongs to type C. The spectroscopic period, from Acker 1976, has never been confirmed by other observations.

References

Acker,A. 1976, Publ.Obs.Strasbourg 4, n.1.
Acker,A. 1980, A.A. 89,33.
Barbaro,G., Pigatto,L. 1984, A.A. 136,355.
Bernacca,P.L., Perinotto,M. 1974, A.A. 33,443.
Bertelli,G., Bressan,A., Chiosi,C., Angerer,K. 1986, Astr.Ap. Suppl. 66, 191.
Cahn,J.H., Kaler,J.B. 1971, Ap.J. Suppl. 22,319.
Cudworth,K.M. 1974, Astron.J. 79,1384.
D'Antona,F. 1987, in "6th European Workshop on White Dwarfs", Monte Porzio, June 1986, Mem.S.A.It., in press.
Greenstein,J.L. 1986, Ap.J. 304,334.
Greig,W.E. 1971, A.A. 10,161.
Greig,W.E. 1972, A.A. 18,70.
Harm,R., Schwarzschild,M. 1975, Ap.J. 200,324.
Herman,J., Habing,H.J. 1985, Phys.Reports 124, 255.
Iben,I.Jr. 1984, Ap.J. 277,333.
Iben,I.Jr., Tutukov,A. 1984, Ap.J. Suppl. 54,335.
Kaler,J.B. 1983, Ap.J. 271, 188.
Kutter,G.S., Sparks,W.M. 1974, Ap.J. 192,447.
Mazzitelli,I. 1987a, in "6th European Workshop on White Dwarfs", Monte Porzio, June 1986, Mem.S.A.It., in press.
Mazzitelli,I. 1987b, in preparation.
Mazzitelli,I., D'Antona,F. 1986a, Ap.J. 308, 706.
Mazzitelli,I., D'Antona,F. 1986b, Ap.J. in press.
Mendez,R.H., Niemela,V.S. 1977, M.N.R.A.S. 178,409.
Paczynski,B. 1971, Acta Astronomica 21,417.
Paczynski,B. 1976, in "Structure and evolution of close binary systems", IAU Symp. 73, ed.P.Eggleton,S.Mitton and J.Whelan (Dordrecht, D.Reidel), p.75.
Peimbert,M. 1978, in IAU Symp. 76 "Planetary Nebulae", ed.Y.Terzian, D.Reidel,Dordrecht.
Peimbert,M., Torres Peimbert,S. 1982, in IAU Symp. 103 "Planetary Nebulae, ed. D.R. Flower (Dordrecht,Reidel)
Pottasch,S.R. 1983, in IAU Symp. 103 "Planetary Nebulae, ed. D.R. Flower (Dordrecht,Reidel) p.391.
Pottasch,S.R. 1987, in this volume.
Renzini,A. 1983, in IAU Symp. 103 "Planetary Nebulae", ed.D.R. Flower (Dordrecht, Reidel), p.267.
Ritter,H. 1987, Catalogue of Cataclysmic Binaries, Low mass X-ray Binaries and Related Objects", 4th edition.
Sabbadin,F. 1986a, A.A. Suppl.Series 64,579.
Sabbadin,F. 1986b, A.A. 160,31.
Sabbadin,F., Minello,S. 1976, A.A.Suppl.Series 33,233.
Sabbadin,F. 1987, in preparation.
Schönberner,D. 1979, A.A. 79,108.
Schönberner,D. 1981, A.A. 103,119.
Schönberner,D. 1983, Ap.J. 272, 708.
Schönberner,D. 1987a, Proceedings of the Calgary Workshop "on the late stages of stellar evolution". (Dordrecht,Reidel)

Schönberner,D. 1987b, A. A. 169,189.
Sweigart,A.V., Gross,P.G. 1978, Ap.J. Suppl. 36,405.
Wood,P.R., Faulkner,D.J. 1986, Ap.J. 307,659.

Discussion:
Viotti: Studies of the circumstellar nebulae around symbiotic stars have revealed the presence of bipolar structure. However the central objects are thought to be detached or semidetached binaries, not common envelope binaries, nor I think that they have passed through this phase in the past. I agree with you that they should be massive objects.

Perinotto: In a recent similar study of bipolarity as function of other parameters including stellar duplicity, Zuckerman and Aller (1986 Ap.J. 301,772) found, if I remember correctly, more evidence than the one you have shown in favour of your point: i.e. there are more close double stars which are associated with bipolar nebulae than with other kinds of nebulae. however they also conclude that the statistics is insufficient for a clear decision.

Renzini: The relevant thermal timescale for post-AGB stars is $\sim GMM_e/RL$ rather than $\sim GM^2/RL$, where M_e is the envelope mass ($\sim 10^{-3} M_\odot$). My impression is that you have overestimated the thermal timescale by perhaps three orders of magnitude.

Considerations about Renzini's comment:
The question was motivated by the fact that in the presentation given at the workshop the thermal time was approximated by the expression GM^2/RL, as, although this is clearly an overestimate of the thermal equilibrium time, nevertheless it is considerably shorter than the nuclear burning timescale, and this was the main point we wished to enphasize.
We are also aware and must remember that the very definition of thermal timescale on a structure with a very complex thermal history, stratified in a complex way, and with several different modalities of substantial energy generation and loss (including neutrinos, coming out from the whole structure) is not so trivial, and, by experience, only numerical computation can give the proper values.
If, to avoid misunderstandings, we decide to take in consideration only the <u>gravitational</u> energy which can be obtained by the contraction of the non degenerate envelope, it seems to us that envelope mass is not only the <u>outer</u> envelope ($10^{-3} M_\odot$, order of magnitude of the hydrogen envelope mass), but also the helium intershell, as it has been consequently assumed in the present written version.
If the timescale to be considered is that of the hydrogen envelope only, (10 to 50yr for the PNe nuclei for which the kinematic age is 500 to 1000 yr) it is necessary to come back to the idea that all these nuclei are in a nuclear burning phase.

Rambling along a Schönberner Track:
Nebular Evolution and NLTE Stellar Atmospheres

J. Köppen
Institut für Theoretische Astrophysik
Im Neuenheimer Feld 561
D-6900 Heidelberg
West Germany

Summary: Based on Schönberner's calculations of the evolution of stars in the post AGB phase, we computed models for the evolution of the hydrodynamical and ionization structure of the planetary nebulae around these central stars. We also computed NLTE model atmospheres for these stars. The results are discussed and compared with observations.

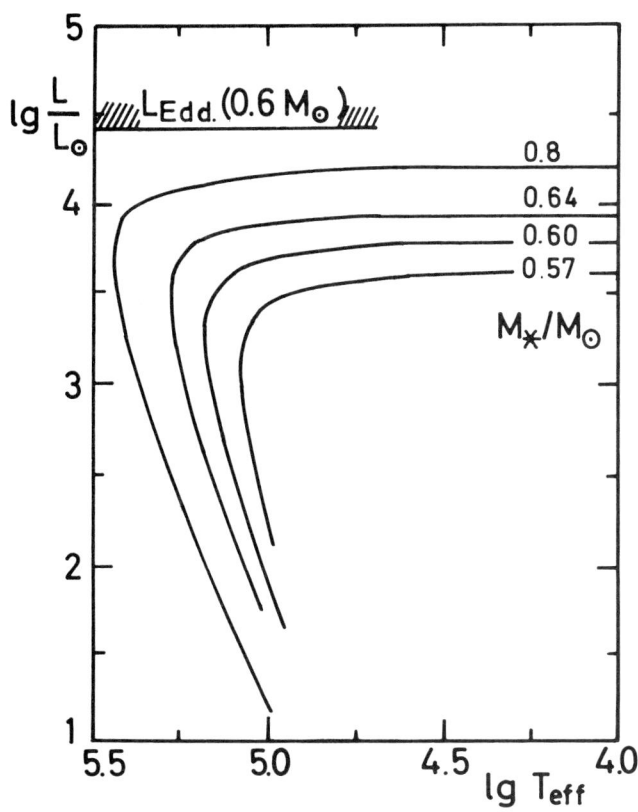

The basic question of the evolution of the central stars of PN is how quickly they cross certain regions in the HR-diagram. The computations of Schönberner (1981) have shown that stars which burn hydrogen quietly in their envelopes are able to explain the observable relation of stellar absolute magnitude and nebular age (see his Fig.7): After the ejection of the nebula the central star first heats up at constant luminosity to reach a temperature maximum just before the energy sources are exhausted. Now the luminosity drops rapidly after which the central star begins to cool very slowly to become a white dwarf. The rapid drop in luminosity at the "knee" in the HR-diagram is found to correspond to a pucity of central stars with M_V of 4 to 5 mag. Since the evolutionary time scale is a very steep function of the mass of the central star, Schönberner finds the average star mass to be about 0.57 $M_\odot$.

Now what happens to the Nebula? Schmidt-Voigt and Köppen (1986) have done hydrodynamical calculations for the evolution of the nebular gas, whose ionization under the influence of the evolving central star is treated time dependently. We consider spherically symmetric models which consist of a sequence of ejecta from the central star: When the star still was on the AGB, it lost mass by a slow (10km/s) wind. We assume the ejection of the nebula to take place as a "superwind", which is a short (1000a) period of enhanced mass loss. After this, the central star has a fast (2000km/s) stellar wind. We have varied the various parameters for the three winds, calculating Three Wind Models which represent models where the nebula was created in a single sudden event. If the AGB wind and the superwind are indistinguishable, we call it a Two Wind Model which represents the "colliding winds" model of Kwok et al.(1978) where the fast central star wind sweeps up the AGB wind to form the nebula.

We have compared our models with various observational data. The most fruitful way is to compare the evolution of the HeII 4686/Hbeta line ratio - which measures the nebular ionization - with the line ratio observed in nebulae of different sizes.

Because the central star's path in the HRD folds back after the temperature maximum, we separate the nebulae in two groups:
The UPs have central stars absolutely brighter than 5 mag in the visual, and are interpreted as objects whose central star still heats UP at constant luminosity.
The DOWNs are objects with faint central stars which are thought to fade DOWN to the white dwarfs.
This distinction is motivated by the "gap" in the M_V distribution (cf. Schönberner).

Looking at the UPs, we notice that the nebular ionization goes up as the stellar temperature increases, until in the nebula all helium is doubly ionized. This first part of nebular evolution - the Ionization Phase - is dominated by the evolution of the central star, and independent of nebular parameters within a reasonable range. This means that by studying the UPs one essentially measures the time scale of the heating up of the central star, which is a sensitive function of its mass. The Figure shows that Schönberner tracks of 0.60 to 0.64 $M_\odot$ are suitable to explain the nebular ionization, rather than the 0.57 $M_\odot$ model which evolves too slowly. This discordance, which we do not want to emphasize too strongly, with Schönberner's (1981) findings could well be due to his analysis being more sensitive to the assumed nebular distance.

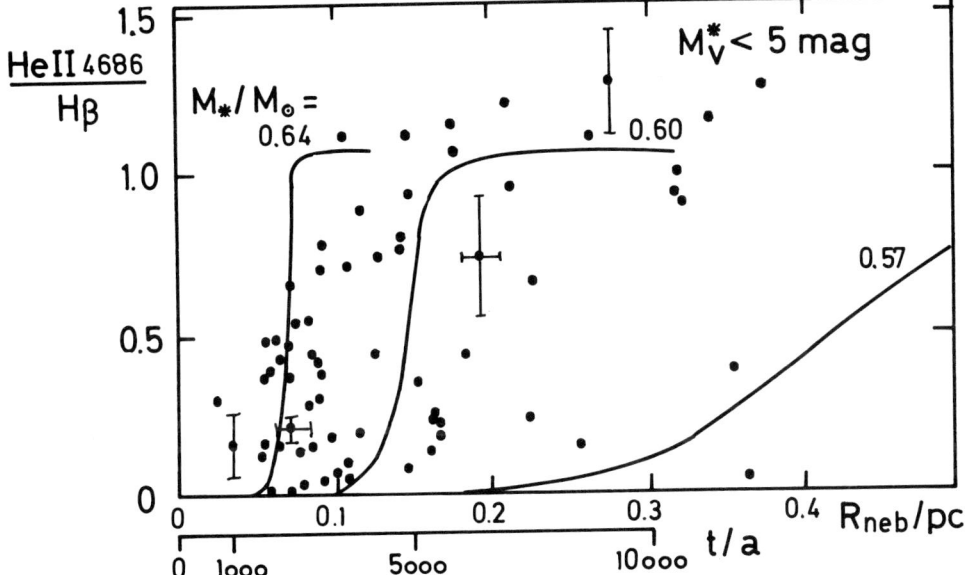

Looking at the DOWNs - the more evolved objects with faint stars - we note: After the temperature maximum, the stellar luminosity and hence the number of ionizing photons drop rapidly. The nebula recombines partially - unless it contains very little mass - and the HeII/Hbeta ratio goes down. After that the nebula enters its Final Phase: Its central star cools off at snail's pace, and the nebula expands into its surroundings while ionized by a radiation field almost constant in time. Observationally, the HeII/Hbeta ratio is low and does not vary much with nebular size, which agrees rather nicely with some of our models:

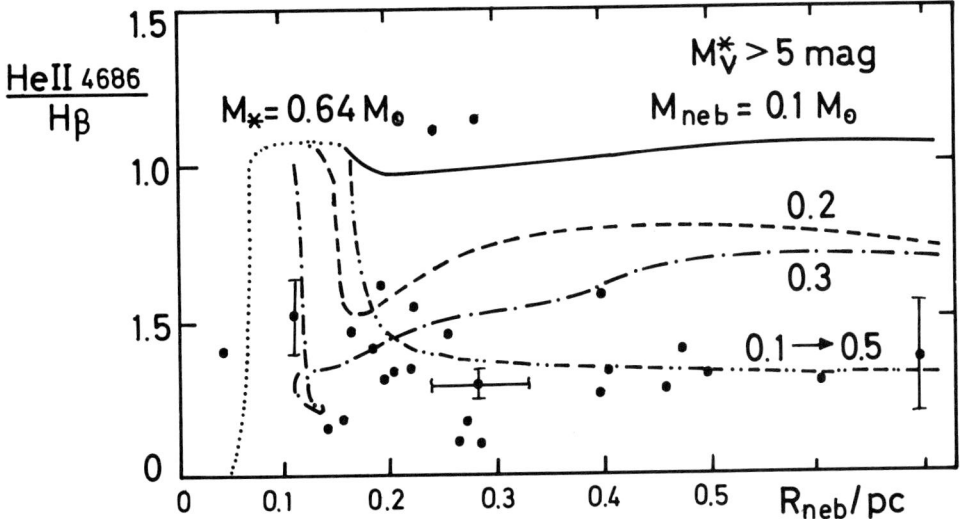

We show results only for the 0.64 $M_\odot$ star, noting that models with the 0.60 $M_\odot$ star have very similar properties except that their stars have the temperature peak when the nebular radius is about 0.3 pc.

An important consequence of the slow stellar evolution is that in evolved objects the nebular properties play a more important role, in particular the initial conditions: Three Wind Models which consist entirely of the matter ejected via the superwind, remain highly ionized during the Final Phase, or their ionization increases again. Furthermore, these models expand too quickly and because the density in the nebula decreases with radius show a rather diffuse outer rim. The models which do reproduce the constant low ionization, belong to a class we call "Accreting Models": Here the nebula runs into the dense remnant of a rather massive AGB wind, and accretes a great amount of material from it. The accretion flow causes the density to increase with radius, and hence the nebulae exhibit rather well defined outer rims. It also causes the nebulae to have lower expansion velocities and to have masses increasing with time. Two Wind Models belong by their very nature to the Accreting Models, but when young they are less massive than Three Wind Models.

Thus by studying the old extended nebulae we gain information about the very early life of planetaries: Some of the nebula material must have been ejected in a single event ("superwind"), but an appreciable portion must have been accreted from rather massive AGB winds. The mass loss rate on the AGB should have been greater than about $3 \cdot 10^{-6}$ $M_\odot a^{-1}$.

If we may now take Schönberner's tracks as a good description of the evolution of the central stars, what are the consequences for the spectral distribution of the ionizing flux from these stars? For the hydrodynamical calculations we had assumed black body spectra. We note that the evolutionary tracks take the stars rather close to their Eddington limits, and therefore we may expect quite strong NLTE effects in the HeII Lyman continuum, which may even be in emission rather than absorption (Husfeld et al., 1984).

We have calculated a number of atmospheric models for stars that follow the 0.60 and 0.64 $M_\odot$ Schönberner and 0.8 $M_\odot$ Paczyński (1971) tracks. Planar LTE and NLTE continuum models with a helium to hydrogen ratio of 0.1 by number were constructed. All lines were assumed to be in radiative detailed balance. In some models, metal opacities were included in LTE. This caused a very slight weakening of the NLTE effects. Between 60000 K and 70000 K, our NLTE model atmosphere program in its present form converged much too slowly to find acceptable solutions.

An useful quantity to characterize the ionizing spectrum is the ratio of the number of ionizing photons beyond 54 eV, the HeII ionization threshold, to the stellar flux at Hbeta. In the Figure we show the logarithm of this ratio (in SI units) versus effective temperature.

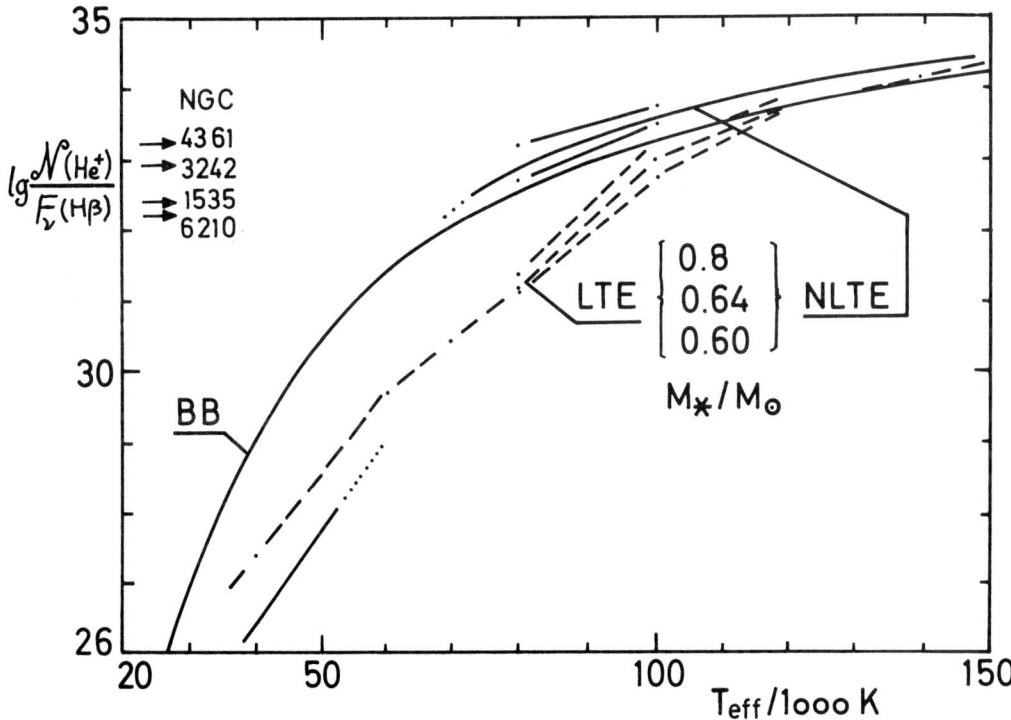

The curve marked BB shows the relation for black body spectra, and for any given ratio the resulting temperature is what is called the HeII Zanstra temperature. LTE models above 100000 K have a ratio quite close to that of BB spectra, but for temperatures below 100000 K their flux distributions show a deep absorption edge at 54 eV, which causes the ratio to be much smaller than that for a BB of the same temperature. NLTE models below 60000 K have an even deeper absorption. This deficit of ionizing photons beyond 54 eV noticed by Heap (1977) induces a discrepancy between Zanstra temperatures (derived from the nebular ionization) and effective temperatures derived from stellar spectral features.

For higher temperatures, however, this discrepancy vanishes, as NLTE effects in the HeII Lyman continuum will weaken the absorption in low surface gravity atmospheres (Husfeld et al., 1984). These gravities are implied from the theoretical evolutionary tracks! The dependence on the star's mass, i.e. on surface gravity is greatest for the 70000 K. NLTE models actually produce more photons harder than 54 eV than even BB spectra. This result is in accord with the findings of detailed analyses of high excitation nebulae with ionization models (e.g. Aller, 1982, Köppen, 1983): In order to achieve the desired fit to the observed nebular emission line spectrum, one often is forced to artificially "bend" the flux distributions (from LTE or high gravity NLTE models) because of their lack of hard photons. Studies of this kind thus tell us that the atmospherical parameters implied from theoretical evolutionary tracks are indeed the correct ones for the central stars!

Acknowledgment: This work was supported by the Sonderforschungsbereich 132 of the Deutsche Forschungsgemeinschaft.

References:

Aller, L.H.: 1982, Astrophys.Space Sci. 83, 225.
Heap, S.R.: 1977, Astrophys.J. 215, 864.
Husfeld, D., Kudritzki, R.-P., Simon, K.P., Clegg, R.E.S.: 1984, Astron. Astrophys. 134, 139.
Köppen, J.: 1983, Astron. Astrophys. 122, 95.
Kwok, S., Purton, C.R. Fitzgerald, P.M.: 1978, Astrophys.J. 219, L125.
Paczyński, B.: 1971, Acta Astron. 21, 417.
Schmidt-Voigt, M., Köppen, J.: 1986, Astron.Astrophys., in press.
Schönberner, D.: 1981, Astron.Astrophys. 103, 119.

PEIMBERT: If you have density fluctuations, the densities are higher and the recombination times are shorter. Does this affect your HeII/Hbeta ratios?
KÖPPEN: Since in the DOWNs we can essentially ignore changes in the stellar flux, there would be enhanced recombination in the dense blobs, resulting in a lower HeII/Hbeta. Though this probably could change the parameter values for our "preferred model", Three Wind Models that do not accrete much AGB wind material would still have to be jedged undesirable because of their rise in ionization and their high expansion velocities.
CLEGG: Where did you put the zero point of your time axis?
KÖPPEN: To be strictly consistent with Schönberner's tracks, we stuck to his zero point, i.e. at the end of the superwind phase.
WEIDEMANN: What is the influence of the stellar He/H abundance on the stellar flux? Henry and Shipman demonstrate that for pure H NLTE model atmospheres the flux beyond 54 eV increases even more.
KÖPPEN: Yes, lowering the helium abundance does push up the HeII continuum. But Henry and Shipman did models with high surface gravity where NLTE effects in HeII are not that strong. So they must use pure hydrogen models to get the same level of flux that models on Schönberner tracks give even with He/H=0.1.
PREITE-MARTINEZ: Would you say that above 70000 K one can safely use blackbodies, or are there other features that make model atmospheres different?
KÖPPEN: If you are interested in quantities such as the ratio of ionizing photons (HeII/H), it may be O.K. to use BB. However, in model atmospheres the spectrum at very energies (70 eV) is flatter than that of a BB, which certainly affects e.g. NeV lines.

EVOLUTION OF HOT SUBDWARFS - AN EMPIRICAL APPROACH

K. Hunger and U. Heber
Institut für Theoretische Physik
der Universität Kiel
Olshausenstraße 40, 2300 Kiel, F.R.G.

Spectral analysis has been making rapid progress in the recent past. On the side of the theory, NLTE-model atmospheres plus NLTE line formation for hot stars have to be mentioned, and also the LTE metal line blanketed atmospheres. On the side of the observations, the role of the IUE telescope as a spectrophotometer has to be stressed, and the great impact of the ESO CASPEC instrument. The quality of the quantitative analyses is such that it may yield narrow constraints which allow to define the evolutionary state of the class of stars under consideration. The class of stars we are dealing with are the hot subdwarfs. How they are related to the central stars of planetary nebulae, will be shown at the end of this talk.

Spectroscopic analyses yield the fundamental atmospheric parameters T_{eff}, gravity g and the abundance ratio n_{He}/n_H. For stars with $T_{eff} < 35000$ K, T_{eff} is determined from calibrated, low resolution IUE spectrograms, while for hotter temperatures, the HeI/HeII ionisation ratio is used. Gravity as well as n_{He}/n_H are determined from the profiles of hydrogen and helium lines. The average error in T_{eff} amounts to $\pm$ 5%, whereas the error in gravity and n_{He}/n_H is $\pm$ 0.2 dex typically.

The results of the analyses are best studied on hand of a (g, T_{eff})-diagram. This diagram is morphologically identical to the conventional HR-diagram (see Fig. 1). The former requires only data which are directly derived from observations, whereas the latter requires the additional knowledge of the distances, in order to arrive at luminosities. Distances are often ill-defined, especially in those cases where the stars in question are located off the galactic plane, and when we are dealing with exotic objects.

Before we discuss the results in detail (see also Hunger et al. 1981, Groth et al., 1985 and references cited therein), let us review the scenario of post horizontal branch evolution. There are two distinct classes of evolutionary tracks, depending on the mass left in the outer hydrogen rich envelope (Fig. 2). If the envelope mass is in excess of 0.02 $M_\odot$, we deal with a true horizontal branch star, with two energy sources, He burning in the center, and H burning in the shell. The

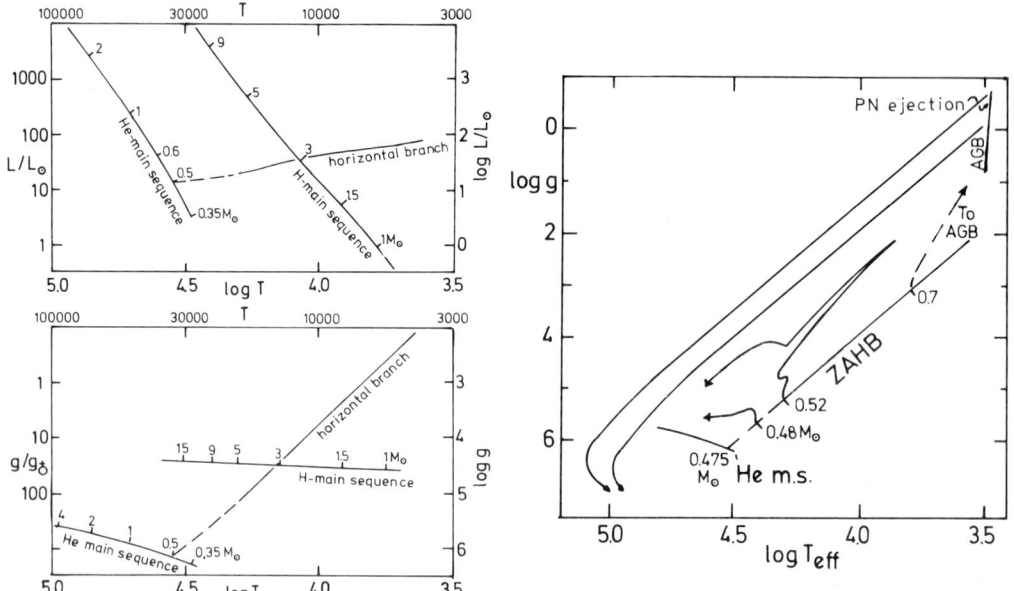

Fig. 1: Conventional HR-diagram (top), (g, T_{eff})-diagram (bottom). The two diagrams are morphologically identical.

Fig. 2: Post-HB evolution, in the g-T_{eff}-plane, schematically (see text)

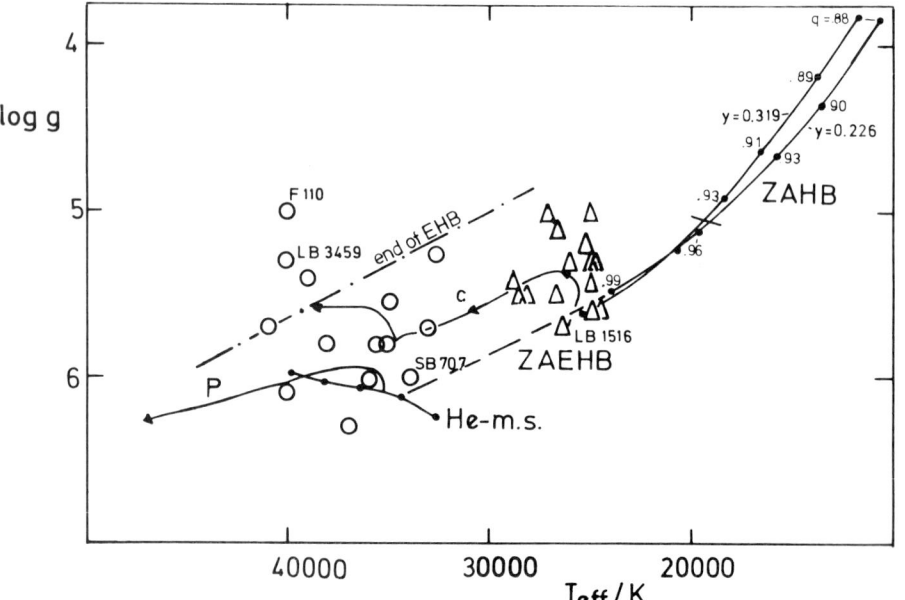

Fig. 3: (g, T_{eff})-diagram of sdB's ($\triangle$) and sdOB's ($\bigcirc$). All sdOB's have a helium content below 9% (by number), while in sdB's, He is reduced even to below 1%.

further evolution proceeds along a "Sweigart-track" (Sweigart et al., 1974), i.e. an evolution towards lower gravities and temperatures before at some point the star reverses its expansion and returns to the hot and high gravity domain. Whether it reaches the asymptotic giant branch, and whether it has enough nuclear fuel to reach the tip of the AGB where it throws off a planetary nebula, depends on the initial mass contained in the shell. Since theoretical tracks are scarce in this domain, the precise critical mass is not known.

When the envelope mass M_e is smaller than $0.02\ M_\odot$, the star is left with only one energy source, the central source which burns helium. These stars are called "extended HB stars". Their further evolution proceeds roughly at constant gravity towards higher temperatures, along a "Castellani-track" (Castellani, private communication). Because of these almost oppositely directed tracks, the region that should be occupied by stars with $M_e \approx 0.02\ M_\odot$ appears depopulated. This is the Newell-Gap II near $T_{eff} = 22000$ K (Newell, 1973).

In the following we discuss the results separately for the two classes of hot post-horizontal branch stars: the sdB/sdOB stars first, and then the sdO stars.

From the (g, T_{eff})-diagram (Fig. 3) it follows that the majority of the sdB/sdOB stars are He-main sequence stars with hydrogen-rich envelopes reduced to almost nothing. Most of them appear slightly evolved, along a Castellani-track. An apparent contradiction is the observed low He content, He being reduced to as little as 10^{-3} in some cases. This contradiction is resolved by the theory of diffusion which predicts the gravitational settling of all elements heavier than hydrogen in high gravity stars. As a consequence, elements such as carbon and silicon may appear reduced by as much as a factor 40000 (Feige 66) and 300000 (Feige 110), respectively.

Subdwarf O stars have effective temperatures above 40000 K. They are all He-rich. Their positions in the (g, T_{eff}) diagram (Fig. 4) can be reached either along a Castellani track or along a Sweigart track. The observed helium abundances will tell us which track is relevant. The border line between the He-poor sdOBs and the He-rich sdOs was thought to be due to He II convection that reaches up to the photosphere when T_{eff} 40000 K. Stars evolving on a Castellani track from the sdOB domain would then suddenly experience mass motions that would inhibit diffusion. This picture, however, is wrong on two grounds:
- the convenction domain does not exactly match the observed border line (Fig. 4)
- convection becomes appreciable only if He is enriched by at least a factor of two with respect to normal composition. A He-poor sdOB star that reaches the potential convection domain stays He-poor.

The conclusion is that the known sdOs cannot have evolved along a Castellani track but along a Sweigart track, where they have reached low gravity regions close to the AG branch. Here motions are likely to

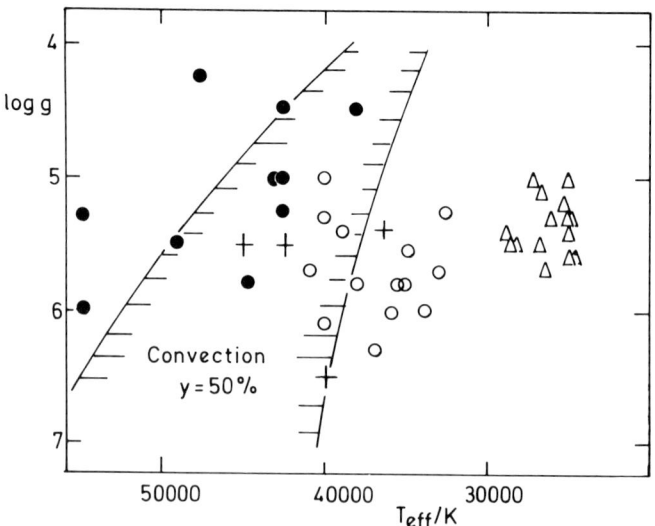

Fig.4: (g, T_{eff})-diagram of hot subdwarfs. ++ designate sdO's which show no hydrogen and ●● those with helium enriched to 50% (by numbers), typically. The other symbols are as in Fig. 3. The region of potential He II convection is indicated (for an assumed He-content of 50% by number).

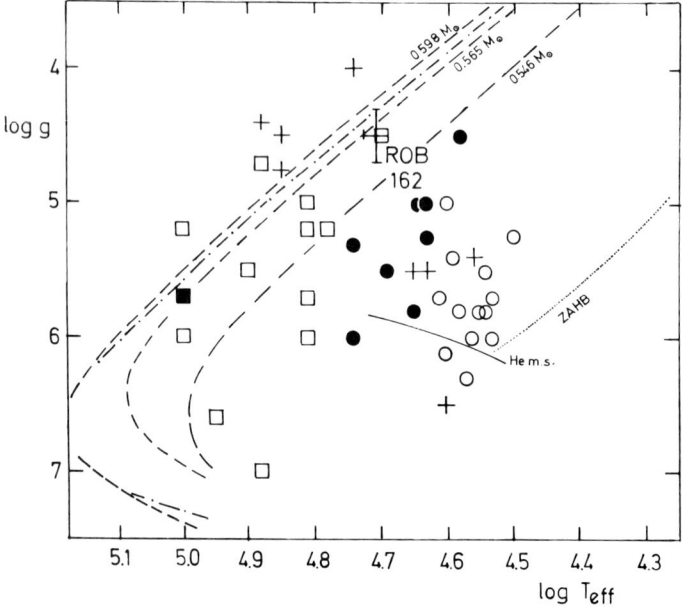

Fig. 5: (g, T_{eff})-diagram of all subdwarfs excluding components of binary systems but including also Central Stars (squares). The other symbols are as in Fig. 3 and 4. Evolutionary tracks for post-AGB stars (Schönberner, 1979, 1983) are also shown.

occur which stop diffusion. This assumption explaines a normal helium
composition. Where the observed He-enrichment has its origin, still
remains an enigma. We come back to this point later.

That a sizeable fraction of sdOs have reached the tip of the AG branch
has to be concluded from their present location in the (g, T_{eff})-
diagram. Five stars lie on a Schönberner track (Schönberner, 1979,
1983, Fig. 5) and hence are suspected to be Central Stars of PN. For
one of them a nebula has already been found. In this respect it may be
speculated whether the planetary nebula phenomenon is restricted to the
canonical cases of central stars, or whether nebula ejection is a more
common feature, shared also by the subdwarf O stars. For whether we
readily see a nebula depends on the proper timing of nebula expansion
and UV-brightening of the central star which leads to the required
ionization of the otherwise invisible nebula. The close relationship
between central stars of PN and subdwarfs may also be anticipated from
the similarity of the masses: from the Schönberner tracks follows that
a typical sdO has a mass that is only by less than 0.1 $M_\odot$ smaller than
that of a typical central star. There remains, however, one important
difference between the two classes: all sdO's are He-rich, while almost
all CSPN's are He-normal or even He-poor. (The CSPNs referred to
comprise only the sdO-type subgroup that has been spectral analysed
directly.)

Last but not least let us now turn to an important question: why are
the subdwarf O stars enriched in helium? We have discussed before, that
the abundances of the progenitors on the (high gravity) horizontal
branch stars are distorted due to diffusion and hence do not reflect
the true atmospheric composition. During the subsequent evolution
towards lower gravities, diffusion comes to an end, and the original
abundances are restored. Does this then mean that the helium is already
enriched in the outer layers of the horizontal branch progenitors of
the sdO's, or have hitherto unknown processes close to the AGB-phase
produced the He-enrichment? In the first case, it would mean that the
progenitors of the sdO's have suffered severe mass losses during the
first He-flash, a hypothesis which has the appeal that it could explain
the small masses of the subdwarf O stars. Spectroscopy, however, cannot
prove this directly, as in the hot high gravity HB-stars diffusion
spoils the analysis, and in the cool low gravity stars HeI cannot be
observed directly as it is not excited.

An attempt has been made to follow up observationally the evolution of
hot post-HB stars towards lower gravities (Heber and Langhans, 1986),
in the hope that there may be found stars which have not reached the
AGB yet where composition changes may take place but which have
gravities small enough to quench diffusion. Such stars if uniquely
defined would reveal the true cmposition of hot HB-stars. Fig. 6 shows
the result of the analysis of a sample of stars which are considered
the best candidates. The enormous difficulties encountered in such a
study becomes manifest immediately: the Sweigart-track crosses the H-
main sequence ZAMS near $\log g = 4$.

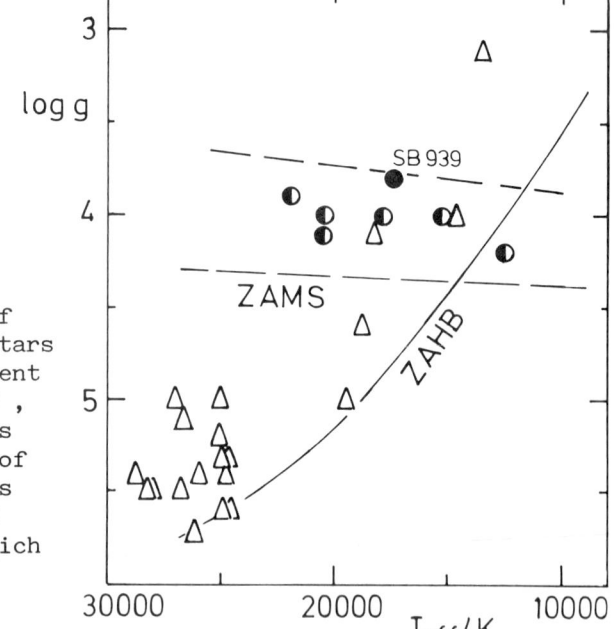

Fig. 6: (g, T_{eff})-diagram of evolved HB stars. Stars with normal He-content are designated by ◐, otherwise symbols as before. The region of the main sequence is indicated by dashed lines. The helium rich object (SB 939) is shown as a filled circle.

Even for stars chosen at high galactic latitude, the probability to find runaway B-type main sequence stars (fainter than 10th magnitude) is high. The 6 normal composition stars are aligned along the ZAMS which strongly suggests that these stars indeed are runaway main sequence stars. However, there is one object (SB 939) with log g = 3.7 and T_{eff} = 17400 K in which helium appears enriched as in subdwarf O stars. If SB 939 is confirmed as a post HB-star, it would mean that hydrogen depletion takes place during the first He-flash. Alternatively it may be conjectured that SB 939 is a runaway intermediate helium star. Whether our key object SB 939 is a true post HB-star and whether our far reaching conclusions as to the He-flash are correct will be checked by a further analysis of the N/C abundance ratio.

References:

Groth, H.G., Kudritzki, R.P., Heber, U.: 1985, Astron. Astrophys. **155**, 33
Heber, U., Langhans, G.: 1986, Proc. 5th IUE conf., London, in press
Hunger, K., Gruschinske, J., Kudritzki, R.P., Simon, K.P.: 1981, Astron. Astrophys. **95**, 244
Newell, E.B.: 1973, Astrophys. J. Suppl. **26**, 37
Schönberner, D.: 1979, Astron. Astrophys. **79**, 108
Schönberner, D.: 1983, Astrophys. J. **272**, 708
Sweigart, A.V., Mengel, J.G., Demarque, P.: 1974, Astron. Astrophys. **30**, 13

EXCITATION OF PLANETARY NEBULAE AND
EVOLUTION OF THEIR CENTRAL STAR

Detlef Schönberner
Institut für Theoretische Physik und Sternwarte
der Universität Kiel
Olshausenstr. 40
D-2300 Kiel, F.R.G.

SUMMARY: The evolution of central stars of planetary nebulae has been
studied by means of a new distance-independent method. Using Kaler's
(1983) sample of large planetary nebulae, it is shown that the
distribution of the nebular excitation, which is distance-independent,
indicates a very fast luminosity decrease of the exciting stars and
can only be explained by post-AGB evolutionary models which burn
hydrogen in a shell. The nebular excitation is also well-correlated
with the luminosity function of their central stars in that low-
excitation nebulae have intrinsically faint central stars, whereas the
bright central stars belong to the highly excited nebulae. These
observations confirm earlier results of Schönberner (1981), according
to which PN formation must occur preferentially during the quiescent
hydrogen burning phase at the tip of the AGB.

I. INTRODUCTION

The question about the timing of the PN formation is of crucial
importance for our understanding of the late phases of stellar
evolution. Schönberner's (1981 and this Workshop) conclusion about CPN
being hydrogen burners was not only based on the well-known time
argument but also on the fact that the observed shape of the
luminosity function of CPN could only be explained by hydrogen-burning
post-ABG models (e.g. Fig. 9 of Schönberner 1981, and discussion in
Schönberner and Weidemann 1983). One special feature of this
luminosity function is a deficit ("gap") of CPN with $M_v \approx 5$. This gap
was already apparent in the classical study by O'Dell (1968), and can
be explained by hydrogen burning post-AGB models of $\approx 0.6\ M_\odot$ because
these models experience a rapid luminosity drop of ≈ 1 dex within only
about 10^3 yrs when hydrogen burning starts to cease. Conversion into a
luminosity function leads to a pronounced dip between $M_v \approx 4.5$ and 6.0,
the exact position depending somewhat on the mass of the models. Such
a luminosity drop is not found in models that leave the AGB whilst
burning helium (cf. Fig. 1 in Iben 1984), and this fact clearly
indicates that at least the majority of CPN must be hydrogen burners.

A gap is also apparent if one uses Kaler's (1983) ensemble of larger PN (cf. Fig. 1).

The reality of the gap has, however, been doubted, probably because of the well-known distance problem of planetaries. We therefore investigate a distance-independent method that can tell us something about the evolution of CPN. Taking into account the fact that nebular and central star evolution are strongly coupled with each other, we consider the nebular excitation, measured by the flux ratio of two appropriate emission lines, as an indicator of the evolutionary state of the central object. If a CPN evolves as outlined in the recent literature, the increasing temperature, combined with a steadily decreasing radius, results in a temporal change of the spectral distribution of the emergent flux which, when coupled with the expansion of the PN, determines the excitation of a PN. An idea of how the number of Lyman photons varies during the evolution of a 0.6 $M_\odot$ model can be obtained from Fig. 6 of Schönberner (1981) or from Fig. 2 of Kwok (1985). The maximum occurs between effective temperatures of 60,000 and 70,000 K, followed by a fast decrease when the stellar luminosity drops rapidly due to a very fast shrinking of the envelope. Such a large variation in the number of emitted ionizing photons must also lead to a temporal variation of the nebular excitation.

2. THEORY

We decided to choose the flux ratio of He II 4686 Å to H_β as a measure of the nebular excitation:

$$E = 100\, F_{4686}/F_\beta$$

This choice for E has the advantage of being only weakly dependent on the interstellar extinction because of the relatively small wavelength separation of both lines. Furthermore, F_{4686}/F_β is known for many PN. Assuming a spherical model PN with constant density, the excitation parameter can be expressed as follows:

$$E = 100\, j(4686)\, N_{He} V_{He^{++}} / j(\beta)\, N_H V_{H^+} ,$$

where j is the emission coefficient, N the particle density and V the ionized volume. Inserting numerical values (cf. Osterbrock 1974), we get for $N_{He}/N_H = 0.1$

$$E = 110\, V_{He^{++}}/V_{H^+}$$

For a further discussion, we consider the following two cases: ionization-bounded PN and density-bounded PN.

2.1. Ionization-bounded PN

In this case, and assuming ionization equilibrium, we have

$$V_{He^{++}}/V_{H^+} = \frac{Q(He^+)}{N_{He^{++}}N_e \alpha(He^+)} \Big/ \frac{Q(H^0)}{N_{H^+}N_e \alpha(H^0)}$$

with Q being the number of ionizing photons per second, N_e the electron density and α the recombination coefficient. Rearrangement and inserting numerical values leads to

$$V_{He^{++}}/V_{H^+} = 1.7\, Q(He^+)/Q(H^0), \text{ or}$$

$$E = 190\, Q(He^+)/Q(H^0).$$

Thus, E is only a function of the central star's effective temperature!

A 0.6 $M_\odot$ post-AGB model reaches at the turn-around point its highest effective temperature of 160000 K which corresponds, for a black body, to $Q(He^+)/Q(H^0) = 0.25$, and, consequently, to $E_{max} \approx 50$. Note further that a more sophisticated study by Stasinska and Tylenda (1986) showed that E approaches only 60 if T_{eff} goes to infinity! We conclude that the nebular excitation of an optically thick PN always remains below 60.

2.2. Density-bounded PN

Now we always have $V_{H^+} = V_{PN}$, but we have to distinguish between $V_{He^{++}} < V_{PN}$ or $V_{He^{++}} = V_{PN}$. In the first case we have

$$(V_{He^{++}}/V_{H^+})_i \leq V_{He^{++}}/V_{H^+} < 1$$

$$E_i \leq E < 110,$$

where the subscript i indicates the ionization-bounded limit. In the second case, we have simply

$$V_{He^{++}}/V_{H^+} = 1$$

$$E = 110.$$

In the density-bounded case, the excitation E depends on the optical depth and is larger than for the optically thick case. The excitation maximum is reached if also all the helium is doubly ionized.

We are now able to estimate the possible influence of the central star's evolution on the excitation of its nebula. Hydrogen-burning post-AGB models show a rapid luminosity and temperature drop at the turn-around point because of the decline of their nuclear burning. For a 0.6 $M_\odot$ model, which may be considered as typical for real CPN, it is found that, within about 2000 yr, $Q(H^o)$ drops by a factor of about 10, this being mainly due to the radius decrease, whereas $Q(He^+)$ drops even faster (i.e. by a factor of 20) because of its higher sensitivity to temperature. Thus, it may happen that recombination leads, at least temporarily, to $V_{He^{++}} < V_{PN}$ or even also to $V_{H^+} < V_{PN}$, i.e. even to an ionization-bounded PN. This means that a very low excitation, $E \approx 10$, can be reached because at $T_{eff} = 100000$ K, $Q(He^+)/Q(H^o) = 0.06$ which gives $E = 10$).

Whether this will happen depends on how fast the PN expands, compared to the evolutionary changes of its CPN. Fast expansion leads to lower densities which, in turn, keep the ionization high even during a luminosity drop of the CPN. The same holds for only slowly evolving CPN. Since the evolutionary timescale of post-AGB models varies as M^{-10} (Iben and Renzini 1983), low-mass CPN (i.e. with ≈ 0.55 $M_\odot$) are expected to evolve too slowly (cf. Schönberner 1983) to be able to induce nebular recombination. This is also true for helium-burning post-AGB models of 0.6 $M_\odot$ since they evolve about 3 times slower than the corresponding hydrogen-burning models. On the other hand, if a CPN evolves too quickly (as is expected for very massive CPN), the PN will remain optically thick - and E rather low - until the star dims to about 10^2 $L_\odot$. Continued nebular expansion will then eventually lead to a high excitation.

From this discussion, it is clear that an investigation of the excitations in PN may shed some light on the evolution of their exciting sources. If, for instance, due to a rapid stellar luminosity drop recombination really occurs, the excitation parameter E should be either high ($E \approx 100$) or rather low ($E \leq 50$), depending on whether the exciting CPN is just before or after the luminosity drop. PN with intermediate excitation should then be rare because of the extremely fast stellar evolution from high to low luminosity. In other words, the gap of the central stars' luminosity function should have an equivalent in the distribution of the excitation parameter E.

3. DATA ANALYSIS AND RESULTS

We have examined observational data for 82 older planetaries compiled by Kaler (1983). These PN have absolute radii (inferred from the angular diameters and distances) of at least 0.18 pc, which, assuming an average expansion velocity of 20 km/s, corresponds to a minimum age of about 8 10^3 yrs. This sample is especially useful for our purposes since it does not contain younger objects which frequently also have low excitation because of their cooler central stars. We can assume that all CPN of Kaler's sample are very hot and cover the evolutionary stages from just above till well below the turn-around point of the evolutionary tracks.

The compilation of Kaler contains many newly measured objects which were not available for the study of Schönberner (1981). Thus, we examined first the luminosity function of the central stars in order to check whether the "gap" between its bright and faint part is also visible in this larger sample. We extracted 66 CPN whose apparent

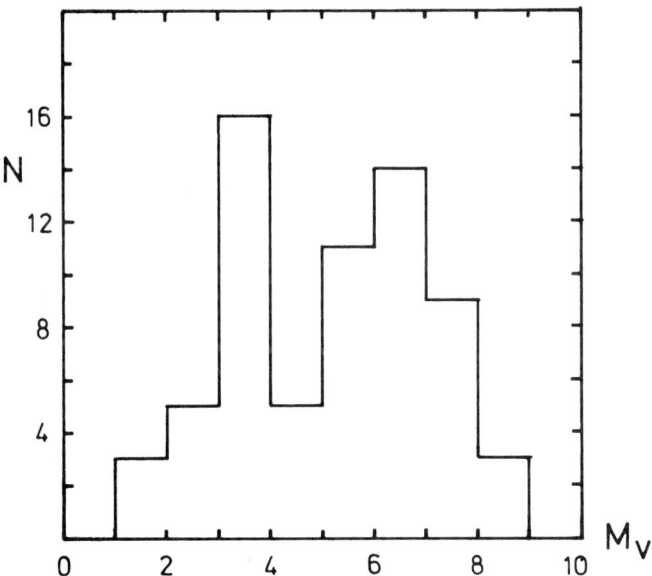

Fig. 1: Central star luminosity function of Kaler's (1983) sample of larger PN. Total number of objects: 66.

magnitudes are known within at least $\pm$ 0.5 mag and, using the distances of Daub (1982) or Cahn and Kaler (1971) and correcting for interstellar extinction, we found the luminosity function as shown in Fig. 1. Between M_v = 4 and 5, a pronounced dip occurs, which is identical with that discovered by Schönberner in 1981 (cf. Fig. 9a therein). In Fig. 1, the decrease towards the brighter end of the luminosity function (M_v < 3) is due to the absence of young planetaries with intrinsically bright central stars in Kaler's sample. Note also that Fig. 1 contains in the interval M_v = 3 to 7 about twice as many objects as Schönberner's (1981) Fig. 9a, namely 46 compared to 26!

Thus, the earlier result of Schönberner (1981) is supported by the much larger Kaler sample and, when interpreting this dip as being caused by a rapid evolutionary phase, the claim that the majority of central stars are hydrogen burners is strengthened. We are, however, still left with the distance problem and will next investigate whether evolutionary effects of this sort will also show up in the nebular excitations, which were also compiled by Kaler (1983). We selected the objects from Kaler's list according to the following criteria: only PN with small errors, $\Delta E \leq 20$, were accepted, the rest rejected. This procedure resulted in a sample of 57 PN, 37 of which have even errors $\Delta E \leq 10$. In most cases the correction due to interstellar extinction is negligible compared to the observational errors ΔE. Fig. 2 shows the distribution of E for the sample of 57 objects together with that of

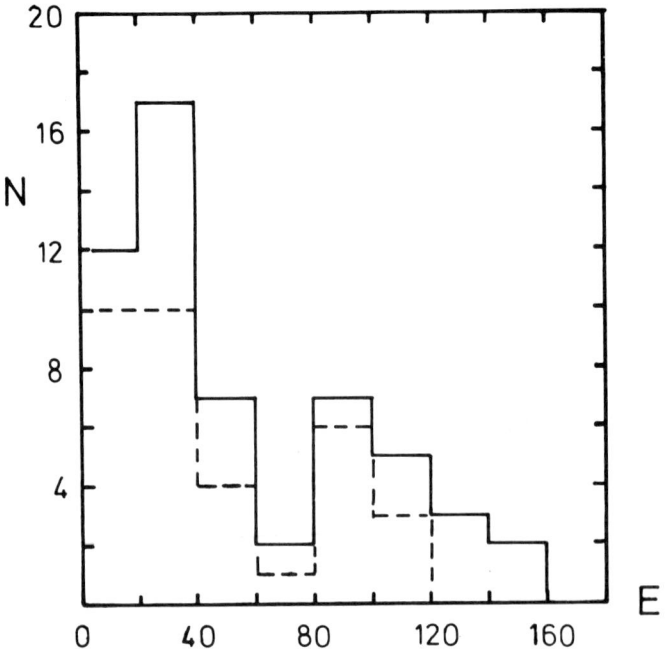

Fig. 2: Distribution of the nebular excitations in Kaler's (1983) sample of larger PN. ——— : Objects with $\Delta E \leq 20$; --- : Objects with $\Delta E \leq 10$.

the smaller subsample with $\Delta E \leq 10$. Both distributions are about equal and appear to be bimodal, with about one third of the PN having high excitation (E > 80) and about two thirds low excitation (E < 60). Only two PN have excitations between 60 and 80. The same "gap" is evident in Fig. 5 of Kaler (1983).

It is admitted that the statistics is not very satisfying: the gap in Fig. 2 might be spurious because of the small number of objects per sampling interval (especially for E > 40) and the relatively large observational errors. Selecting only objects with, for example $\Delta E \leq 5$, does not help either because then only 18 objects remain, of which 16 have E < 60 and 2 E > 80. It is, however, safe to conclude that the majority of the old planetaries of Kaler's sample have a low excitation parameter E. The distribution in Fig. 2 is what we would expect if planetaries go through a phase of high excitation (E ≈ 100) which is followed by a rapid recombination to low excitation (E $\leq$ 50).

Accordingly, we interpret the deficit of PN with intermediate excitations, and also the large number of low excitation PN, as shown in Fig. 2, in the same way as in the case of the luminosity function, namely: by a rapid drop in the luminosity of the central star as it occurs in hydrogen-burning post-AGB models! If this scenario is true,

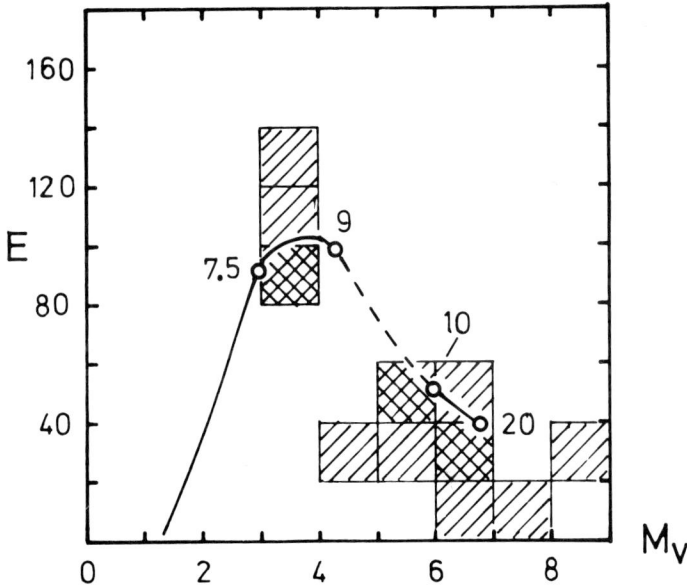

Fig. 3: Nebular excitation E vs. absolute magnitude M_v of the central stars, for the same sample as in Fig. 2. Dashed squares contain 3 or 4 objects, crossed squares more than 4 objects. Squares containing only one or two object are not marked. The curved line illustrates the evolution of a model PN as a function of its CPN absolute magnitude. Some ages (in 10^3 yr) along the evolutionary track are also indicated. The fast luminosity decline of the model is emphasized by the dashed part of the evolutionary track. See text for further details.

we expect a correlation between the intrinsic brightness of the central stars and the excitation of their nebulae, in the sense that the old PN with low excitation also belong to intrinsically faint CPN. We plotted therefore in Fig. 3 the excitation E vs. M_v, the absolute magnitudes of the central stars. In order to reduce the scatter introduced by the distance uncertainties, we binned the data as shown in the figure and marked only squares that contained at least 3 data points, thereby loosing about one third of the sample (we used all 57 objects of Fig. 2).

The observed correlation between E and M_v is exactly what we expect from the above discussion. The old, low excitation nebulae preferentially have intrinsically faint, and the high excitation nebulae brighter central stars. We have 29 CPN with $M_v \geq 5$, the mean excitation of their PN being $\bar{E} = 39 \pm 6$; the mean excitation for the 12 CPN with $3 \leq M_v \leq 4$ is $\bar{E} = 101 \pm 7$. Both groups are clearly separated and the "gap" in the M_v-distribution is obviously correlated to the deficit of old PN with intermediate excitations. We also find that the brightest CPN in Fig. 3 have on the average smaller, i.e. younger, PN than the faint CPN. The differences between the mean radii of the 15 PN with central stars of $5 \leq M_v \leq 6$ ($\overline{E} = 50 \pm 10$) and of the 12 PN with central stars of $3 \leq M_v \leq 4$ is $\Delta \overline{R}_N \approx 0.05$ pc. Thus, the mean nebular radii of both groups are about equal and their very small difference corresponds, if one assumes an expansion velocity of 20 (30) km/s, to a mean age difference between both groups of about 2500 (1700) yr. Regardless of the exact amount of the mean expansion velocity, these observations clearly indicate that the decrease of the nebular excitation occurs on an extremely short timescale of some 10^3 yr, demanding also a significant luminosity drop of the central-star luminosities within the same period of time!

Also shown in Fig. 3 is an "evolutionary" track of a model PN with $M_{PN} = 0.3\ M_\odot$ that expands with $V_{exp} = 20$ km/s whilst coupled to a $0.6\ M_\odot$ post-AGB model of Schönberner (1979, 1981) as the exciting source. This track is extracted from Vilkoviskii et al. (1983 and private communication), to which the reader is referred for more details. The evolution proceeds from left to right, and the dashed portion emphasizes the rapid luminosity drop of the model by about 1 dex within only 10^3 yrs when hydrogen burning starts to cease about 10^4 yrs after the model has left the AGB (Schönberner 1981). This model not only reproduces the observed correlation of nebular excitations with the visual luminosities of the exciting stars astonishingly well but also the fast temporal decrease of the nebular excitations.

Finally, we would like to add that helium burning post-AGB models of $\approx 0.6\ M_\odot$ do not fade below $M_v \approx 4$ within the typical life time of a PN! Some of the high excitation PN might well have such helium burning CPN, or, alternatively, may have hydrogen-burning CPN of lower mass ($0.56\ M_\odot$). In both cases, the excitation stays at a high level during the entire lifetime of the nebula shell.

3. CONCLUSION

We have shown by means of a sample of older PN that the distribution of their excitation is strongly correlated with the luminosity function of the corresponding central stars. There is also some evidence for an excitation "gap" between $E = 60$ and 80, which appears to be related to a distinctive "gap" in the central star luminosity function. It would be very important to make more observational efforts to improve the statistics of the excitation distribution of PN since we would then have a distance-independent tool for the study of later evolutionary phases of their central stars.

We conclude that only post-AGB models which burned hydrogen quietly when they left the AGB, and which continue to do so, depict such a rapid luminosity drop as is necessary to account for the observations. This, in turn, is another confirmation of the conclusion already made earlier by Schönberner (1981), namely that the PN-formation is <u>not</u> initiated by a thermal pulse, but that it occurs preferentially during the quiescent hydrogen-burning phase. We would like to emphasize that this conclusion does not depend on the existence of the excitation dip in Fig. 2 but follows from the fact that practically all intrinsically faint CPN have nebular shells with E < 60.

Our conclusion, of course, does not exclude the occurrence of helium burning CPN, either because a thermal pulse may coincide quite by chance with the star's departure from the AGB, or because a late thermal pulse may convert a hydrogen burning CPN into a helium burner (cf. Iben 1984, also Schönberner 1979). The fraction of helium-burning CPN depends directly on the (unknown) details of the PN formation on the AGB, i.e. on the wind strength (the superwind), as a function of the evolutionary state on the AGB. An estimate of the fraction of helium-burning CPN, which is only based on observational arguments, is given by Schönberner (1987). We will only mention here that helium-burning CPN seem to be rather rare, their fraction of the total CPN population being certainly less than 25%, maybe even as small as 15%.

ACKNOWLEDGEMENTS. The author thanks Prof. V. Weidemann for many fruitful discussions.

REFERENCES

Cahn, J.H., Kaler, J.B.: 1971, Astrophys. J. Suppl **22**, 319
Daub, C.: 1982, Astrophys. J. **260**, 612
Iben, I.Jr.: 1984, Astrophys. J. **277**, 333
Iben, I.Jr., Renzini, A.: 1983, Ann. Rev. Astron. Astrophys. **21**, 271
Kaler, J.B.: 1983, Astrophys. J. **271**, 188
Kwok, S.: 1985, Astrophys. J. **290**, 568
O'Dell, C.R.: 1968, IAU Symp. No. 34, "Planetary Nebulae", D.E. Osterbrock and C.R. O'Dell Eds., Reidel, Dordrecht, p. 361
Osterbrock, D.E., 1974, "Astrophysics of Gaseous Nebulae", Freeman and Co., San Francisco, p. 66
Schönberner, D.: 1979, Astron. Astrophys. **79**, 108
Schönberner, D.: 1981, Astron. Astrophys. **103**, 119
Schönberner, D.: 1983, Astrophys. J. **272**, 708
Schönberner, D.: 1987, Astron. Astrophys. in press
Schönberner, D., Weidemann, V.: 1983, IAU Symp. No. 103, "Planetary Nebulae", R.D. Flower Ed., D. Reidel, Dordrecht, p. 359
Stasinska, G., Tylenda, R.: 1986, Astron. Astrophys. **155**, 137
Vilkoviskii, E.Y., Kondrateva, L.N., Tambovtseva, L.V.: 1983, Sov. Astron. **27**, 194

DISCUSSION.

<u>Rodriguez</u>: You divided planetary nebulae in density bounded and ionization bounded. There is, however, good evidence that some PN (in particular the bipolar ones) are density bounded in some direction and ionization bounded in others. Will this affect your analysis significantly?

<u>Weidemann</u>: Certainly. The effects are even more important than that of the filling factor.

<u>Peimbert</u>: Is it possible from the I(4686)/I(H-beta) vs. radius diagram to improve on the observational mass distribution of the central stars?

<u>Weidemann</u>: I don't think so. This is too approximate. I presented a viewgraph displaying in the H-R diagram the expected locations of CPN's distributed according to the white dwarf mass distribution and within the framework of Schonberner and Wood and Faulkner tracks. It demonstrates that the few high mass nuclei are close together at low luminosities and equal ages (or radii) with the 0.6-0.7 $M_\odot$ CPN's; then it is hard to separate them.

IONISATION AND DYNAMICAL STRUCTURE OF PLANETARY NEBULAE

L. Bianchi(1), M. Grewing(2), C. Falcetta(1), M.Baessgen(2)
(1)Osservatorio Astronomico di Torino
I-10025 Pino Torinese, Italy
(2)Astronomisches Institut der Universitaet
D-7400 Tuebingen, W.Germany

ABSTRACT. The ionisation structure of planetary nebulae (PNe) is investigated by means of monochromatic images in the strongest optical emission lines from the most abundant ionic species, and by means of optical and ultraviolet spectroscopy. These data are supplemented by high-resolution ($R \sim 10^5$) spectroscopic observations which allow to map out the velocity fields in the nebular envelopes. This is particularly interesting since several nuclei show evidence for conspicuous supersonic winds, which could have a significant effect on the material ejected at much lower velocities during earlier evolutionary phases of the central stars. As this is an ongoing study, we present here preliminary results for two objects : NGC 3918 and NGC 2440.

1. THE OBSERVATIONS

As part of a long-term program we obtained direct images of several planetary nebulae through narrow-band interference filters in front of the ESO-CCD cameras at the Danish 1.5m and the ESO 2.2m telescopes on La Silla.

Also, we obtained high resolution ($R \sim 10^5$) slit spectra in the neighborhood of the H-alpha and [N II] 654.8nm and 658.4nm lines with the CES spectrograph at the ESO 1.4m CAT. The actual resolution measured from unblended lines in the Thorium calibration spectrum corresponds to about 0.005nm or 3 km/s. The wavelength calibration, performed with the ESO IHAP routines, allows an accurate pixel-to-wavelength conversion but with a zero-point uncertainty. As a consequence, the nebular expansion velocities derived from the line profiles should be rather precise, whereas the systemic velocities remain somewhat uncertain.

Though the instrument is designed for high resolution observations of bright point sources (stars), we were able in a few observing runs to obtain good S/N spectra for several planetary nebulae also in off-center positions. The exact location of the spectrograph slit is, however, a problem, given the facts that the TV camera used for target aquisition and guiding is usually not sensitive enough to bring up the nebulae emission, and that the field is rotating. As the latter effect can not be compensated for, the spectra suffer from spatial smearing despite the narrow slit (3 arcsec) chosen to match the spectral resolution.

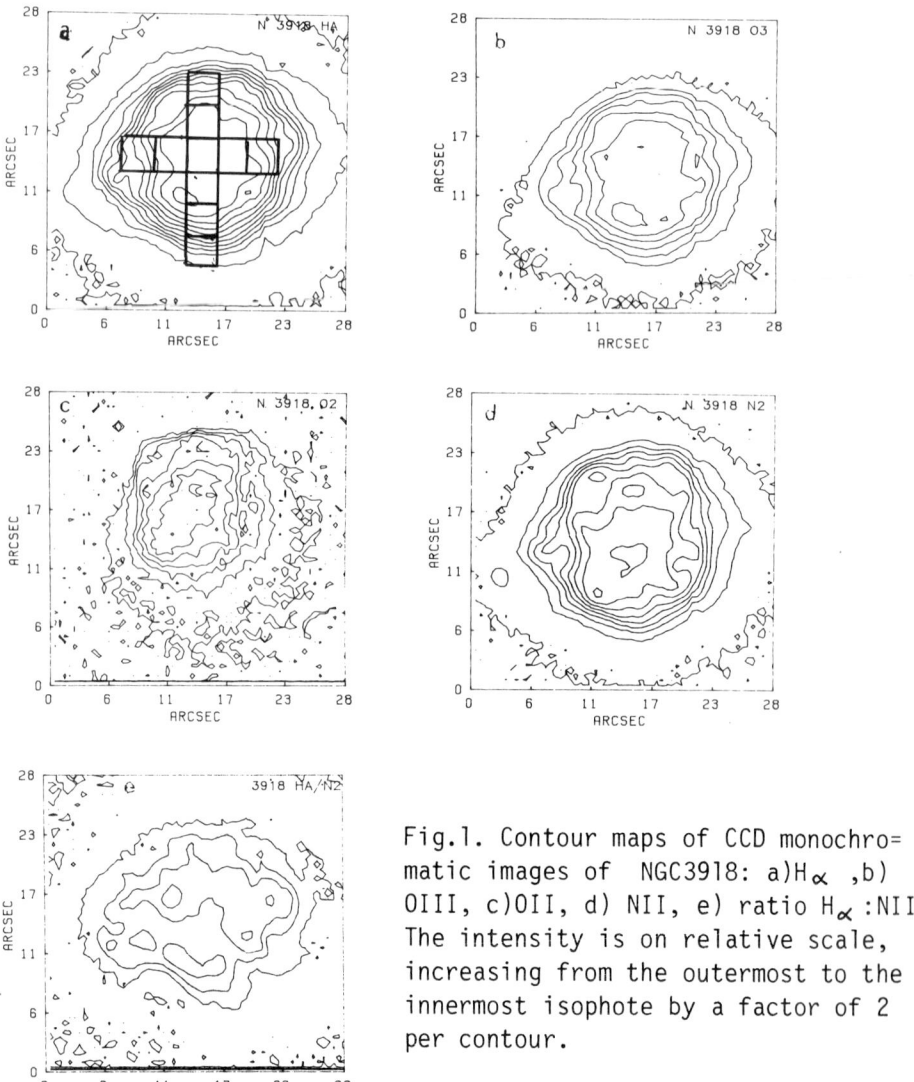

Fig.1. Contour maps of CCD monochromatic images of NGC3918: a)H_α, b) OIII, c)OII, d) NII, e) ratio H_α:NII. The intensity is on relative scale, increasing from the outermost to the innermost isophote by a factor of 2 per contour.

In the next two subsections we discuss the results obtained from the imaging and the high resolution observations of NGC 3918 and NGC 2440. They are discussed in this sequence because NGC 3918 looks the morphologically simpler object.

2. NGC 3918

NGC 3918 has a rather smooth surface brighness distribution and is almost circularly symmetric. A photo-ionization model for the object has been constructed by Clegg et al.(1986), based on UV, optical and radio observations. In order to reproduce these data, these authors had to assume, however, that the nebula consists of two thick cones at the front and rear, with an optically thin equatorial region. In an earlier paper, Clegg and Harrington (1985) had studied the velocity field of NGC 3918 with a resolution of 20 km/s.

In Figures 1a-d we show contour maps of the object derived from an H-alpha , [O III], [O II], and [N II] image of NGC 3918, taken with the CCD at the Danish 1.5m telescope on La Silla. Fig. 1a also shows the positions in which high resolution spectra were obtained.

Such spectra are shown in Fig.2 which displays the H-alpha and [N II]658.4nm profiles on a velocity scale, centered on the systemic velocity. The corresponding [N II]654.8nm emission is not shown here because of its lower S/N but it generally confirms the shape of the stronger [N II] line.

The first thing to notice is the narrower width of the [N II] lines as compared to H-alpha. This is often found in PNe (see e.g. Pottasch 1984). The average line widths correspond to about 18km/s ([N II]) and 30-35km/s (H-alpha), respectively. These values clearly exceed the thermal line widths of 6km/s and 22km/s, respectively, that one calculates for the physical conditions in the nebula. The excess can be caused by a velocity gradient within the nebular shells and/or turbulence. Indeed, Clegg and Harrington (loc.cit.) suggest that the velocity is increasing outwards in this object.

The expansion velocities measured in various locations are compiled in Fig.3. As is to be expected, the separation of the peaks decreases from the centre towards the edge of the object due to the projection effect along the line of sight. Surprisingly enough, however, the lines remain double-peaked out to a radial distance of 9 arcsec and beyond, i.e. out to a distance beyond the nebular radius that is normally quoted (7.7 arcsec; see e.g. Pottasch 1984). The CCD images confirm indeed that NGC 3918 has a faint outer halo (cf.Fig.1).

From Fig.2 one can read a maximum expansion velocity of $\sim$26km/s from the [N II] and $\sim$18km/s from the H-alpha line profiles. This is consistent with the expansion velocity increasing outwards and the fact that the H-alpha/[N II] intensity ratio is much higher in the inner regions of the nebula than in the outskirts (cf. Fig.1e).

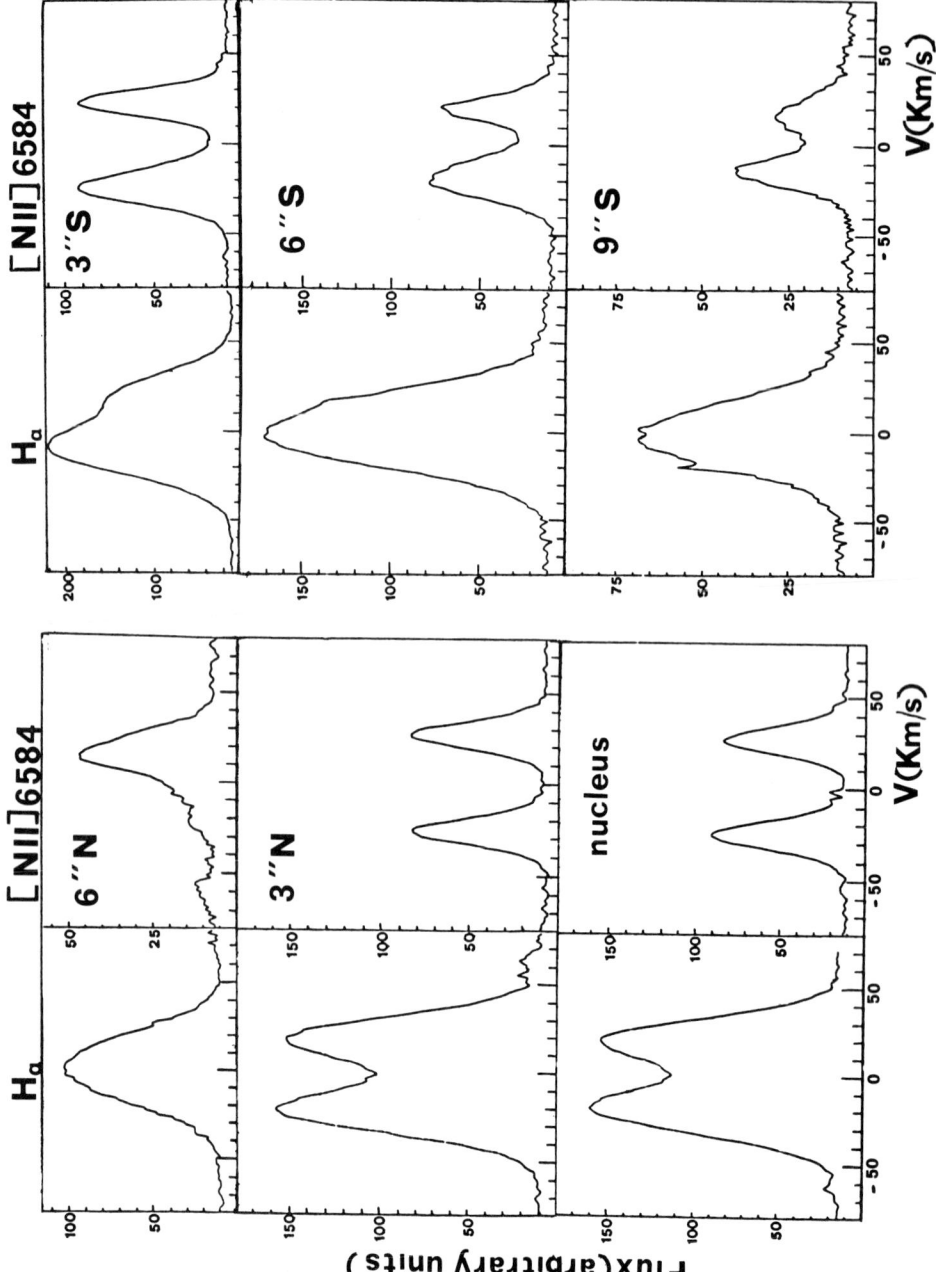

IONISATION AND DYNAMICAL STRUCTURE OF PLANETARY NEBULAE

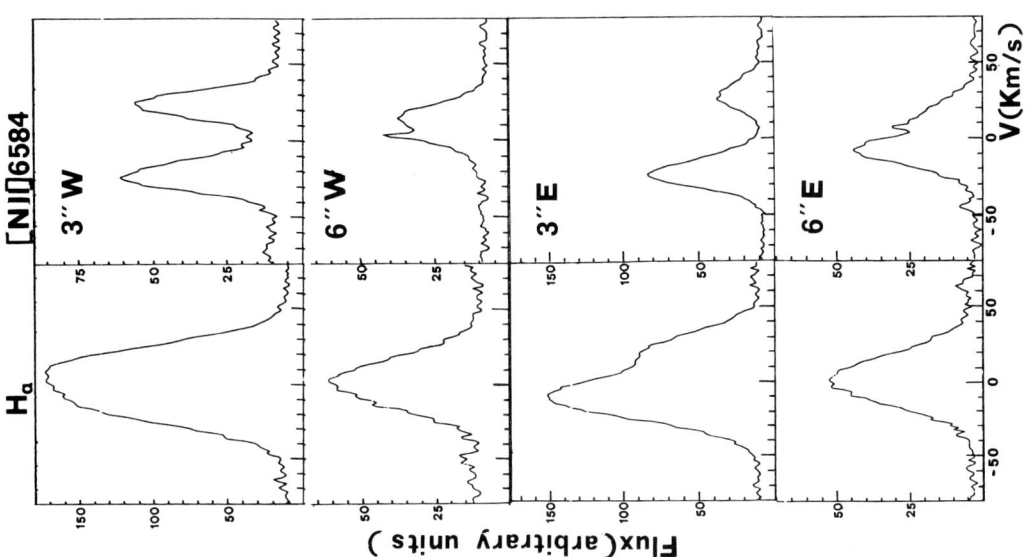

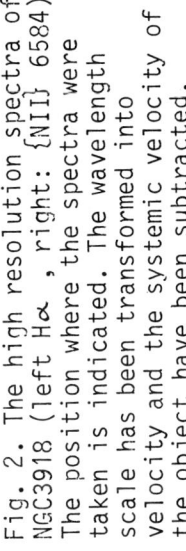

Fig. 2. The high resolution spectra of NGC3918 (left Hα, right: [NII] 6584). The position where the spectra were taken is indicated. The wavelength scale has been transformed into velocity and the systemic velocity of the object have been subtracted.

Fig.2 also shows that the rather regular morphological structure suggested by the direct images is not supported by the kinematical data. Considering e.g. the spectra taken to the east of the central star, the component which approaches the observer is much stronger than the receeding one whereas in other locations the two are similar. Further differences are seen in the shape and the width of the lines some of which seem to be blends of different velocity components (see e.g. the profile observed 6 arcsec south of the nucleus). Clearly, a detailed modelling of the density and velocity structure within the object will be needed to fit the spectra quantitatively. This work is in progress.

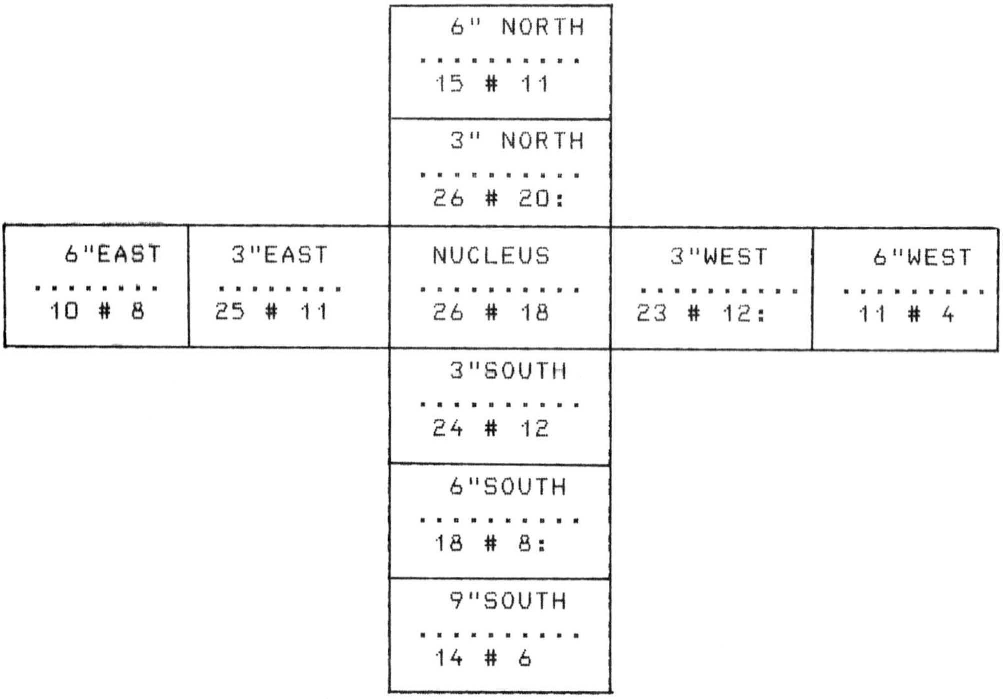

		6" NORTH .... 15 # 11		
		3" NORTH .... 26 # 20:		
6"EAST .... 10 # 8	3"EAST .... 25 # 11	NUCLEUS .... 26 # 18	3"WEST .... 23 # 12:	6"WEST .... 11 # 4
		3"SOUTH .... 24 # 12		
		6"SOUTH .... 18 # 8:		
		9"SOUTH .... 14 # 6		

Fig.3. The measured expansion velocities in NGC3918. Each box represents a position where a high resolution spectrum was taken. In each box the position on the object and the measured expansion velocities (H_α =[NII]) are given.

3. The velocity field in NGC 2440

NGC 2440 has a much more complicated morphological structure as compared to NGC 3918. This can be seen from Fig.4 which shows monochromatic images taken through narrow band interference filters with the ESO-CCD attached to the 2.2m telescope on La Silla. The nebula consists of a rather thick bar and two large thinner ansae with a rather filamentary structure.

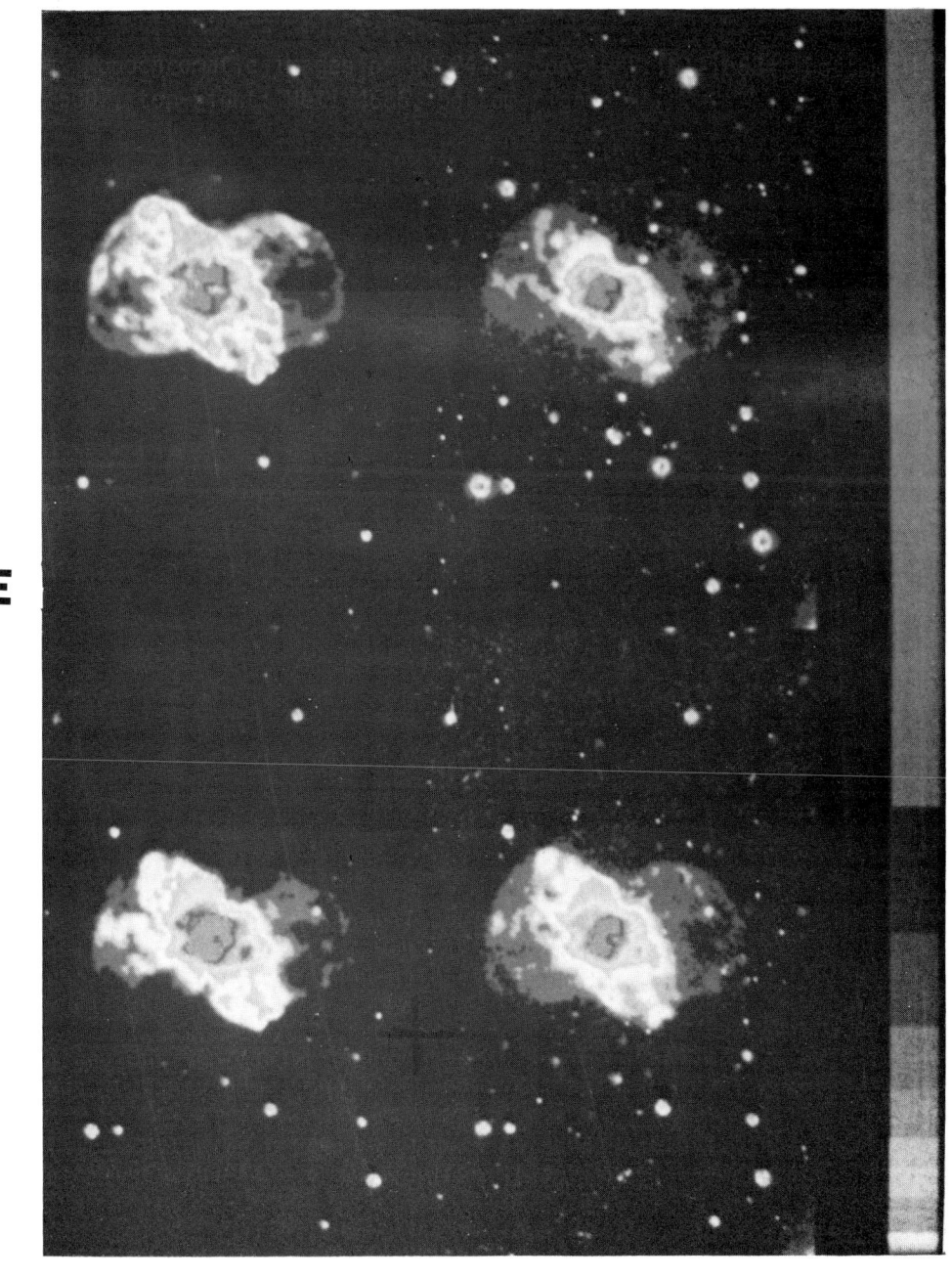

Fig.4a. Monochromatic images of NGC2440: top-left:[NII]6584, bottom-left: [OIII]5007, top-right: HeII 4686, bottom-right: H_β .

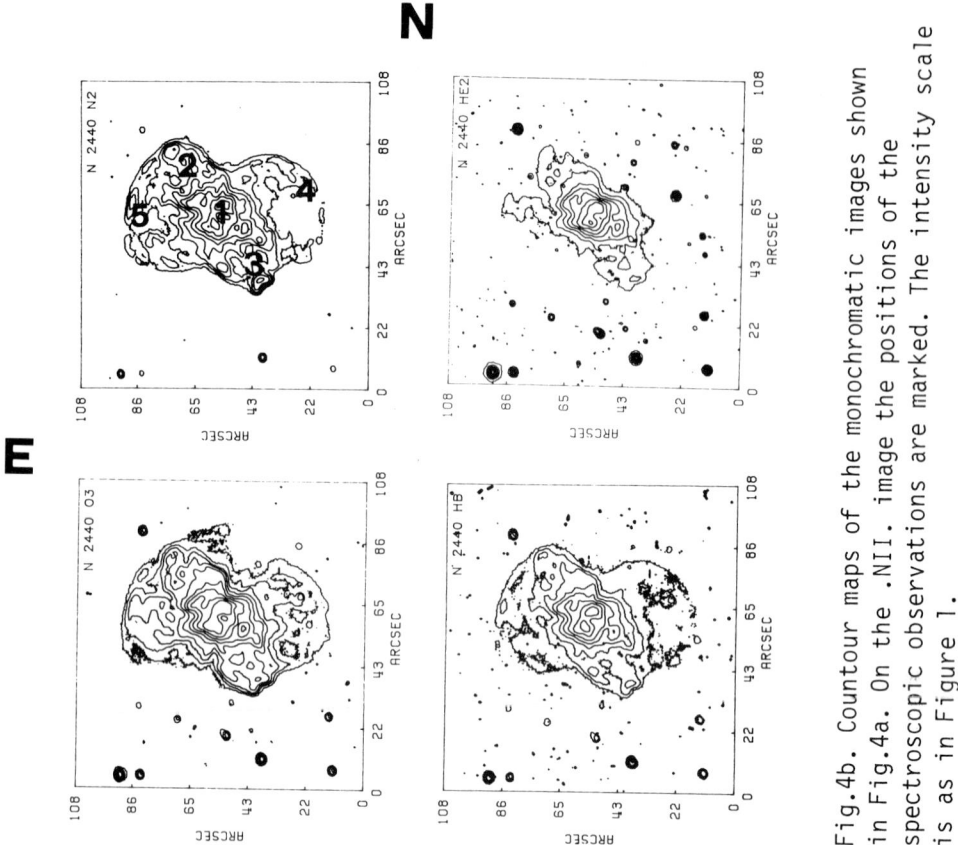

Fig.4b. Countour maps of the monochromatic images shown in Fig.4a. On the .NII. image the positions of the spectroscopic observations are marked. The intensity scale is as in Figure 1.

In Fig.5 we display H-alpha and [N II] line profiles from high resolution CAT-CES spectra observed in the locations indicated in Fig.4. From the [N II] lines which are clearly separated into two components, we find the following expansion velocities for positions 1-5 : 16.0, 17.0, 19.0, 16.3, and 16.5 km/s, respectively, i.e. almost constant values. By contrast, the relative intensities of the approaching and receeding components of the [N II] profiles vary widely, indicating major density fluctuations. Special attention should be drawn to the profile observed in position 2 : here one of the main components (the one arising from the near side of the object) is very weak while a strong additional component appears at V$\sim$-60km/s. Clearly, a more detailed velocity mapping of this nebula is required to reveal its 3-dimensional density and velocity structure.

REFERENCES

Clegg,R.,Harrington,J.,1985:in 'Workshop on model nebulae',ed.D.Pequignot, publication de l'Observatoire de Paris, 1986
Clegg,R.Harrington,J.,Barlow,M.,Walsh,J.,1986,Astrophys.J.,preprint
Pottasch,S.,1984; 'Planetary Nebulae',Ap.and Space Sci.Lib. Vol.107,p.296

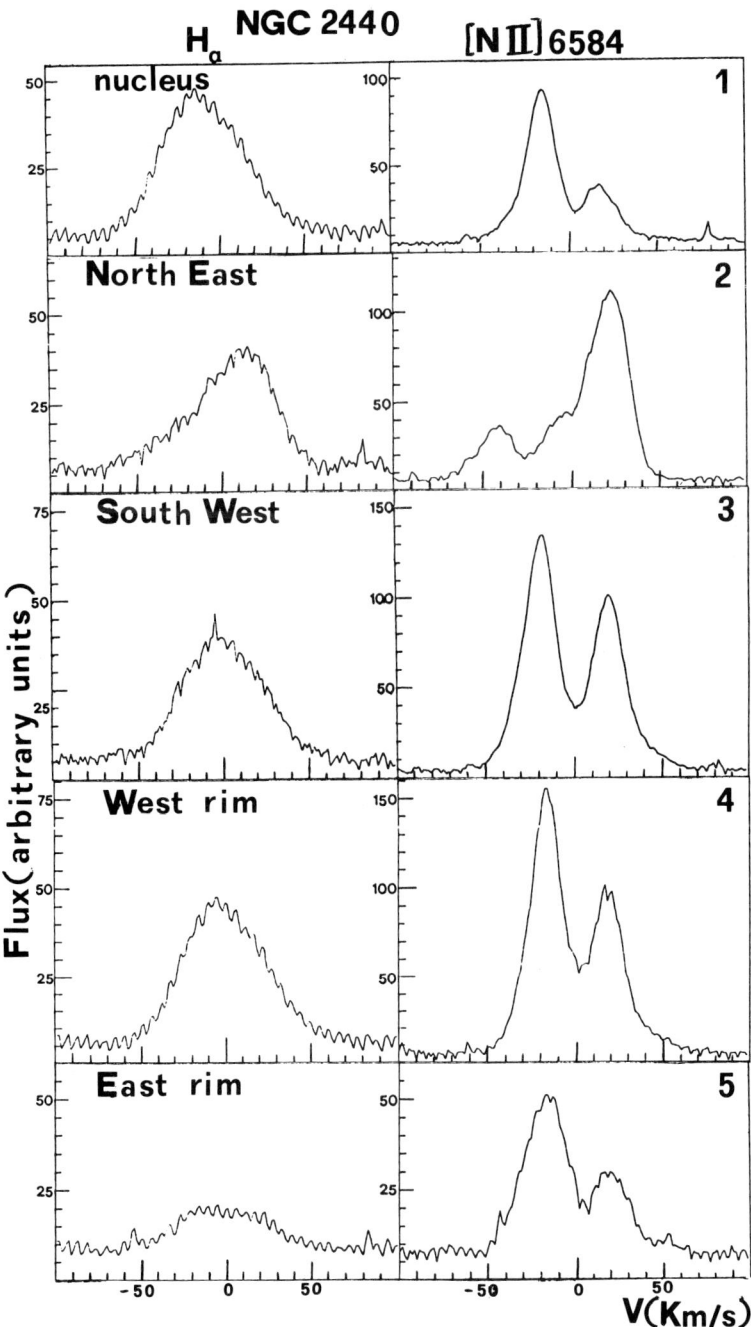

Fig.5. The line profiles of Hα (left) and [NII]6584 in some positions of NGC 2440. The positions are shown in Figure 4.

DISCUSSION.

<u>Clegg</u>: Have you any idea what the 3-dimensional structure of NGC 3918 might be, from your velocity data?

<u>Grewing</u>: Not yet, we just finished the reduction of the data. Some of the computations made to interpret the data by folding a density distribution with a velocity field has, however, already been finished.

<u>Clegg</u>: For NGC 40, what would the mass of the carbon "curtain" be?

<u>Grewing</u>: The mass would be $2-4 \times 10^{-3} M_\odot$.

<u>Perinotto</u>: If the carbon curtain is located far from the star and it is spherically simmetric, i do not see clearly why it is not producing an emission component in the observed profile.

<u>Grewing</u>: It may, but since it would be in a very thin shell it would be very difficult to resolve within a broaded emission profile.

<u>Preite-Martinez</u>: Did you try to compare the temperature of the central star you derive from the UV with other determinations, as that from the energy-balance method?

<u>Grewing</u>: We compared our results with published values derived by a variety of methods. Clearly, most of them imply a temperature for the central star much lower than determined from UV data.

FORMATION AND STRUCTURE OF NEBULAE AROUND SYMBIOTIC OBJECTS BASED ON RADIO TO X-RAYS OBSERVATIONS (A REVIEW)

R. Viotti
Istituto Astrofisica Spaziale, CNR
Frascati, italy

ABSTRACT. One of the defining characteristic of symbiotic stars is the presence of prominent emission lines typical of a nebular spectrum. High spatial resolution radio observations have recently given direct evidence of the presence of small nebulae around symbiotic stars, with a variety of structures, such as shells, jets, bipolar nebulae and extended halos. The origin and physical structure of the nebulae in the most interesting objects, V1016 Cyg, HM Sge, CH Cyg, AG Peg and R Aqr, is discussed, and compared with information in other wavelength ranges, especially infrared (for the presence of c.s. dust) and X-ray (very hot regions), and with current theoretical models. It is shown that the sudden appearance of the nebular structures is not associated to violent ejection of matter, but rather to the increase of the flux of ionizing photons from the central hot source, with the formation and propagation of a ionization front through the circumstellar nebula produced by the cool star wind.

1. INTRODUCTION

More than fifty years ago Merrill and Humason (1932) called attention to a group of stars characterized by a peculiar spectrum, showing TiO absorption bands typical of a late type star, together with prominent emission lines, such as [OIII] and HeII which pertain to emission nebulae. These stars, later called by Merrill "Symbiotic Stars", thus display a spectrum which is a combination of the spectra of a cool star and of a planetary nebula. To explain this association of two apparently conflicting features, we need a source of ionizing photons. According to the different models, this source could be identified with either strong active regions and extended chromosphere-corona around the cool star, or a dwarf companion with high surface temperature and/or with an accretion disk.

The latter binary hypothesis is the one most commonly accepted, although direct evidence of binarity has been found in a very few objects from the observation of periodic eclipses (e.g. CI Cyg and AR Pav) or of periodic variation of the cool star radial velocity (e.g. EG

And and AG Dra, Garcia 1986). It is true however that energy balance considerations favour the binary hipothesis, since in many cases the total amount of energy emitted outside the red, near-IR is comparable, or at least is a consistent fraction of the cool star luminosity, and cannot be supported by it.

Whatever be the nature of the symbiotic systems, it is clear that we are dealing with objects in the late stages of stellar evolution, and that in many cases at least one component is in the post-AGB and pre-WD stage, i.e. close to the CPN one.

A number of symbiotic stars, the Symbiotic Novae, are known for having undergone in recent times a nova-like outburst, characterized by a large, but very slow brightening up to several magnitudes in one or a few years, associated to a drastic change of their optical spectrum, with a spectral evolution similar to that of novae. Typical objects are: RR Tel, HM Sge, and V1016 Cyg, which are also known to have a strong IR excess which is attributed to thermal emission from extensive circumstellar dust clouds (the so-called D-type symbiotic stars, see Allen 1982). Intense radio emission has also been found and high resolution (VLA) radio observations recently led to the discovery of compact nebulae around most of the symbiotic novae, well probably appeared only in recent times. Thus these events might be the formation and first evolution of planetary nebulae in (probably) binary systems. In any case the symbiotic novae may provide us precious information about the role of the cool star wind and of the radiation and wind of the hot source in the PN problem.

2. HIGH RESOLUTION RADIO OBSERVATIONS OF SYMBIOTIC NOVAE

So far radio observations represent the ideal mean to investigate the structure of gaseous nebulae at the highest spatial resolution. In the following we shall therefore mostly concern ourselves with the radio observations of the symbiotic novae. Table I summarizes the main data on symbiotic novae collected from several sources (Viotti 1987).

2.1 V1016 Cyg. This object is one of the most studied symbiotic novae. It rose from B=15mag to 11 during 1964-68. Its visual spectrum changed from Me (before outburst) to a P Cygni one (during the first luminosity rise), and finally to a rich, high excitation emission line spectrum (e.g. Mammano and Ciatti 1975).

This spectral and photometric evolution led different authors to conclude that V1016 Cyg could be considered a good candidate for an object in the transition phase from AGB to the hot subdwarf stage, with the formation of a planetary nebula (e.g. Baratta et al. 1974). This star is also known for its large mid-IR excess. In the framework of the above model, this can be explained by emission from dust grains which are present in the extended nebula produced by the continuous mass outflow from the cool giant. The dust is hot, but it not clear which is the heating mechanism, the hot star or cool star radiation.

High resolution radio map of V1016 Cyg at 1.3 cm using the largest configuration of VLA (35 km) shows two peaks separated by about 0.1

arcsec in the NE-SW direction (figure 1, from Newell 1982). This structure reminds one of that of the bipolar nebulae and may be associated with objects surrounded by equatorial disks, ejecting matter to the polar direction.

TABLE I. Basic data on Symbiotic Novae

	(1)	(2)	(3)	(4)	(5)	(6)	(7)
AG Peg	1855	S		10.6	13	1"5	core,jet,halo
RR Tel	1944	D	X	14.5	10	0"5	unknown
V1016 Cyg	1964	D	X	10.6	85	0"3	bipolar
V1329 Cyg	1966	S		10.6	14/0v	<2"8	unknown
HM Sge	1975	D	X	10.5	40/100	1"5	bipolar, halo
Related objects:							
R Aqr	1974	D	X	14.9	17.9	7"	5 sources
CH Cyg	1984	S	X	14.9	9/26	0"4	bi/tripolar

Notes to the table.
1. Date of beginning of the optical "outburst".
2. IR type (D: dust, S: stellar; see Allen 1982).
3. X-Ray flux detected.
4. Radio frequency in GHz.
5. Radio flux in mJy. Radio variability is indicated.
6. Maximum radio size in arcsec.
7. The radio structure.

It would be important to study the structure of the nebula also at other spectral region in order to have more information on the physical parameters in the different parts of the nebula. Howvwer, in the case of V1016 Cyg and of the other symbiotic stars the size of the c.s. nebulae is too small to be detected at optical wavelengths with ground telescopes. However, Solf (1983) was able to observe the nebula of V1016 Cyg using a special technique which combines high spatial with high spectral resolution. His observations in the light of the |NIII| 6583 A line suggested the presence of an elongation of the nebula of about 0"4 with a velocity difference of the extremes of the nebula of 51 km/s, which shows the existence of a slow expansion of the nebula.

2.2 HM Sge. This symbiotic nova is in many aspects very similar to V1016 Cyg. Both are D-type symbiotics with a high excitation emission line spectrum. They were also found X-ray sources with the Einstein Observatory. HM Sge "appeared" in 1975 after a luminosity increase of more than 6 mag. Radio observations made since 1977 disclosed a gradual flux increase of about a factor two in three years. The radio spectrum was always optically thick. This radio brightening implies an increase of the angular diameter at a rate of 0.024 arcsec/year, i.e. 56 km/s (if the distance is 1 kpc), consistent with the line broadening (Kwok 1982).

The radio observations at 1.3 cm at the highest resolution (0.07 arcsec) show a central double peak with a separation of 0.08 arcsec, while the lower resolution observations at 2 and 6 cm indicate the presence of a spherically symmetric structure, or halo with a maximum extension of 0.5 arcsec.

In the southern sky RR Tel presents the same features and was found to be a radio source. The expected extension should be around 0.3 arcsec, but the small sensitivity and angular resolution of the southern radio telescopes did not allowed so far to study the radio structure of this object.

2.3 CH Cyg. This is the brightest symbiotic star with a visual magnitude of 6. In July 1984 the star underwent a luminosity decline to V=8 followed by the appearence of a rich emission line spectrum, especially in the ultraviolet (Selvelli and Hack 1985). An X-ray source was recently discovered with the EXOSAT satellite by D.A. Leahy and others (not yet published). Both facts are probably associated with the 1984 event.

VLA observations were made November 1984 and 75 days later in January 1986 by Seaquist et al. (1986). At the first epoch they found two sources about 0.2 arcsec apart (figure 1), while two months later they were three, with a stronger overall flux. This was interpreted by Taylor et al. as a rapid expansion of the two sources and the subsequent formation of a third source in the central position. If this interpretation is correct, the rate of elongation of the double source should correspond to a projected velocity of $v\sin i=1050$ km/s. There is however an alternative interpretation: the two sources observed in November 1984 correspond to the central and SE sources in the January 1985 map, while the NW source is a new one. In such a hypothesis the apparent motion of the radio structures is much lower and probably, within the errors, could also be zero. Thus their appearance could be explained by a sudden increase of the ionizing photons from the central object, well probably at the time of the light fading of July 1984. (Actually, this fading could have been only apparent, but the result of a change of the temperature of the object at constant bolometric luminosity, with flux redistribution at shorter wavelengths). This is a kind of "illumination" of the circumstellar nubeculae by light beams from the central star, like the scenery in a theater. The further appearance of the third source could be either due to the propagation of the ionizing front through the circumstellar environment, or, more probably, the result of a new "beam" of light. In any case the important point is that the three lobes are aligned, which again suggests a polar structure.

2.4 AG Peg. This is the oldest among the known symbiotic novae for having "exploded" after 1850 (Lundmark 1921). Radio observations have revealed the presence of an extended (1.5 arcsec) spherical nebulosity or halo, and a central point source (<0.1 arcsec). In addition there is a jet-like feature extending from the central source to SW with a total extension of about 0.8 arcsec. It should be noted that the IR spectrum of this star is S-type, that it there is no evidence for dust emission,

FORMATION AND STRUCTURE OF NEBULAE AROUND SYMBIOTIC OBJECTS 167

contrary to the provious symbiotic novae V1016 Cyg, HM Sge and RR Tel. Could this fact related to the longer time elapsed since the outburst? That is, there is a gradual dissipation of the c.s. dust after the outburst?

2.5 R Aqr. This Mira variable is surrounded by a filamentary nebula extending about one arcmin to the East and West. The ionization of the nebula and the symbiotic spectrum of the central star both suggest the presence of a hot source near the red giant, which could be the inner edge of an accretion disk or a hot subdwarf. The central core of the nebula is variable. In particular, very recently Wallerstein and Greenstein (1980) discovered a short tail oriented to the NE which was not present a few years before (see Herbig 1980). This could be the direct evidence for ejection of matter from a stellar source. However, radio observations made two years apart in 1982 and 1984 did not show evidence for a rapid expansion. More recently, Kafatos et al. (1986) made an accurate radio map of R Aqr and established the presence of five separate sources, placed about on the same line. Source B is diffuse and is coincident with the jet. The central part consist of two point sources separated by about 0.4 arcsec. In addition, Hollis et al. (1986) found a SiO source not coincident with the expected position of the Mira, but about one arcsec offset, probably originating by maser action in the circumstellar nebula.

Are these structures the result of successive and collimated eruptions from the central star? Let me first recall that a kinematic study of the outer R Aqr nebula led Solf and Ulrich (1984) to find a general expansion of the filaments which has been interpreted as the result of two metter ejection occurred some 180 and 400 years ago. Thus the R Aqr system was subject to violent phenomena also during its less recent history. The radio monitoring from 1982 to now has not revealed any significant change of the structure (Hollis et al. 1985). That is the radio sources do not expand, as it would have been expected if they were ejected from the central object a few years ago. Thus, we are forced to find another explanation for the sudden appearence of the jet in R Aqr.

R Aqr was recently found by Viotti et al. (1985, 1986) to be a weak X-ray source. This result is not unexpected for such a peculiar object; may be the observed flux is too weak. The ultraviolet observations could be of help to understand the origin of the X-rays. IUE spectra of R Aqr and its jet in fact indicate a steady increase of the excitation of the jet, which in 1982 appeared lower in excitation with respect to the central source (Michalitsianos and Kafatos, 1982), while since 1985 the high excitation lines of NV 1240 A and HeII 1640 A have intensified in the jet and are stronger than in the UV spectrum of the star. One possible explanation of this effect is that the flux of ionizing photons from the central hot source has increased in recent times. This could be associated with the recurrent anomalies of the Mira light curve, with an almost complete disappearence of the pulsation, which could be due to the tydal effects during the periastron passage of a close binary system, moving in an eccentric orbit (Willson et al. 1981). The last anomaly occurred during 1974-77 and this could be the time of the last close passage of the two stars of the R Aqr system. This could be followed by

an increase of the mass transfer in the system with the heating of the
hot star (or of the accretion disk) and subsequent production of an
excess of UV and X-ray photons. Because of the presence of opaque matter
in the orbital (=equatorial) plane, most of the radiation is emitted in
a cone towards the poles. The matter, in form of clouds, which is
present within the cone is "illuminated" by the radiation and suddenly
appeared to our eyes. Hence, we associate the recent appearence of the
jet in R Aqr to the recent periastron passage of the system followed by
the ionization of the surrounding matter by the increased flux of ionizing photons. Matter could be ejected during these phase, but at rather
low velocity. What is however still to be investigated is the role of
the cool star, and the interaction of the asymmetric radiation field and
wind of the hot star with the gas and dust in the cool star wind.

3. INTERPRETATION.

Bipolar structures like those observed in V1016 Cyg and HM Sge are
typical of young objects, such as MWC 349, and are commonly associated
to the presence of an equatorial disk, which is shielding the stellar
radiation and wind in the equatorial plane. Jets, chains of nubeculae,
cometary structures are common in Herbig-Haro objects. Another interesting case is the nearby cool supergiant a Sco. The radio map revealed
the presence of two separate source, an optically thick point like one
centered on the M star, which is associated to the extended chromosphere
of the star, and a thin, diffuse source near the B-type companion
(Hjellming and Newell 1983). This latter emission has a bow structure
and is the outer part of the cool star wind ionized by the hot star
radiation. Radio emission was also found in all the VV Cep binary systems closer than 2.5 kpc (Hjellming 1985). The components are much
closer than in a Sco, so that the UV photons of the hot stars penetrate
into the inner and denser parts of the cool star winds. This is confirmed by the radio spectrum which appears optically thick with a=0.8 to
1.2.
  In the case of the symbiotic systems we are dealing with a cool
star supporting an intense wind, which is filling the circumstellar
space with gas and dust. Which is the role of the radiation and a high
velocity wind from the stellar companion, or from the accretion disk?
And their time variability?
  It should be noted that the D-type symbiotic stars, i.e. those
symbiotics showing a strong IR excess which can be attributed to dust
emission, generally present a stronger radio flux. In fact Wright and
Allen (1975) found a correlation between the 2 cm and 10 um fluxes. D-
type systems are probably surrounded by extended gas and dust regions
produced by massive outflows from the Mira type secondary. At the time
of the outburst, the origin of which is not essential in this picture,
there is a sudden production of ionizing photons, and a ionization front
propagates through the circumstellar matter. This explains the gradual
rise of the optical and radio luminosity. Such a model has for instance
been proposed to explain the evolution of V1016 Cyg by Baratta et al.
(1974) and Ahern et al. (1977), and of HM Sge by Kwok et al. (1984).

To account for the radio properties of symbiotic stars, Taylor and Seaquist (1984) developed a model in which the wind from the cool star is ionized by the hot star radiation. They have computed the shape of the ionized nebula for different binary separation, wind density and ionizing photon luminosity, and found that most of the radio emission arise in the portion of the cool star wind facing the hot companion. This might explain the periodic variation of the UV spectrum in some symbiotic stars, such as AG Dra (Viotti et al. 1984). Most of the UV continuum is produced in a bow shaped ionized region in the cool star wind which is periodically occulted during the orbital motion of the binary system. An important element to be considered is the wind from the hot star which may collide with the cool star wind and produce shock fronts. Interactive winds models have been developed by Kwok et al. (1978) for planetary nebulae. The interaction of the two winds may produce an expanding shell structure which explains the observed shape of the nebula around HM Sge (Kwok et al. 1984). A shocked region is formed between the two stars, closer to the red component. As suggested by Willson et al. (1984), this region is probably the place where the highest ionization species (e.g. lFeVIIl) and the X-rays are formed. A similar model was also proposed by Kwok and Leahy (1984) to explain the X-ray emission from symbiotic novae and its time evolution.

4. CONCLUSIONS.

Small nebulae have been observed around symbiotic stars known to have undergone explosions in recent years (the so-called symbiotic novae and related objects). These objects contain a cool Mira-type giant which has for a long time lost a large amount of gas and dust, presently forming an extended cool nebula. The innermost parts of this nebula have been ionized by the flux of UV photons emerging after the outburst.

Bipolar and jet-like structures have been identifed which are associated to the presence of matter, in form of disks or streams, in the equatorial plane. Asymmetry in the ionized region is also evident from the study of the optical and UV light curve of a number of symbiotic stars. The wind from the hot source probably interacts with the cool wind, and a shock front is produced where high ionization species and X-rays are possibly emitted.

This picture applies both to the single star and to the binary system model for symbiotic star. In the former case it is the cool giant which is evolved to the blue part of the H-R diagram which produces the ionizing photons (see e.g. Figure 4 in Baratta et al. 1974). While, in the interactive binary hypothesis, the mass transfer onto a dwarf, possibly degenerate companion is the main responsible of the outburst and the following appearence of the nebula. These objects should be surrounded by extended regions of neutral gas, molecular clouds and dust clouds, which should be put in evidence by radio and IR mapping. It should be possible from these observations to derive the structure and chemical composition of the furthest parts of the envelope, uneffected by the hot radiation, which should give precious information on the past history of the cool giant.

The author is grateful to Dr. Michael Friedjung for discussions.

REFERENCES

Ahern,F.J., FitzGerald,M.P., Marsh,K.A., Purton,C.R.: 1977, Astron. Astrophys. 58, 76.
Allen,D.A.: 1982, in The Nature of Symbiotic Stars, M.Friedjung and R.Viotti eds., Reidel, Dordrecht, p.27.
Baratta,G.B., Cassatella,A., Viotti,R.: 1974, Astrophys.J. 187, 651
Herbig, G.: 1980, IAU Circular No. 3535.
Hjellming,R.M.: 1985, in Radio Stars, R.M.Hjellming and P.M.Gibson eds., Reidel, Dordrecht, p.151.
Hjellming,R.M., Newell,R.T.: 1983, Astrophys.J. 275, 704.
Hollis,J.M., Kafatos,M., Michalitsianos,A.G., McAlister,H.A.: 1985, Astrophys. J. 289, 765.
Hollis,J.M., Michalitsianos,A.G., Kafatos,M., Wright,M.C.H., Welch,W.J.: 1986, Ap.J. (Letters) in press.
Kafatos,M., Michalitsianos,A.G., Hollis,J.M.: 1986, Astrophys.J.Suppl. Ser. in press.
Kwok,S., Bignell,R.C., Purton,C.R.: 1984, Ap.J. 279,188.
Kwok,S., Leahy,D.A.: 1984, Astrophys. J. 283, 675.
Lundmark,K.: 1921, Astr. Nachr. 213, 93.
Merrill,P.W., Humason,M.L.: 1932, P.A.S.Pacific 44, 56.
Michalitsianos,A.G., Kafatos,M.: 1982, Ap.J.Lett. 262, L47.
Solf, J.: 1983, Astrophys. J. (Letters) 266, L113.
Solf, J., Ulrich, H.: 1983, Astron. Astrophys. 148, 274.
Taylor,A.R., Seaquist, E.R.: 1984, Astrophys. J. 286, 263.
Taylor,A.R., Seaquist, E.R., Mattei, J.A.: 1986, Nature, 319, p.38.
Viotti,R.: 1987, The Symbiotic Stars, in Cataclysmic Stars, M.Hack ed., Monograph Series on Nonthermal Phenomena in Stellar Atmospheres, in preparation.
Viotti,R., Piro,L., Friedjung,M., Cassatella,A.: 1985, IAU Circ.No.4083
Viotti,R., Rossi,L., Piro,L., Cassatella,A.: 1986, IAU Circ.No.4169
Wallerstein, G., Greenstein, J.L.: 1980, PASP 92, 275.
Willson,L.A., Garnavich,P., Mattei,J.A.: 1981, Inf.Bull.Var.Stars,1961
Willson,L.A., Wallerstein,G., Brugel,E.W., Stencel,R.E.: 1984, Astron. Astrophys. 133, 154.
Wright,A.E., Allen,D.A.: 1978, M.N.R.A.S. 184, 893.

DISCUSSION.

Rodriguez: The radio maps of the symbiotic stars frequently show a bipolar geometry. What is the cause of this geometry?
Viotti: It should reflect the presence of matter in the equatorial plane (of a single or binary object), which is shielding the stellar radiation and possibly the stellar wind, so that only clouds above the poles are ionized (or, alternatively, matter is preferentially ejected at right angles with respect to the orbital/equatorial plane. For more details you may see the papers of Solf and Ulrich and of Kafatos et al. on R Aqr.

THE PECULIAR PLANETARY NEBULA NGC 2346 AND ITS NUCLEUS

L.Rosino, T. Iijima , S.Ortolani
Astrophysical Observatory of Asiago

A. Mammano
Department of Mathematics of the University of Messina

ABSTRACT. A preliminary account is given of a research on NGC 2346 and its central nucleus V 651 Mon. An acceptable mean light curve of this star has been drawn with the elements: $2445010.32 + 15^d.882$ E. The eclipses of V651 Mon ceased at the beginning of 1986. The spectrum and the distribution of nebulosity around the nucleus of NGC 2346 are briefly discussed.

I. INTRODUCTION

The bipolar planetary nebula NGC 2346 and its visible nucleus, a relatively cool star type A5-V, have been thoroughly investigated after Kohoutek (1982) discovered that the A5-V star (later called V 651 Mon) presented periodic eclipses of relatively large amplitude with a period of about 16 days.

This discovery raised some problems. NGC 2346 shows a high excitation spectrum which cannot be due to the A-type star, too cool to ionize the nebula. So it was assumed that the A-type star had a collapsed O-type companion responsible for the nebular spectrum. On the other hand this hot star of very small radius cannot produce the eclipses. And moreover it must be considered that the minima occur when the A-type star reaches its higher radial velocity and therefore is in a position of quadrature and not of conjunction with its companion (Mendez et al.,1982).

Two other points are significant:a) The A-type component is a single-lined spectroscopic binary with a period of $15^d.995$ (Mendez and Niemela,1981) . b) From 1899 to November 1981 the central star of this planetary nebula did not show any trace of eclipses (Schaefer, 1985) . The eclipses began abruptly in December 1981 and ceased in 1986, the light curve having changed several times its form and amplitude during this interval. In addition infrared observations (Roth,al. 1984) have shown the presence of dust clouds therefore suggesting

(Schaefer,1985) that the eclipses were produced by an obscuring cloud moving in front of the system. Several models, mostly based on occultation by dusty envelopes have been hitherto proposed to explain the periodic eclipses of the central star of the nebula.

The present communication deals with the visual light curve of this star from 1981 to 1986 and also reports some new data on the spectrum of the nebula and the distribution of gas and dusts around the nucleus of NGC 2346.

2. ECLIPSES OF THE A5-V STAR (V 651 Mon)

In order to obtain the mean light curve of the central star of NGC 2346 we have collected all of the available visual observations of the star which we were able to find in the literature. Most of the observations were due to Brian Marino and H.O.Williams of Auckland (New Zealand), who used a 53 cm cassegrain telescope with Kodak Tri-X plus a yellow filter, but also to Kohoutek, Mendez, Gauthier, Niemela et al. Some plates were occasionally obtained at Asiago with the 67 cm Schmidt and the 182 cm telescope of Ekar (Table 1). All the observations reduced to the V system and partly corrected in order to eliminate the contribution of the nebular radiation, have been used to draw the light curve of V 651 Mon from 1981 to 1986 (Fig.1).

TABLE 1 . Asiago observations of V 651 Mon

JD 244	V	Phase	JD 244	V	Phase
5706.4	14.2	0.83	6114.5	11.75	0.52
6033.5	13.0	0.42	6114.6	11.3	0.53
6035.5	12.5	0.55	6117.4	12.05	0.71
6061.5	14.2	0.19	6119.4	14.0	0.83
6090.5	15.7	0.01	6435.5	10.9	0.74
6095.5	12.1	0.33	6444.4	11.0	0.30
6096.4	11.8	0.38	6447.5	10.9	0.49
6096.5	11.3	0.39	6448.6	11.0	0.56
6110.3	12.75	0.27	6477.4	11.15	0.37

There are several gaps in the light curve due to the period of seasonal invisibility of the star from May to September-October and the lack of observations. Moreover, in many cases the observers were unable, for the presence of the nebula, to follow the star when it was fainter than magn. 14-14.5. In this circumstance the only available data were those giving an inferior limit for the brightness of the variable, writing for instance : " V weaken than 14.2 ", which can mean everything from 14.3 to, say, 16 . In addition, the estimates of magnitude near minimum, or below 14 , were always rather uncertain.

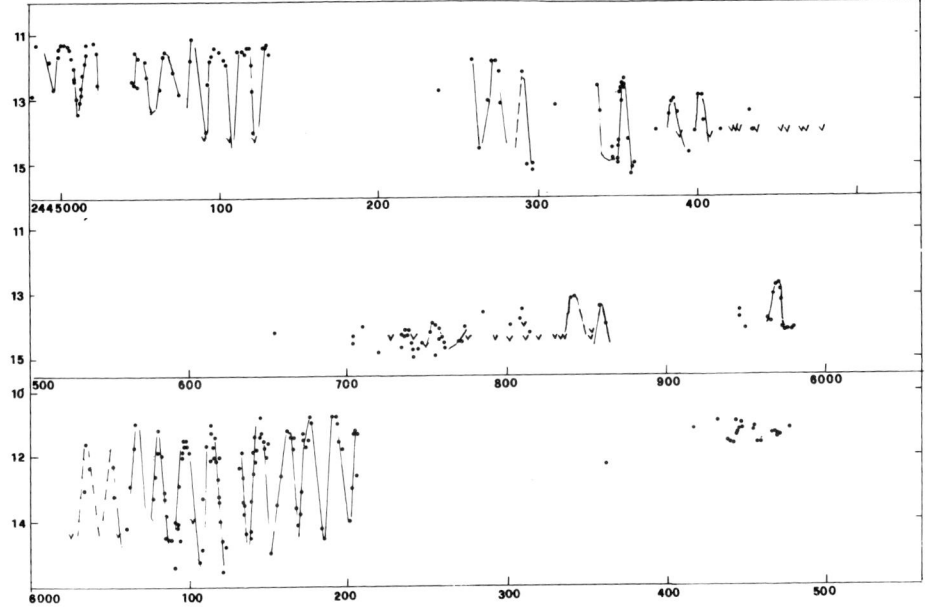

FIG. 1 Visual light curve of V 651 Mon.

A quick look to the light curve reproduced in Fig.1 is sufficient to show that the star is a sort of eclipsing binary,with a period near 16 days; however, a strange eclipsing binary. During 1983 was observed a progressive weakening of the variable which at maximum scarcerly reached magn.13,while at minimum was generally invisible,weaker than 14.4 . At the same time was noticed a broadening of the minima and a drastic diminution of the amplitude. Some periodicity, however, was maintained,although the maxima,now narrower than the minima, seem to have shift in phase by about 0.2. The situation was still more critical in 1984 when the reconstruction of an acceptable light curve became somewhat arbitrary for lack of effective observations near minimum. The picture changed in 1985. The amplitude became again as large as in 1982 or even larger and the variable reached at maximum visual magnitude 11 . Finally the star at the beginning of 1986 suddenly ceased to be variable, becoming nearly constant at maximum, as it was prior to 1982. Recent observations carried out at Asiago (Sept.-Oct. 1986) confirm that the star is still constant with visual magn. 11.2-11.4.

Kohoutek (1983) discussing the observations available at that

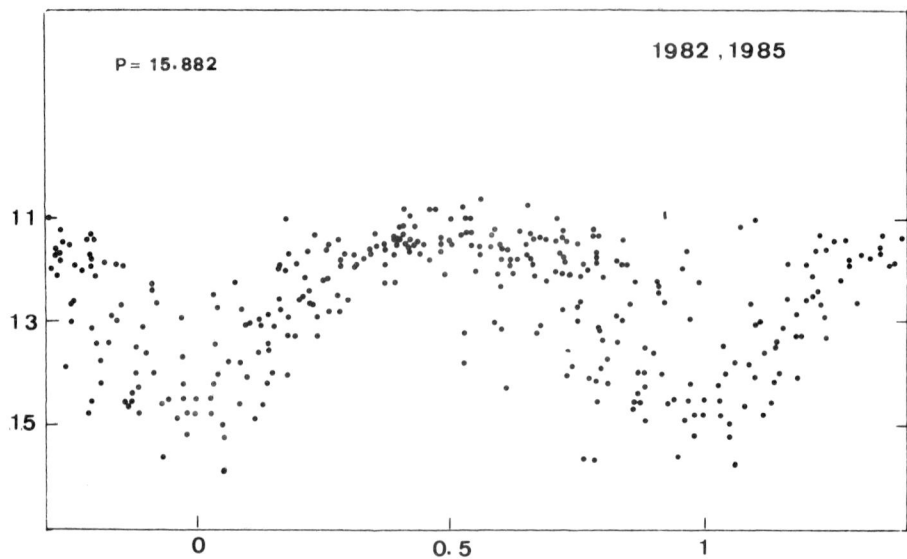

FIG.2 . Mean visual light curve of V 651 Mon in the years 1982 and 1985.

time derived a period of $15^d.957$. Later, the period was slightly changed to $15^d.991$ by Acker and Jasniewicz (1985) . These Authors remarked that there was a jump in the phases during 1983-84.

To draw the light curves reproduced in the Figs.2 and 3 , after discussing all the available observations, including those made in 1985, we

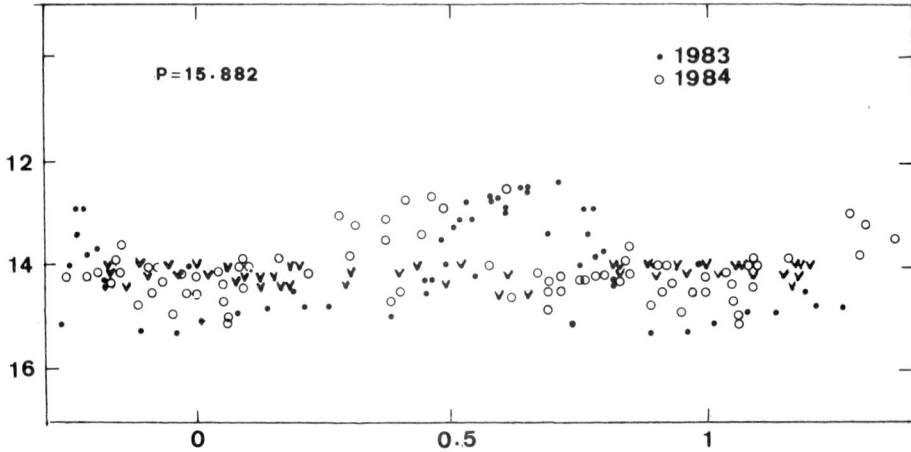

Fig.3 . Mean light curve during the years 1983-84.

have adopted the following elements, which seem to give the best possible representation of the points in order of phase:

$$T_{Min} = 244\ 5010.32 + 15^d.882\ E$$

These elements satisfy well the data obtained in the years 1982 and 1985 which appear perfectly in phase, but they give a rather flat light curve plotting the magnitudes observed in 1983-84. This is certainly due, at least in part, to the small amplitude of the variable in that period and to the lacking of good estimates near minimum. A shift of the phase is possible, but it is not apparent.

Several models have been proposed in order to explain the strange behaviour of the central star of NGC 2345, the temporary occurring of the eclipses and the weakening of the light curve in 1983-84. Ultraviolet and infrared observations have proved the presence of dark matter in front of the star. So that the eclipses have been generally attributed to the occultation of the A5 star by a clumpsy dust cloud ejected by the collapsed O-type component. Schaefer (1985) has suggested that the ejecta by the hot subdwarf condense in dust grains at a distance of some hundred A.U., forming dust clouds. When the A-type star in its orbital motion passes behind these obscuring clouds along the line of sight we observe an eclipse. The passage of dark matter of different consistence can explain the peculiarities above remarked.

Apparently the episode which has taken place from December 1981 to the last months of 1985 went to its end at the beginning of 1986 when the variability of V 651 Mon abruptly ceased, as confirmed by the observations carried out at Asiago and those published by Huruhata (1986). Evidently the dusty material in front of the star moved out from the line of sight or dissipated.

The star is still kept under control at Asiago for a possible re - prise of variability.

3. SPECTRA AND GAS DISTRIBUTION

The spectrum of NGC 2346 prior to the eclipse episode has been described by Kaler et al.(1976) and by Sabbadin (1976). After 1982 several other spectra of the nebula and its nucleus were obtained at Asiago with the purpose of ascertain whether some changes had occurred in consequence of the variability ot the central star. The conclusion has been negative, in the sense that , as previously remarked by Mendez et al. (1982) neither the A-type star nor the nebula showed substantial differences in the spectrum before and after the eclipse phenomenon.

Fig.4 reproduces the intensity tracing of the spectrum of the nu - cleus of NGC 2346 obtained on Jan.20,1986 at the 182 cm telescope of A- siago with a Reticon applied to the Boller and Chivens spectrograph.The

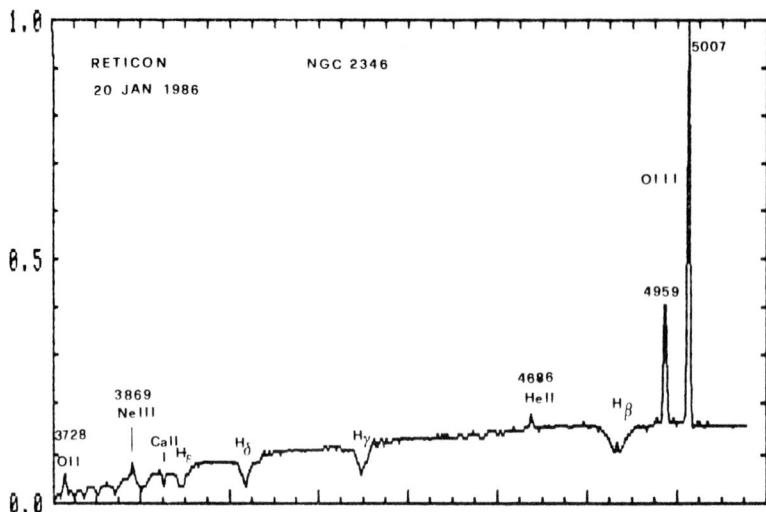

FIG.4 . Reticon tracing of the spectrum of V 651 Mon.

broad absorption lines of H and CaII are typical of an A5-V spectral class. Superposed to the absorption spectrum are visible the narrow emission lines of H_β, HeII λ 4686 , NeIII λ 3869,3968 and of the blended pair of OII λ 3727-29. Outstanding are the forbidden lines of OIII λ 4959,5007. These emissions are obviously due to nebular gas.

Another reproduction of the spectrum of the nebula and its central star is shown in Fig.5 . The spectrum was obtained with the prism spectrograph at the cassegrain focus of the 1.22 m telescope of Asiago, equipped with a Carnegie-RCA image-tube.

Through the courtesy of E.Budding we had also the possibility of examining an excellent intensity tracing of the spectrum of NGC 2346 obtained at the 1.86 m telescope of the Mount Stromlo Observatory in Australia on April 17, 1984 (Fig.6) in the spectral range 3800-4050. Table 2 gives the list of the emission lines recorded and their relative intensity, having assumed for the line λ 3868 of NeIII intensity 30. The spectrum looks alike to that of planetary nebulae of medium -high excitation, with a clear evidence of stratification proved by the presence of lines due to ions of largely different I.P. Some relatively strong lines (λ 3875, 3957, 3983, 3988) have not been identified.

The density distribution of the gaseous envelope surrounding the central A5-V star is shown in Figs. 7 and 8 , reproduced from CCD frames obtained on Jan 23 ,1986 by one of us (S.O.) with the 2.2 m telescope of ESO. Fig.7 represents the image of the field taken through a B filter.Since there are no strong lines in that spectral region, the nebula

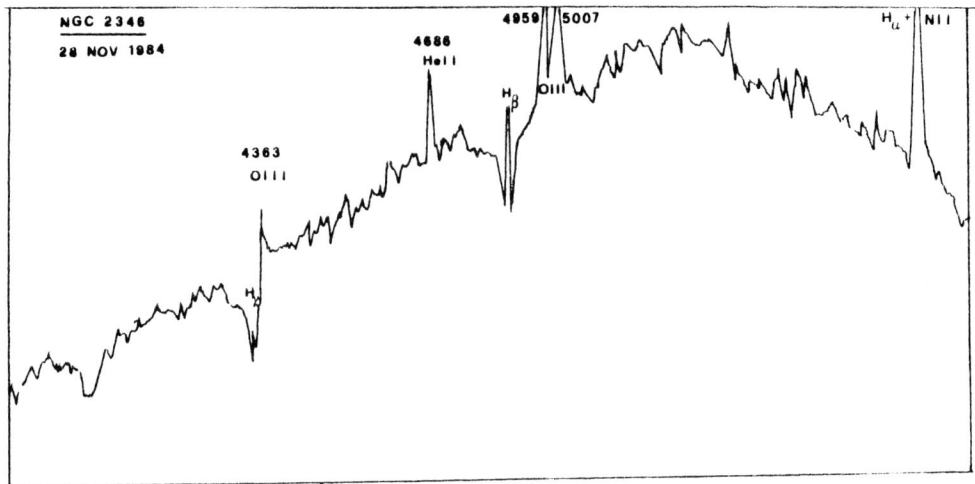

FIG.5 . Microphotometer tracing of the spectrum of the nucleus and the nebula from H_α to H_δ.

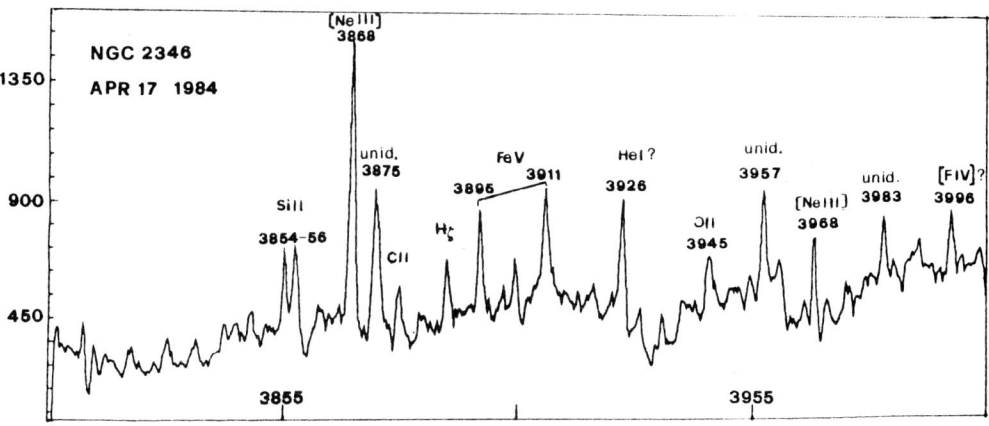

FIG.6 . Tracing of the spectrum of NGC 2346 from λ 3800 to 4050 . In the ordinates photon counts.

TABLE 2. Identification and relative intensities of emission lines between λ 3800 and λ 4050 in NGC 2346

Ion	λ	Mult.	Rel. Int.	Notes
H_9	3835.4	2	2.8	
SiII	3853.7	1	10.7	
SiII	3856.0	1	10.9	
NeIII	3868.7	1F	30.0	Possibly blended with HeI, mult.20.
....	3875	..	15.6	Auroral ? Possibly blended with CII λ 3876, mult.23.
CII	3880	33	5.5	
H_8+HeI	3889	2,2	6.9	
FeV	3895.7	1F	11.3	
....	3902	..	5.5	Unidentified
FeV	3911.1	3F	12.3	
HeI ?	3926.5	58	11.6	The identification is uncertain; the line seems to be too strong
OII	3945.0	6	5.1	
CII	3946.4	31,32	2.8	
CII	3949-52	36,32	2.0	
OII	3954.4	6	3.1	
....	3956.8	..	12.0	Unidentified. This line is recorded also in the spectra of other planetary nebulae.
....	3960.2	..	4.6	Unidentified
HeI	3964.7	5	2.8	
NeIII	3967.5	1F	9.2	
H_ε	3970.0	1	3.3	
....	3982.7	..	7.0	Auroral
....	3988	..	4.0	Unidentified
FIV	3996.3	1F	6.6	
HeI	4009.2	55	3.7	

appears rather weak with a mean diameter of about 35 arcsec; while, in Fig.8, which shows an image obtained through an OIII interference filter, the nebula is much more extended, with a clear bipolar structure. The central region of the nebula, with a mean diameter of about 35 arcsec, has an uneven density, showing the maximum brightness along a semicircular zone immediately south of the nucleus, extended from SE to SW. A stratification of the OIII ions, with decreasing density outwards, is evident. Out of this central region of higher intensity, the CCD frames in OIII light show an extended halo, of much lower brightness, which displays

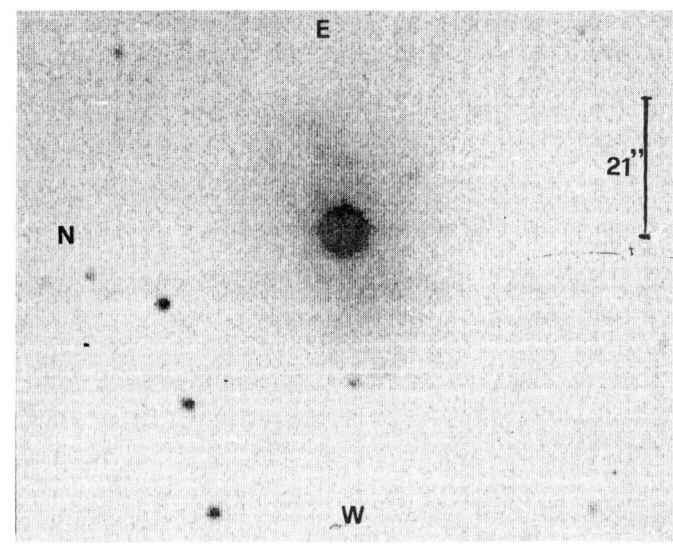

FIG.7 .Intensity distribution of the nebula around the central star of NGC 2346 from CCD frames obtained at La Silla with a B - filter

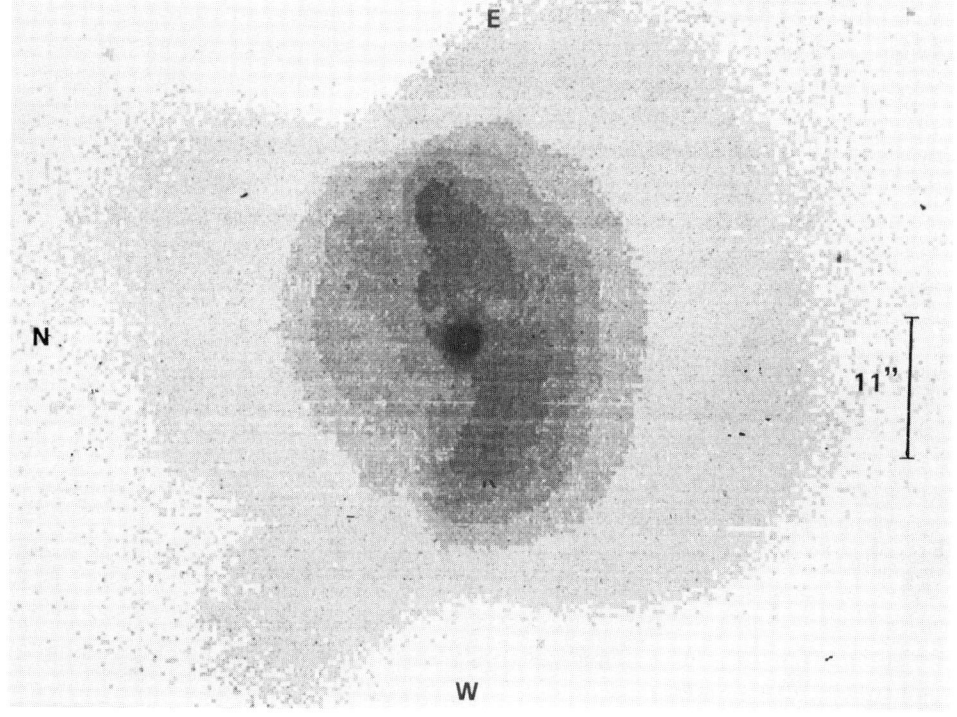

FIG.8. Distribution of the OIII ions in NGC 2346. Note the central roundish region of higher brightness and the outwards extended halo with a bipolar dumb-bell like structure. From a CCD OIII frame.

a dumb-bell like bipolar form, with an elongation of about 75 arcsec in the N-S direction and a W-E elongation of about 40 arcsec.

The reduction of these CCD frames and their interpretation in order to obtain a representative picture of this strange nebula and the study of the radial velocities of the central star are still under way.

4. CONCLUSIONS

We conclude summarizing the principal facts hitherto ascertained.
a) The central star of the planetary nebula NGC 2346, a single-lined spectroscopic binary of spectral type A5-V, apparently normal under any point of view, which prior to 1981 had a constant brightness, displayed, beginning from December 1981, deep eclipsing-like minima which ceased four years later.
b) The visual amplitude of the star during the eclipsing episode was at first (1982) of the order of three-four magnitudes. Then, in 1983-1984 the maxima drastically weakened, the amplitude decreased to one or two magnitudes or even less and the minima broadened occupying a substantial part of the period. The star recovered its primitive amplitude in 1985. Finally the eclipses disappeared and in 1986 the star appeared constant, as it was before 1981, at visual magnitude $\sim$ 11.
c) By computing the phases with the photometric period $P = 15^d.882$, very near to the orbital one ($15^d.995$), we have obtained a mean light curve of β Lyr-type which represent fairly well the observations of 1982 and 1985 (Fig.2). The light curve , however, becomes flat when the 1983-84 magnitudes are taken into consideration (Fig.3). A shift of the maxima is somewhat apparent, but the minima, although badly defined, seem to maintain their phase.
d) There is no doubt, in our opinion, that the "eclipses", observed during this transient episode, were due to an obscure cloud of gas and dust, as proposed by Schaefer (1985) et al., ejected by the hot collapsed companion and slowly passing in front of the system . The eclipses ceased when the dark cloud dissipated or went out of the visual to the A-type star.

Since the facts have proved a certain activity of the hot subdwarf similar episodes very likely will occur also in the future. The central star of the nebula should be kept therefore under control.
e) The spectra of the A-type component and the nebula have not shown substantial changes during and after the eclipsing episode. We have given a list of lines, in the near ultraviolet, which previously were not been recorded. Other spectra in this spectral region, in the far ultraviolet and in the infrared should be desirable.
f) Finally we observe that the distribution of density of the nebular envelope in the central region, around the nucleus, as shown

in Figs. 7-8,looks rather peculiar for a concentration of the gas in a semicircular zone south of the nucleus. The external parts of the envelope are more symmetric relative to the nucleus and typical of a bipolar nebula.

REFERENCES

Acker,A.,Jasniewicz,G. 1985, Astron.Astrophys. 143,L1.
Huruhata,M. 1986 , Inf.Bull.Var.Stars No.2923.
Kaler,J.B.,Aller,L.H., Czyzak,S.J. 1976, Astrophys.Journal 203,636.
Kohoutek,L. 1982, IAU Circ. No.3667.
Kohoutek,L.1983, Mon.Not.R.astr.Soc. 204,83P.
Marino,B.,Williams,H.O. Inf.Bull.Var.Stars Nos. 2266,2467,2583,2774.
Mendez,R.H., Niemela,V.S. 1981, Astrophys.Journal 250,240.
Mendez,R.H.,Gathier,R.,Niemela,V.S.1982,Astron.Astrophys. 116,L5.
Roth,M.,Echevarria,J.,Tapia,M., Carrasco,L., Costero,R., Rodriguez,L.F. 1984,Astron.Astrophys. 137,L9.
Sabbadin ,F.1976, Astron.Astrophys. 52,291.
Schaefer,B.E. 1985, Astrophys.Journal 297,245.

DISCUSSION.

Viotti: What is the spectrum of NGC 2346 at the light minima? Is the A-type spectrum still present?
Rosino: Yes, it is. No changes in the spectrum of the A5-V central star have been observed from maximum to minimum .

d'Hendecourt: In the first figure the spectrum of the nebula seems to show a very broad emission. Is this real?
Rosino: No. This spectrum was obtained with a prism spectrograph. The hump in the 5500-6200A region is simply due to an effect of dispersion and sensitivity of the plate.

Panagia: Has anybody tried, or is anybody planning, to make polarization measurements? It seems to me that if the apparent magnitude variations are due to occultation by a dust cloud, then appreciable polarization should be detectable and be especially strong at minimum light.
Rosino: As far as I know a polarization study of the nebula has not yet been made.

NGC2346:
VISIBLE AND INFRARED OBSERVATIONS OF SEVERAL MASS-LOSS EPISODES?

R. Costero[1], M. Tapia[1,3], R. Méndez[2], J. Echevarría[1], M. Roth[1,3], A. Quintero[1] and J. Barral[1].
[1]) Instituto de Astronomía, UNAM, México.
[2]) CONICET (Argentina) and IAA, Munchen, F. R. Germany.
[3]) Guest Observer, Cerro Tololo Inter-American Observatory, operated by AURA for the National Science Foundation.

ABSTRACT. New observations of the eclipses of the central star of NGC2346 are presented; these, together with other data available in the litterature allow us to elaborate on a detailed model for the obscuring dust cloud.

NGC2346 is a well studied planetary nebula with a binary central object which became an attractive target after the central object began to present eclipses in late 1981 (Kohoutek, 1982; Méndez et al., 1982). It was shown by Méndez et al. (1982) that the eclipses are due to an obscuring dust cloud. Warm dust (T $\cong$ 1200K) was detected prior to the eclipse phase by Cohen and Barlow (1975) and Whitelock (1985). The origin and nature of the cold eclipsing cloud has raised considerable controversy over the past years (Roth et al., 1984 (hereafter Paper I); Schaefer, 1985, Méndez et al., 1985).

In Paper I we presented a model of the obscuring cloud and predicted the possible behaviour of the eclipses. In this paper we present new observations at visible and infrared wavelengths which confirm our predictions. Our previous model is revised and detailed calculations of the physical parameters and morphology of the obscuring cloud are presented.

Our new results can be summarized as follows:

1.- The obscuring cloud is most probably a self-gravitating fragment of a toroid surrounding the central object. It's dust density is
$$n_d \cong 1.2 \times 10^{-2} \text{ cm}^{-3},$$
and it's mean density (dust only) is
$$<\rho> \cong 3.5 \times 10^{-17} \text{ g cm}^{-3}.$$

The total mass (dust and gas) is

$$M_{cl} \cong 2 \times 10^{-10} \, M_\odot$$

which implies that only a small fraction of the total emitting dust in the planetary nebula is present in the obscuring cloud.

2.- The shape of the obscuring cloud which best explains the evolution of the eclipses is that of an elongated eclipse or a disk seen edge-on. The lines of equal absorption (isocrypts) are roughly of the same shape. The obscuring cloud has a velocity component parallel to the major axis of the planetary nebula.

3.- It is unclear wether the warm dust cloud and the obscuring cold cloud have a simultaneous origin or have been formed in more than one mass-loss episode.

The present results are the first evidence of a dense cloud of mass similar to that of a minor planet orbiting an evolved object and which is probably, the result of the fragmentation of a disk or toroid around the central object of NGC2346.

The details of the present work will appear in the *Revista Mexicana de Astronomía y Astrofísica*.

REFERENCES

Cohen, M. and Barlow, M. J. 1975, *Ap. J. (Letters)*, 16, 165.
Kohoutek, L. 1982, *Inf. Bull. Var. Stars*, No.2113.
Huggins, P. J. and Healy, A. P. 1986, *M. N. R. A. S.*, 220, 33p.
Méndez, R. H., Gathier, R. and Niemela, V. 1981, *Astr. and Ap.*, 116, L5.
Méndez, R. H., Marino, B. F., Clariá, J. J. and van Driel, W. 1985, *Rev. Mexicana Astron. Astrof.*, 10, 187.
Roth, M., Echevarría, J., Tapia, M., Carrasco, L., Costero, R. and Rodríguez, L. F., 1984, *Astr. and Ap.*, 137, L9.
Schaefer, B. 1985, *Ap. J.*, 297, 245.
Whitelock, P. 1985, *M. N. R. A. S.*, 213, 59.

COMMENTS

D'HENDECOURT: "What can you say about the very high temperature of the warm (1200K) dust? How can you explain it?"

ROTH: "We believe that this material has been ejected recently by the central system and is the hottest material condensed in grains".

PANAGIA: "You showed the variations of the extinction as a function of orbital phase. How was this determined?"

ROTH: "Simply by subtracting the reported magnitude prior to the eclipsing stage from the value determined during eclipses".

INFRARED AND OPTICAL SPECTROSCOPY OF THE SUSPECTED PLANETARY NEBULA He2-77*

M. de MUIZON
Sterrewacht Leiden, Postbus 9513, NL-2300 RA Leiden,
and Observatoire de Paris, F-92195 Meudon.
A. PREITE MARTINEZ
Instituto di Astrofisica Spaziale, C.N.R., C.P. 67, I-00044 Frascati, Italy.
M. HEYDARI-MALAYERI
European Southern Observatory, Casilla 19001, Santiago 19, Chile,
and Observatoire de Paris, F-92195 Meudon.

ABSTRACT: We present infrared and optical spectroscopic data of the suspected planetary nebula He2-77 (PK 298-0.1). The object has been selected from the IRAS-LRS database because it has a bright LRS spectrum with many emission lines and features. The observations include the IRAS Low Resolution Spectrum and IRAS far-infrared photometry, a ground-based 2-4 μm spectrum and 1-5 μm photometry, and a ground-based optical spectrum between 4300 and 6800 Å. A number of emission fine-structure lines from ionized atomic species are measured in the infrared and optical spectra. Also emission features between 3 and 13 μm, attributed to polycyclic aromatic hydrocarbon molecules, are detected. Properties of the gas and dust components of the nebula are discussed and they are used in order to clarify the nature of He2-77.

1. INTRODUCTION.

We selected the nebula He2-77 from the IRAS-LRS database because it is a strong far-infrared source and because of the presence in its Low Resolution Spectrum (LRS) of several emission lines from ionized gas and emission features from hot dust. The visual aspect of the nebula on the ESO-SERC Sky Survey Plate is that of a bright, very red, irregular patch, about 12 arcsec in diameter and with no apparent central star. It is located in the galactic plane, at very low declination ($-63°$). He2-77 is included in the Perek-Kohoutek catalogue under the number PK 298-0.1. In earlier observational studies of planetary nebulae, it has been classified as a non-stellar planetary nebula (Allen and Glass 1974), or a faint suspected planetary nebula (Sanduleak 1976). It appears clearly as a small but strong source on radio surveys at 5GHz by Goss and Shaver (1970) and Milne (1979); it is located about half a degree from the very strong and extended radio source-giant HII region complex G298.2-0.3, but not necessarily associated with it. The integrated 5GHz

---

Based on observations collected at the European Southern Observatory, La Silla, Chile. The Infrared Astronomical Satellite (IRAS) was developped and operated by the Netherlands Agency for Aerospace Programs (NVIR), the US National Aeronautics and Space Administration (NASA) and the UK Science and Engineering Research Council (SERC).

flux of He2-77 is 2.3 Jy. Cohen and Barlow (1980) obtained eight points of narrow-band infrared photometry from 8 to 20 μm, together with some optical spectrophotometric data. They estimate a visual extinction of $A_v=8.9$, assign the predicted optical depth at 10 μm to silicate absorption and suggest that He2-77 is a heavily obscured compact HII region. Finally, Whitelock (1985) obtained J, H, K photometry of He2-77 in a study of eighty planetary nebulae. She also suggests that He2-77 could be fundamentally different from the PNe in her sample. The distance to He2-77 was estimated to be 0.6 kpc by Maciel (1984).

We present in this paper a set of additional infrared and optical, IRAS and ground-based, spectroscopic observations of He2-77. These data are discussed in order to derive new elements about the nature of He2-77 and about its gas and dust composition and properties.

2. OBSERVATIONS AND RESULTS.

Several sets of observations are reported here:
i) The IRAS 8-23 μm LRS spectrum, and the IRAS far-infrared photometry.
ii) Near-infrared J, H, K, L, M photometry and 2 to 4 μm spectrophotometry, obtained at the ESO-1m telescope.
iii) Optical spectroscopy from 4300 to 6800 Å obtained at the ESO-2.2m telescope.

The LRS spectrum was obtained in 1983 by the Low Resolution Spectrometer (Wildeman et al. 1983) onboard IRAS. Two wavelength ranges, 7.7-13.5 μm (band 1) and 11-22.5 μm (band 2), were recorded simultaneously, with respective fields of view of 6' × 5' and 6' × 7.5'. However the instrument is basically an objective prism and therefore good quality spectra are obtained only for sources less than 20 to 30 arcsec in diameter. In each wavelength band, the resolving power increases with wavelength and varies from 10 to 40. The LRS data have been calibrated using the Survey calibration (IRAS Explanatory Supplement 1985). The absolute flux calibration of the LRS spectra is thought to be accurate to ±30%.

The LRS spectrum of He2-77 is presented in Figure 1. It is the average of six individual spectra in band 1 and four in band 2. On each individual spectrum, the baseline at both edges is symmetric; this ensures the good quality of the spectrum, and shows that the source was pointlike enough for the LRS. Several lines and features can be identified on the LRS spectrum of He2-77. These are the atomic fine-structure lines of ionized gas such as: [ArIII]9.0 μm, [SIV]10.5 μm, [NeII]12.8 μm, [NeIII]15.5 μm and [SIII]18.7 μm. There is also a doubtful presence of [ClIV] lines at 11.7 and 20.3 μm.

Three emission features at 7.7, 8.6 and 11.3 μm are prominent on this spectrum. They are part of the family of infrared emission features between 3 and 13 μm which have been attributed to Polycyclic Aromatic Hydrocarbons – PAHs – molecules by Léger and Puget (1984) and Allamandola et al. (1985). The apparent dip at 10 μm cannot be attributed to silicate absorption since it appears to be simply the dip between two strong emission lines or features. There is no sign of the broad silicate absorption feature at 17 μm either. The slope of the spectrum is positive from 10 to 22 μm. Far-infrared photometry from the IRAS Survey instrument shows that the peak of the FIR spectrum, either in Inband flux or in energy (λI_λ), is in band 3 of the IRAS Survey Instrument (60 μm).

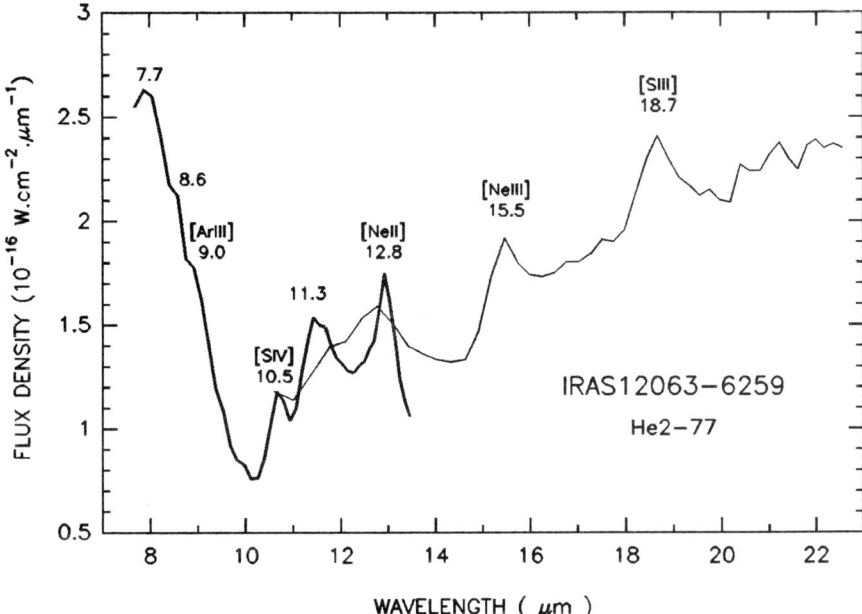

Figure 1: *IRAS-LRS spectrum of He2-77.*
Thick line: band 1 (7.7-13.5μm); thin line: band 2 (11-22.5μm).
In the overlapping region, the resolution is ≈40 in band 1 and ≈10 in band 2.

The near-infrared 2-2.4 and 3-4 μm CVF (Circular Variable Filter) spectrum of He2-77 and J, H, K, L, M photometry were obtained in March and August 1985 at the European Southern Observatory 1m telescope. The data were calibrated by measuring standard stars close in airmass to the object. An entrance aperture of 15 arcsec was used. The total time spent on the source to obtain the CVF spectrum is about five hours.

Several lines and features are also present on this spectrum (Figure 2): the Brγ, Pfγ and Brα lines of atomic hydrogen at 2.16, 3.74 and 4.05 μm, and the [HeI] line at 2.06 μm. Also prominent are features at 3.3, 3.4 and 3.54 μm. The first one is part of the same family as those at 7.7, 8.6 and 11.3 μm seen in the LRS spectrum, attributed to PAHs. It includes, however, a contribution from the the Pfδ line (3.29 μm) of hydrogen which can be estimated from the Pfγ line and is about 20% of the total feature intensity. The near and far-infrared photometric data are summarized in Table I. We have estimated the total infrared flux (1-100 μm) to be 1.6×10^{-10} W.m^{-2}. At a distance of 0.5 to 0.6 kpc, this leads to a total infrared luminosity between 1250 and 1800 $L_\odot$.

The optical spectrum of He2-77 was obtained at the European Southern Observatory 2.2m telescope in June 1986, using a CCD detector and a Boller and Chivens spectrometer at a resolution of 172 Å/mm, in the wavelength range 4300-6800Å. The spectrum is presented in Figure 3. Besides the Hα and Hβ Balmer lines of hydrogen, a number of emission lines from various elements are present, such as: [OIII]4959 and 5007Å, [NII]5755, 6548 and 6584Å, [HeI]5876 and 6678Å, [SII]6717 and 6731Å, [SIII]6310Å.

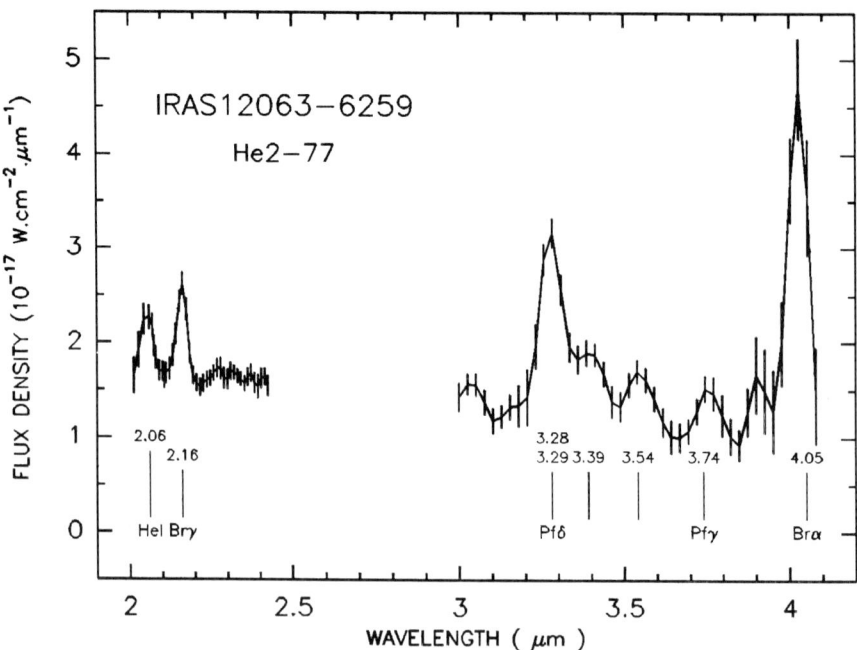

Figure 2: Near-infrared CVF spectrum of He2-77 obtained at the ESO-1m telescope. Spectral resolution is ≈70; aperture is 15 arcsec.

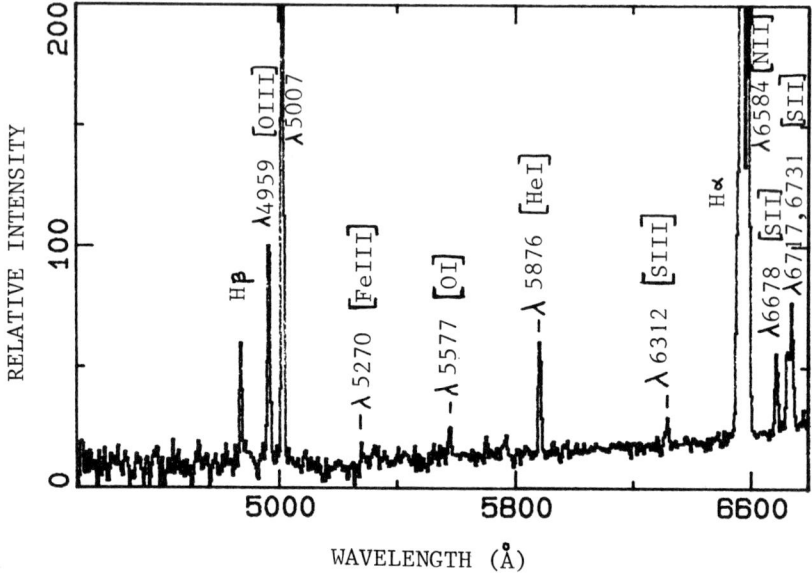

Figure 3: Optical spectrum of He2-77 obtained at the ESO-2.2m telescope, using a Boller and Chivens spectrograph (172Å /mm) and a CCD detector.

Table I: Near and far infrared photometry of He2-77

Band	λ (μm)	Mag.	I_λ (W.m^{-2}.μm^{-1})	Inband Flux (W.m^{-2})	$\lambda.I_\lambda$ (W.m^{-2})
J	1.25	10.9	1.71×10^{-13}		2.14×10^{-13}
H	1.65	9.8	1.39×10^{-13}		2.29×10^{-13}
K	2.2	8.5	1.61×10^{-13}		3.54×10^{-13}
L	3.6	6.2	2.10×10^{-13}		7.57×10^{-13}
M	4.8	5.0	2.09×10^{-13}		1.00×10^{-12}
IRAS Band 1	12			7.21×10^{-12}	1.43×10^{-11}
IRAS Band 2	25			2.26×10^{-11}	5.43×10^{-11}
IRAS Band 3	60			4.82×10^{-11}	9.48×10^{-11}
IRAS Band 4	100			2.23×10^{-11}	6.63×10^{-11}

3. THE GAS COMPONENT

He2-77 is a very reddened nebula, and previous observations estimated an E(B-V) of the order of 3 (Cohen and Barlow 1980). Our optical spectrum, averaged over the entire nebula, is at higher resolution and allows a more accurate determination of the reddening. As can be seen in Figure 3, Hα is much stronger than [NII], so that the intensity of the [NII]6548Å line, blended with the hydrogen line, will be inferred from the intensity of the [NII]6584Å line. The resulting intensities are listed in Table II. E(B-V) can then be derived from the Balmer decrement, giving E(B-V)=2.6$\pm$0.1 or, equivalently, c=3.8, and A_v=8. The intensities of the optical lines and of the near-IR lines were corrected accordingly. In absence of an absolute calibration for the optical spectrum, we are not able to compare the value derived from the Balmer decrement method with the extinction that can be derived using radio and Hβ fluxes. However, we can compare the Hβ flux predicted from the radio with that predicted from near-IR hydrogen lines, corrected for an E(B-V) of 2.6. The predicted fluxes are listed in the last column of Table III. Although errors are of the order of 15-20%, Hβ fluxes from near-IR hydrogen lines are on the average lower (5.5×10^{-10}) than the flux predicted from the radio (7.4×10^{-10}). A possible explanation is that the nebula is slightly more extended than the beam used for the near-IR observations.

In order to derive abundances from optical and infrared ionic lines, we need an estimate of electron density and temperature. The only available ratio of forbidden lines sensitive to electron density is [SII]6731/[SII]6717. Its value is 1.60, indicating an electron density of about 3100 cm^{-3}. The optical spectrum is too noisy shortward of Hβ to allow the determination of the electron temperature from the usual [OIII] ratio. Using only the optical forbidden lines, an upper limit to T_e can be set from the ratio [NII]5755/([NII]6548+6584), although the uncertainties on the intensity of the 5755Å line are too high to set a good upper limit: we can only say that T_e is most probably lower than 20.000 K. Therefore we have to use another temperature dependent ratio, involving both optical and infrared lines: the ratio [SIII]6310 Å/[SIII]18.7 μm. In order to compute this ratio we need to normalize the two lines independently, assuming that:

$$I(6310\text{Å})/I(18.7\mu m) = I(6310\text{Å})/I(H\beta) * F\beta(\text{Radio})/F(18.7\mu m)$$

where $F\beta$(Radio) is the Hβ flux derived from the flux at 5GHz. The value of the ratio is found to be 0.035, corresponding to an electron temperature of about 9000 K, as derived from the theoretical curves shown in Figure 4. The curves were computed by solving the system of balance equations for a 5-level atom, using the atomic data compiled by Mendoza (1983). The error can be estimated of the order of 1000 K.

Table II: Intensity of the optical lines

Ident.	λ (Å)	I_0	I_c
Hβ		100.	100.
OIII	4959	225.	181.
OIII	5007	680.	520.
NII	5755	10:	1.9:
HeI	5876	112.	17.
SIII	6310	18.	1.4
NII	6548	135.	7.8
Hα		5065.	283.
NII	6584	480.	26.4
HeI	6678	76.	3.7
SII	6717	83.	3.84
SII	6731	133.	6.15

Table III: Intensity of the infrared lines (10^{-11} erg.cm^{-2}.s^{-1})

Ident.	λ (μm)	F_0	F_c	Predicted F_β
HeI	2.06	0.38	1.09	
Brγ	2.16	0.42	1.07	40.
Pfγ	3.74	0.50	0.70	77.
Brα	4.05	2.8	3.7	47.
S_ν	5Ghz	2.3Jy		74.
ArIII	9.0	4.0		
SIV	10.5	7.8		
ClIV	11.7	< .52 :		
NeII	12.8	22.		
NeIII	15.5	29.		
SIII	18.7	29.		
ClIV	20.3	< 3.3 :		

Table IV: Ionic abundances ($H^+ = 1$)

Ion	Line	I/I_β	Ionic abundance		Total abundance (LogH=12)
He^+	5876		0.12		11.04
	6678		0.10		
He^+	<>		0.11		
He^{++}			0.0	(e)	
N^+	6548+6584	0.342	6.73×10^{-6}		7.8:
N^{++}			5.50×10^{-5}	(e)	
O^+			3.40×10^{-5}	(e)	8.5:
O^{++}	4959+5007	7.01	2.74×10^{-4}		
S^+	6717+6731	0.10	4.74×10^{-7}		6.8
S^{++}	6310		4.94×10^{-6}		
	18.7μm	0.392	4.94×10^{-6}		
S^{3+}	10.5μm	0.105	3.50×10^{-7}		
Ne^+	12.8μm	0.297	3.50×10^{-5}		7.7
Ne^{++}	15.5μm	0.392	1.76×10^{-5}		
Ar^{++}	9.0μm	0.054	7.70×10^{-7}		
Cl^{3+}	11.7μm	< 0.007	$< 1.6 \times 10^{-7}$		

(e)=estimated

Table V: Estimated line intensities ($H\beta = 1$)

Line	λ	I
OII	3727Å	0.37
	7330Å	0.01
OIII	51.8μm	0.88
	88.3μm	0.17
SIII	9069Å	0.31
	9535Å	0.80
NeIII	3869Å	0.14
	3968Å	0.04

In Table IV, we list the ionic abundances computed assuming that the electron density and temperature derived from the forbidden lines of [SII] and [SIII] are constant throughout the nebula and applicable to zones where ionic species of higher ionization potential are formed. As one can see, S and Ne are present with more than one ionization stage. Nevertheless, we can try to estimate total elemental abundances for He, N, O, Ne and S.

Helium: the optical spectrum is too underexposed shortward of Hβ to search for the [HeII]4686Å line. On the other hand, from the overall behaviour of S and Ne ionic abundances, we can infer that He2-77 is a medium-excitation nebula, with a central ionizing source not able to fully ionize the surrounding gas above $\approx 40eV$. Indeed, the abundance ratios S^{++}/S^{3+} and Ne^+/Ne^{++} are both larger than 1. Thus we expect the strength of the [HeII]4686Å line to be negligible. We estimate a total helium abundance, by number, of 0.11 (+0.02, −0.01).

Nitrogen and oxygen: the only way to estimate the total abundance of these elements is to assume further that $N^{++}=O^{++}/5$ and $O^+=5\times N^+$. For what we said above we do not expect significant contribution from other ions. Total abundances are then 6×10^{-5} and 3×10^{-4} respectively.

Sulphur and neon: we just add up the abundances of the ions seen in the spectra. Total abundances are: $S=5.6\times 10^{-6}$, and $Ne=5.3\times 10^{-5}$.

From the available spectra and the derived ionic abundances we can classify He2-77 as a medium-excitation nebula. We can try to estimate the temperature of the central star using the energy-balance method (Preite-Martinez and Pottasch 1983). In order to apply the method we basically need another quantity, the ratio ρ of the sum of the forbidden line intensities to Hβ. We compute ρ from the observed intensities listed in Tables II and III, and use the ionic abundances of Table IV to estimate the intensities of other unobserved lines of the most abundant ions. These estimated line intensities are listed in Table V. We end up with a value of $\rho \geq 11.5$, which directly transforms into a minimum temperature of the central star of 35.000 K.

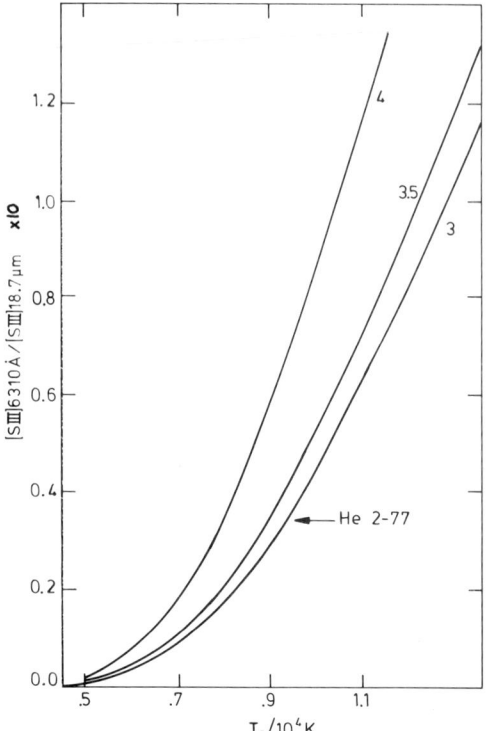

Figure 4: Temperature dependence of the ratio of the forbidden line intensities: [SIII]6310Å /[SIII]18.7μm.

4. THE DUST COMPONENT.

The strong far-infrared flux and the heavily obscured optical aspect of He2-77 are the first indicators of the presence of a significant amount of dust in the nebula. The visual extinction inferred from the ratio of optical lines Hα/Hβ is A_v=8. However, using the ratio of infrared lines Brα/Brγ leads to A_v=19$\pm$4. Finally, combining the radio flux and infrared lines, i.e. 5GHz/Brβ leads to A_v=22$\pm$2. These decrepancies in the value derived for the visual extinction could be due partly to some uncertainty in the measurements of the infrared line intensities. A more likely explanation is that there is a significant inhomogeneity in the local dust distribution, as modelled by Natta and Panagia (1984) who applied their method to the heavily obscured compact HII region K3-50.

The infrared data of He2-77 clearly show the presence of two populations of dust particles. A population of cold dust is responsible for the far-infrared emission. Colour temperatures can be estimated from the ratio of the far-infrared fluxes, assuming various laws for the grain absorption efficiency: $Q_{abs} \propto \lambda^{-1}$, $Q_{abs} \propto \lambda^{-2}$, or Q_{abs} taken from Mathis's model (Mezger et al. 1982). The results are given in Table VI. They refer to a population of big grains, typically 0.1 μm in diameter, thermally heated, with an average temperature of about 50 K. Since there is no obvious silicate absorption in the infrared spectrum of He2-77, the main constituent of these grains is most likely graphite.

The near-infrared emission features at 3.3, 7.7, 8.6 and 11.3 μm account for a totally different population of dust particles. Together with a feature at 6.2 μm, not observable from the ground, they are observed as a set and in a large variety of objects (see d'Hendecourt and Léger, this volume). They have been attributed to Polycyclic Aromatic Hydrocarbons molecules, such as coronene ($C_{24}H_{12}$), by Léger and Puget (1984). The idea was later supported by Allamandola et al. (1985). Each of the features can be identified with a fundamental vibration frequency: C–H stretch (3.3 μm), C=C stretch (6.2 μm), C=C skeletal mode (7.7 μm), C–H bending in-plane (8.6 μm) and C–H bending out-of-plane (11.3 μm). The features at 3.4 μm and 3.54 μm are not clearly identified. They are not always systematically present with the other features of the family. Together with some other features between 3.4 and 3.6 μm, observed in two IRAS sources (Muizon et al. 1986), they may account for a mixture of PAHs molecules: the precise frequency of the C–H stretching vibration depends on the nature of the molecules considered. In an unsaturated hydrocarbon such as an aromatic, the mode occurs at $\approx$3030 cm^{-1} (3.3 μm). In a saturated hydrocarbon such as an alkane, the C–H vibration lies in the frequency interval 2962-2853 cm^{-1} (3.38-3.51 μm). The features at 3.4 and 3.54 μm could be due to PAHs, on which some peripheral H–atoms have been replaced by molecular subgroups such as $-CH_3$ or $-C_2H_5$. However, no exact identification has been obtained so far in laboratory spectra (d'Hendecourt and Léger, this volume).

Table VI: Colour temperatures from IRAS Survey fluxes

	$Q_{abs} \propto \lambda^{-1}$	$Q_{abs} \propto \lambda^{-2}$	Q_{abs} (Mathis Model)
T(60/25)	72	61	64
T(100/60)	46	37	45

These near-infrared emission features can be explained by the presence of a population of very small grains — large molecules, typically 5 to 10 Å in diameter, and which are transiently heated by absorption of a single UV photon (Sellgren 1984).

In order to estimate the average temperature and size of the molecules, a simple method has been proposed by Muizon et al. (1987). It is based on the value of the intensity ratio of the 11.3 to the 3.3 μm features. These two bands originate in two modes of the same molecular subgroup (aromatic C–H stretching and bending, respectively) and it is possible to relate their intensity ratio to an average temperature of the emitting molecule. The number of C–atom per molecule is then only dependent on the energy of the incident UV photon. The application of this method is rather limited however since: i) it would be more realistic to talk about a temperature distribution instead of an average temperature, but this is much more complicated to model. ii) it would be necessary to know or to determine the energy distribution of the incident UV photon flux. iii) the spectra at 3 μm and at 11 μm have been taken with different instruments and telescopes, therefore different apertures; in the case of He2–77, the value of the intensity ratio I(11.3 μm)/I(3.3 μm) is only an upper limit if the infrared emission is more extended than the 15 arcsec used for our 3 μm observations. With all the above restrictions, applying this method to He2–77 gives an upper limit for the intensity ratio I(11.3 μm)/I(3.3 μm) of 25±8, thus a lower limit for the average temperature of the molecules $<T> \geq$ 550 K, and an upper limit for the number of C–atoms per molecule $N_C \leq 60$. Although not accurate it indicates anyway that these small grains–large molecules are much hotter than those responsible for the 60–100 μm emission.

Estimating the mass of gas and dust in He2–77 leads to $M_{gas} \sim 6 \times 10^{-2} M_\odot$ and $M_{dust} \sim 7 \times 10^{-4} M_\odot$, assuming a distance of 0.5 kpc. The dust-to-gas mass ratio is therefore of the order of 10^{-2}, a value which is similar to the average value in the interstellar medium. It is quite high, however, but not exceptional, for planetary nebulae, in which this ratio is commonly between 10^{-2} and 10^{-4}, even 1.8×10^{-2} in the case of BD+30°3639 (Pottasch 1984, p.199).

5. CONCLUSION.

The data presented in this paper contain some observational evidences in favour of the nature of He2–77 being a planetary nebula as opposed to a compact HII region:
i) the presence in the LRS spectrum of the [SIV] 10.5 μm and [NeIII]15.5 μm lines. All LRS spectra of compact HII regions only show [SIII] and [NeII], but no higher ionization stages.
ii) the peak of the far-infrared spectrum of He2–77 is around 60 μm. All far-infrared spectra of compact HII regions peak beyond 100 μm and have a 100 μm flux much higher than their 60 μm flux.
iii) the lower limit of the temperature of the central star is estimated to be 35.000 K. If the total far-infrared luminosity (1 to $2 \times 10^3 L_\odot$) accounts for the luminosity of the central star, this puts He2–77 close to BD+30°3639 in the H–R diagram for planetary nebulae (Pottasch 1984, p.218).

REFERENCES:

Allamandola, L. J., Tielens, A. G. G. M., Barker, J. R.: 1985, *Astrophys. J. (Letters)* **290**, L25.
Allen, D. A., Glass, I. S.: 1974, *M. N. R. A. S.* **167**, 337.
Cohen, M., Barlow, M. J.: 1980, *Astrophys. J.* **238**, 585.
Goss, W. M., Shaver, P. A.: 1970, *Aust. J. Phys. Astrophys. suppl* **14**, 1.
Léger, A., Puget, J.-L.: 1984, *Astron. Astrophys. (Letters)* **137**, L5.
Maciel, W. J.: 1984, Astron. Astrophys. Suppl. Ser. 55, 253.
Mendoza, C.: 1983, in *"Planetary Nebulae"*, IAU Symposium No. 103, ed. D.R. Flower, (Dordrecht:Reidel), p.143.
Mezger, P. G., Mathis, J. S., Panagia, N.: 1982, *Astron. Astrophys.* **105**, 372.
Milne, D. K.: 1979, *Astron. Astrophys. Suppl. Ser.* **36**, 227.
Muizon, M. de, Geballe, T. R., d'Hendecourt, L. B., Baas, F.: 1986a, *Astrophys. J. (Letters)* **306**, L105.
Muizon, M. de, d'Hendecourt, L. B., Geballe, T. R.: 1987, in *"Polycyclic Aromatic Hydrocarbons and Astrophysics"*, Proceedings of the NATO workshop held in Les Houches (F), 17-22 Feb. 1986, eds. A. Léger, L. B. d'Hendecourt, (Dordrecht: Reidel), p. 287.
Natta, A., Panagia, N.: 1984, *Astrophys. J.* **287**, 228.
Pottasch, S. R.: 1984, *"Planetary nebulae"*, Dordrecht: Reidel, vol.107.
Preite-Martinez, A., Pottasch, S. R.: 1983, *Astron. Astrophys.* **126**, 31.
Sanduleak, N.: 1976, *Publ. Warner and Swasey Obs.* **2**, 56.
Sellgren, K.: 1984, *Astrophys. J.* **277**, 623.
Whitelock, P. A.: 1985, *M. N. R. A. S.* **213**, 59.
Wildeman, K. J., Beintema, D. A., Wesselius, P. R.: 1983, *J. British Interplanet. Soc.* **36**, 21.

DISCUSSION

Pottasch: How was the distance estimated?

Muizon: A distance of 0.6 kpc is given by Maciel (1984). It has been obtained using a method to determine the distances of planetary nebulae proposed by Maciel and Pottasch (1980), on the basis of a mass-radius relationship established from selected electron densities and distances.

Panagia: It seems to me that the combination of a low dust temperature (≈ 50 K), high dust-to-gas ratio ($\approx 10^{-2}$), a relatively cool star (≈ 35.000 K) and the high, possibly highly inhomogeneous, extinction in front of it makes He2–77 similar to an HII region (such as K3–50 for example) than a proper PNe. Why is it considered a planetary nebula ?

Muizon: A dust-to-gas ratio of $\approx 10^{-2}$ and a central star of at least ≈ 35.000 K are encountered in planetary nebulae (e.g. BD+30°3639), so is a high extinction (e.g. NGC 7027). What makes He2–77 very different from a normal HII region is the presence in its LRS spectrum of emission lines of ionization stages commonly observed in planetary nebulae, [SIV] and [NeIII], but not seen in HII regions, and also the peak of the far-infrared spectrum which occurs much shortward of 100 μm.

Roche: Normal HII regions always show evidence of silicate dust in their infrared spectra, as expected from the interstellar medium. He2–77 shows no evidence for silicates and so it is very unlikely an HII region and is much more likely to be a planetary nebula.

Peimbert: It is very important to have an accurate distance to decide if it is a PNe or an HII region. If it is a PNe, the ionized mass would be $\approx 0.2 M_\odot$, while if it is an HII region it would be about $20 M_\odot$. From the visual and radio extinction you presented, it is obtained that the ratio of total to selective absorption, R, is about seven. I would not be too worried for this result; if it is confirmed it could be explained with normal dust but with an absorption varying strongly across the face of the nebula.

CONTRIBUTION OF NEBULAR EMISSION LINES TO IRAS
PHOTOMETRIC SURVEY FLUXES

A. Preite Martinez(^), S.R. Pottasch
Kapteyn Astronomical Institute
Postbus 800
9700 AV Groningen, The Netherlands

ABSTRACT. In this paper we discuss the contribution of nebular emission lines to IRAS in-band fluxes. The most affected nebulae are medium-, high-excitation nebulae. An analytical fit to nebular line emissivities is presented.

1. INTRODUCTION

In a recent paper (Pottasch et al., 1984) far-infrared IRAS measurements of 46 planetary nebulae were presented and discussed. It was also pointed out that line emission could contribute significantly to the observed fluxes, especially in the 12μm band. In this paper we examine in detail this problem, presenting also results of a best-fit of line emissivities as a function of electron density and temperature.

Observational evidence and theoretical considerations indicate the presence, in the far-IR spectral region, of line emission originated from levels of the ground term of the most abundant ions.

Three major fine-structure lines have been observed from the ground in the 8-13μm window: AIII 9.0μm, SIV 10.5 μm, and NeII 12.8μm (Grasdalen, 1979; Dinerstein, 1980; Beck et al., 1981; Aitken and Roche, 1982; Roche and Aitken, 1983). Other lines (OIII and OIV) have been observed with airborne telescopes (Forrest et al., 1980) in spectral regions unaccessible from the ground. The low-resolution spectrometer (LRS) onboard the IRAS satellite has confirmed the presence of emission lines in the 8-13μm region, broadening the observable wavelength range from about 7 up to about 22μm (Pottasch et al., 1984; 1986), covering also part of the 25μm survey band.

2. LINE CONTRIBUTION

About 20 emission lines fall in the wavelength range of the IRAS Survey band B1 (8.5-15μm), B2 (19-30μm), B3 (40-80μm), and B4 (83-120μm). The
----
(^) On leave from Istituto di Astrofisica Spaziale, Frascati, Italy

elements responsible for the lines are N, O, Ne, Na, Mg, S, Cl, A, and Ca, but we restricted our analysis to 11 lines of six elements on the basis of elemental abundance and expected emissivity. The selected lines are listed in the first column of Table I.

Using the atomic data compiled by Mendoza (1983) and the more recent data on Ne++ (Butler and Mendoza (1984) we can compute the emissivity of all the selected lines as a function of electron density and temperature. In order to simplify computations we have fitted our emissivity tables with the analytical formula :

$$R(\lambda) = E(\lambda)/E(H\text{-beta}) = A\, t^{\alpha} n^{\beta}/(n + B) \qquad (1)$$

where: $R(\lambda)$ is the emissivity of line λ in units of the H-beta emissivity (Ferland, 1980); t is the electron temperature in units of 10^4K; n is the electron density in cm^{-3} ; A, B, α, and β are the parameters of the fit, shown in Table I.

Table I. Parameters of the analytical fit to line emissivities.

line		$S_k(\lambda)$	A	α	B	β
12µ band						
AIII	9.0	0.75	1.30+5	0.63	31.	0.926
SIV	10.5	0.93	7.44+9	0.66	2.10+4	0.0
NeII	12.8	1.0	3.48+9	0.508	3.88+5	0.0
AV	13.1	1.0	1.24+5	0.772	52.	0.892
NeV	14.3	0.8	2.30+9	0.644	1.72+4	0.018
25µ band						
AIII	21.8	0.9	2.75+8	0.668	4.94+4	0.0
NeV	25.2	1.0	8.97+8	0.713	3.66+3	-0.066
OIV	25.9	0.9	2.17+8	0.721	3.83+3	0.0
60µ band						
OIII	51.8	0.7	1.75+7	0.742	2.07+3	0.0
NIII	57.3	0.8	9.25+6	0.720	1.16+3	0.0
100µ band						
OIII	88.3	1.0	7.70+6	0.829	3.25+2	-0.142

Equation (1) reproduces the computed line emissivities always better than about 10% in the temperature range 6000-16000K and in the density range 100-100,000 cm^{-3}, and better than 2% in the (n , t) region close to (1000cm^{-3} , 10^4 K).

The contribution to the in-band flux detected in the IRAS survey band k is then given by

$$C(k) = S_k(\lambda) \, F(H\text{-beta}) \, N(X)/N(H) \, R(\lambda) \qquad (2)$$

where: $S_k(\lambda)$ is the relative system spectral response in band $\underline{k}$ at wavelength λ (Neugebauer et al., 1984); the value of $S_k(\lambda)$ is listed in Table I. F(H-beta) is the observed H-beta flux corrected for extinction (Pottasch, 1984), and N(X)/N(H) is the abundance of the ion emitting the line. Ionic abundances were taken from Aller and Czyzak (1983), Barker (1983, 1984), and Pottasch et al. (1981).

When the N++ abundance was not directly available from UV data it was estimated to be 1/3-1/4 of O++, as derived from the OIII optical lines. The abundance of S^{3+} was either derived from IRAS/LRS data (Pottasch et al., 1985), or from the O++/O+ ratio, following the ionization correction scheme suggested by Natta et al. (1980). When not directly derived from UV data, the abundance of O^{3+} was evaluated according to the ionization correction formula by Seaton (1968): $O^{3+}/(O^+ + O^{2+}) = He^{2+}/He^+$.

Table II. Observed (o) and corrected (c) IRAS Survey fluxes (Jy).

Nebula	12μ (o)	12μ (c)	25μ (o)	25μ (c)	60μ (o)	60μ (c)	100μ (o)	100μ (c)
NGC 40	15.7	14.4	79.2	79.1	78.7	78.5	36.3	36.1
NGC 1535	1.5	1.15	9.7	7.7	20.4	17.9	13.1	12.3
NGC 2022	0.9	0.35	12.8	6.7	10.0	9.6	8.1	7.7
NGC 2371	0.6	0.19	6.3	2.7	10.1	9.0	11.8	10.6
NGC 2440	4.0	2.7	33.8	14.8	49.3	44.1	33.4	30.0
NGC 3242	4.4	2.9	38.0	34.6	63.0	56.1	36.0	31.8
NGC 6210	2.0	0.54	27.0	23.4	40.0	37.1	21.0	19.7
NGC 6543	7.8	5.3	118.0	111.0	158.0	144.0	80.0	72.4
NGC 6751	4.0	3.1	21.8	20.6	28.7	27.3	11.3	8.9
NGC 6790	15.5	15.2	41.5	39.3	15.6	14.3	4.1	3.6
NGC 6803	0.9	0.05	12.6	12.3	14.9	13.9	3.3	2.8
NGC 6818	1.8	0.6	19.4	11.0	23.7	21.6	14.4	12.9
NGC 6826	5.5	5.1	46.7	45.8	56.8	53.1	27.1	24.6
NGC 6853	5.55	2.2	40.1	6.1	130.0	84.9	151.2	64.3
NGC 6884	1.4	0.85	15.6	13.3	20.3	18.9	6.8	6.1
NGC 6905	0.5	0.14	7.0	6.0	10.2	9.5	8.7	8.1
NGC 7009	7.4	2.84	65.4	62.9	114.0	104.6	58.1	53.0
NGC 7662	4.3	2.8	41.7	32.9	39.8	37.4	17.6	16.1
IC 418	41.4	39.5	242.0	241.8	129.0	127.4	40.5	39.9
Hu 1-2	0.5	0.0	4.2	3.0	4.9	4.8	2.3	2.2

IR line emissivities are practically independent on electron temperature. The temperature dependence (α) mainly comes from the H-beta emissivity. Electron density plays a more important role: collisional deexcitation can drastically reduce the emissivity of a particular line if n is much greater than the critical density for that line. Critical densities for the lines listed in Table I range from <1000 to >10,000 cm^{-3}, so that collisional de-excitation will decrease line contribution to in-band fluxes for high-density nebulae.

Equation (2) has been used to compute the contribution of nebular lines to in-band survey fluxes for a sample of 20 planetary nebulae selected on the basis of great variety in excitation conditions, brightness, size, and electron density. Results are shown in Table II.

An alternative method is to derive the intensity of IR lines contributing to survey fluxes from the the intensity of line(s) of the same ion observed in other wavelength ranges. The same set of atomic data is needed to compute line ratios, and knowledge of the H-beta flux is also necessary.

Subtraction of observed line intensities (from LRS spectra and/or directly observed from the ground) is also possible for the best studied nebulae.

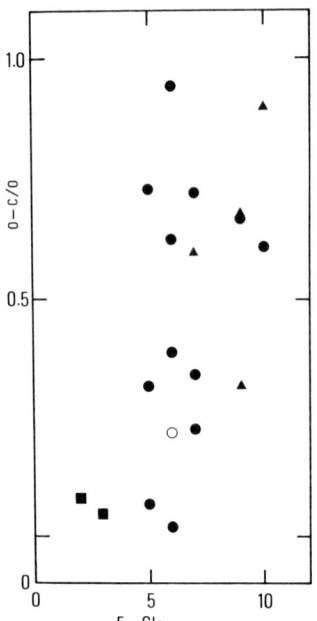
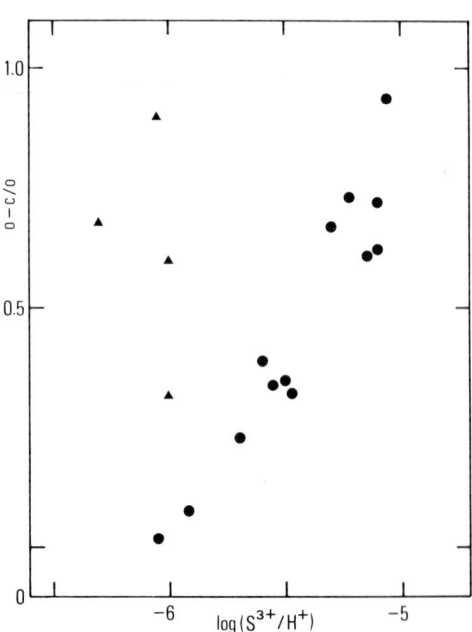

Figure 1. The relative strength of the correction for the 12μm band is plotted against the excitation class of the nebula (Fig.1a) and S^{3+} (Fig.1b).

3. DISCUSSION

As expected, the 12μm survey band is the most affected by nebular line emission. In Figure 1 we plot the relative strength of the correction (o-c)/o for the 12μm band versus the excitation class of the nebula (Fig. 1a), and S^{3+} abundance (Fig. 1b). We indicate with filled circles, squares, and triangles the nebulae where the strongest line in this band is the SIV, NeII, and NeV line, respectively. In NGC 6751, indicated with an open circle, line contribution is dominated by the AV line because of the abnormally high abundance of Argon.

Figure 1 shows one obvious feature: SIV produces the strongest line only in medium-, high-excitation (ME, HE) nebulae, NeII in low-excitation (LE) nebulae, and NeV only in very HE nebulae. No significant correlation is present in Figure 1a, although the correction is negligible in LE nebulae, while it can be tremendously important in ME, HE nebulae.

Infrared line emission can also contribute significantly to the determination of the temperature of the central star, when the energy-balance method is used. Adding up the line contribution to the four IRAS Survey bands we recomputed ρ for the nebulae of our sample. The updated values of ρ and the corresponding temperature T(EB) are listed in Table III.

Table III. Revised temperatures of central stars (T in 10^3 K).

Nebula	ρ_{IR}	$\rho(\hat{\ })$	ρ_{Tot}	T(EB)
NGC 40	1.6	9.5	11.6	39
NGC 1535	4.4	22.4	26.8	78
NGC 2371	14.1	24.3	38.4	92
NGC 2440	8.4	70.8	79.2	205
NGC 3242	3.2	20.7	23.9	60
NGC 6751	11.5	19.7	31.2	94
NGC 6818	10.7	45.7	56.2	130
NGC 6826	1.9	11.3	13.2	43
NGC 6853	>10.	39.4	50:	145:
NGC 6905	6.5	24.6	31.1	72
NGC 7009	4.0	20.0	24.0	69
NGC 7662	4.6	36.1	40.7	105
IC 418	0.6	8.5	9.1	31.5
Hu 1-2	4.7	37.3	42.0	92

($\hat{\ }$) From Preite-Martinez and Pottasch (1984).

It is worth noting that now the Energy-Balance temperature of NGC 6853 is in very good agreement with the helium Zanstra temperature.

References

Aitken,D.K., Roche,P.F. 1982, Mon.Not.R.astr.Soc. 200,217
Aller,L.H., Czyzak,S.J. 1983, Astrophys.J.Suppl. 51,211
Barker,T. 1983, Astrophys.J. 267,630
Barker,T. 1984, Astrophys.J. 284,589
Beck,S.C., Lacy,J.H., Townes,C.H., Aller,L.H., Geballe,T.R., Baas,F. 1981, Astrophys.J. 249,592
Butler,K., Mendoza,C. 1984, Mon.Not.R.astr.Soc. 208,17p
Dinerstein,H.L. 1980, Astrophys.J. 237,486
Ferland,G.J. 1980, Publ.Astron.Soc.Pac. 92,596
Forrest,W.J., McCarty,J.F., Houck,J.R. 1980, Astrophys.J. 240,L37
Grasdalen,G.L. 1979, Astrophys.J. 229,587
Mendoza,C. 1983, IAU Symp.103, 'Planetary Nebulae', ed.D.R.Flower (Reidel,Dordrecht),p.143
Natta,A., Panagia,N., Preite-Martinez,A. 1980, Astrophys.J. 242,596
Neugebauer,G., Habing,H.J., van Duinen,R., Aumann,H.H., Baud,B., Beichman,C.A., Beintema,D.A., Boggess,N., Clegg,P.E., de Jong,T., Emerson,J.P., Gautier,T.N., Gillet,F.C., Harris,S., Hauser,M.G., Houck,J.R., Jennings,R.E., Low,F.J., Marsden,P.L., Miley,G., Olnon,F.M., Pottasch,S.R., Raimond,E., Rowan-Robinson,M., Soifer,B.T., Walker,R.G., Wesselius,P.R., Young,E. 1984, Astrophys.J. 278,L1
Pottasch,S.R., Gilra,D.P., Natta,A., Preite-Martinez,A., Wesselius, P.R. 1981, Proc.Second Eur.IUE Conf., p.185
Pottasch,S.R., Beintema,D.A., Raimond,E., Baud,B., van Duinen,R., Habing,H.J., Houck,J.R., de Jong,T., Jennings,R.E., Olnon,F.M., Wesselius,P.R. 1983, Astrophys.J. 278,L33
Pottasch,S.R. 1984, 'Planetary Nebulae' (Reidel,Dordrecht)
Pottasch,S.R., Baud,B., Beintema,D., Emerson,J., Habing,H.J., Harris,S., Houck,J.R., Jennings,R.E., Marsden,P.L. 1984, Astron. Astrophys. 138,10
Pottasch,S.R., Preite-Martinez,A., Olnon,F.M., Jing-Er,M., Kingma,S. 1986, Astron.Astrophys. 161,363
Preite-Martinez,A., Pottasch,S.R. 1983, Astron.Astrophys. 126,31
Roche,P.F., Aitken,D.K. 1983, Mon.Not.R.astr.Soc. 203,9p

INFRARED CHARACTERISTICS OF POLYCYCLIC AROMATIC HYDROCARBONS AND THE INTERPRETATION OF IR ASTRONOMICAL SPECTRA

L.B. d'Hendecourt, A. Léger
Groupe de Physique des Solides de l' ENS
Université de Paris VII - Tour 23
2, Place Jussieu - 75251 Paris Cedex 05 - FRANCE

ABSTRACT. Recently, the interpretation given to the so-called "unidentified" infrared emission bands has drastically improved. These bands are attributed to the fundamental vibrations of large molecules known in organic chemistry as Polycyclic Aromatic Hydrocarbons (PAH's). In this paper, we review the main arguments which led to this identification and focus our attention to some recent IR spectroscopic studies which allow us to interpret further the main characteristics of IR astronomical spectra. We show that a mixture of compact and partially dehydrogenated PAH's can be the carrier of the main emission features.

1 - INTRODUCTION

Mid infrared spectroscopy has proven to be a useful tool for the identification of molecular species in the spectra of astronomical sources. The fact that molecular vibration frequencies fall in the mid infrared region and that each molecular subgroup yields characteristic vibration frequencies over most of the mid IR spectrum (Allamandola, 1984), has been widely used for the interpretation of astronomical spectra. IR absorption spectroscopy of various molecular ices in the laboratory has been very successful not only for the identification of the numerous absorption bands observed towards IR sources embedded in dense molecular clouds (Lacy et al., 1984; Tielens et al., 1984) but also for the understanding of the role dust grains can play in the overall chemistry of molecular clouds (Tielens and Hagen, 1982; d'Hendecourt et al., 1985, 1986).
   In emission, apart from an important feature around 10 micron

attributed to silicates, several bands at 3.3, 6.2, 7.7, 8.6 and 11.3 micron, although discovered as early as 1973 (see Allamandola, 1984 and Willner, 1984 for a review), have resisted identification until recently. To summarize, we can say that these bands are observed simultaneously in a wide variety of astronomical objects : reflection nebulae, bipolar nebulae, planetary nebulae and whole regions of active galaxies such as M82 (figure 1). A prerequisite condition for their observation seems to be the presence of a strong UV field so that the energy emitted in the IR results from the conversion of absorbed UV photons. Any explanation to account for the observed features must explain both the <u>position and respective intensities</u> of the bands. In other words, this explanation should lead to a precise explanation of the emitting species in terms of their physico-chemical properties and simultaneously account for the emission mechanism (d'Hendecourt, 1987).

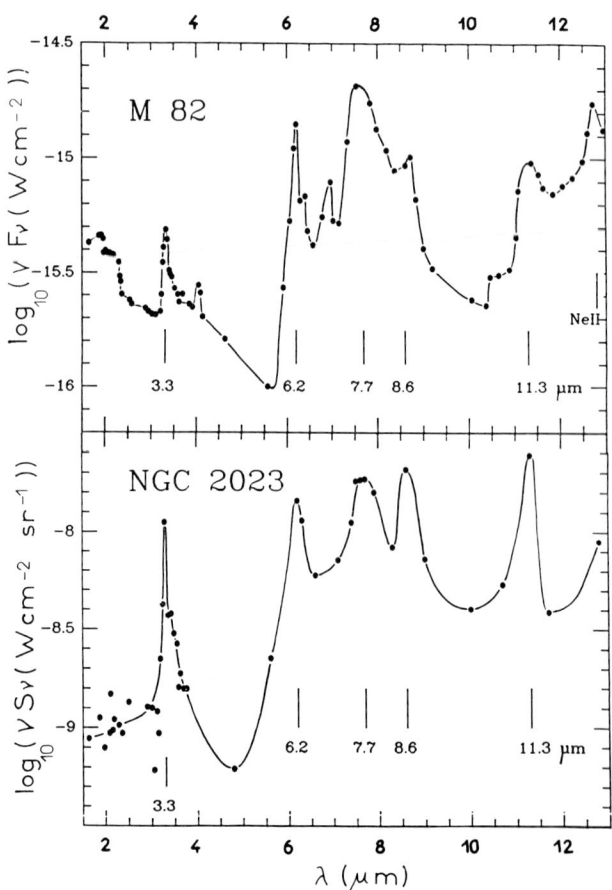

Figure 1. Mid IR spectra of the central part of the active galaxy M82 (top) adapted from Willner et al. (1977) and Gillett et al. (1975) and of the reflection nebula NGC 2023 from Sellgren et al. (1985).

2 - THE PAH's HYPOTHESIS

In 1983, Sellgren et al. made a decisive observation in measuring the IR spectrum of some reflection nebulae. They showed that, in addition to the prominent emission bands, a strong continuum, not explained by scattered light from the illuminating star, was present and that this continuum could be fitted by a diluted blackbody emission spectrum of rather high temperature ($\neq$1000 K). Later on, Sellgren (1984) showed that such a high temperature could be achieved by the quantum heating of very small grains upon the absorption of one UV photon. For a temperature increase of about 1000 K, the number of atoms involved in such small grains is only about 50. Leger and Puget (1984) suggested that only graphitic material could survive such temperature spikes and considered the optical properties of very small graphite grains. They showed that the use of bulk optical properties for very small graphite grains is incorrect because the strong electronic transitions that arise in a large grain and completely screen the lattice modes, are in fact forbidden in very small, molecule sized grains (for a simple example, this is why benzene is transparent in the visible while graphite is black). Going along further, they showed that a sphere of graphite containing 50 atoms and subjected to high temperature spikes will split along the graphitic planes because the interplane binding energy is quite small (0.06 ev.) compared to the binding energy of the carbon atoms in the plane (7.5 ev.). In an astronomical environment, because of the overwhelming abundance of hydrogen, the unsaturated carbon atoms at the periphery of the planar cluster will acquire a full hydrogen coverage. Figure 2 reproduces this situation, leading to the concept of large PAH molecules which are well known in organic chemistry (Clar, 1964).

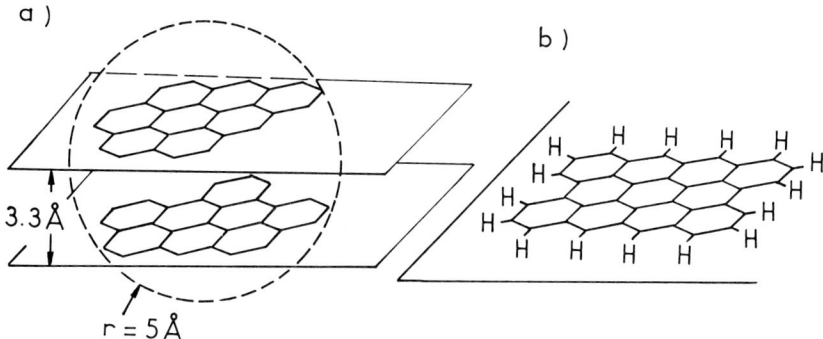

Figure 2. Schematic representation of a 50 atoms graphite "sphere" (a) and proposed structure of a hydrogenated graphitic cluster (b).

IR spectra of some aromatic molecules are available in the literature and a general description of the IR properties of aromatic molecules can be found in Bellamy (1966). Each observed emission line can be assigned to a particular fundamental frequency of a molecular subgroup: The 3.28 μm band is due to an <u>aromatic</u> C-H strech, the 6.2 and 7.7 μm ones are attributed to a C=C strech and a skeletal carbon mode respectively and the 8.6 and 11.3 μm lines are attributed to a C-H bending in and out of the plane of the molecule. Moreover, as stated by Bellamy, in analytical IR spectroscopy, these frequencies are highly characteristics of the presence of aromatic compounds in the substance studied. The emission spectrum of a particular molecule, coronene $C_{24}H_{12}$, has been computed by Léger and Puget (1984) and this spectrum is reproduced on figure 3 together with two typical astronomical spectra.

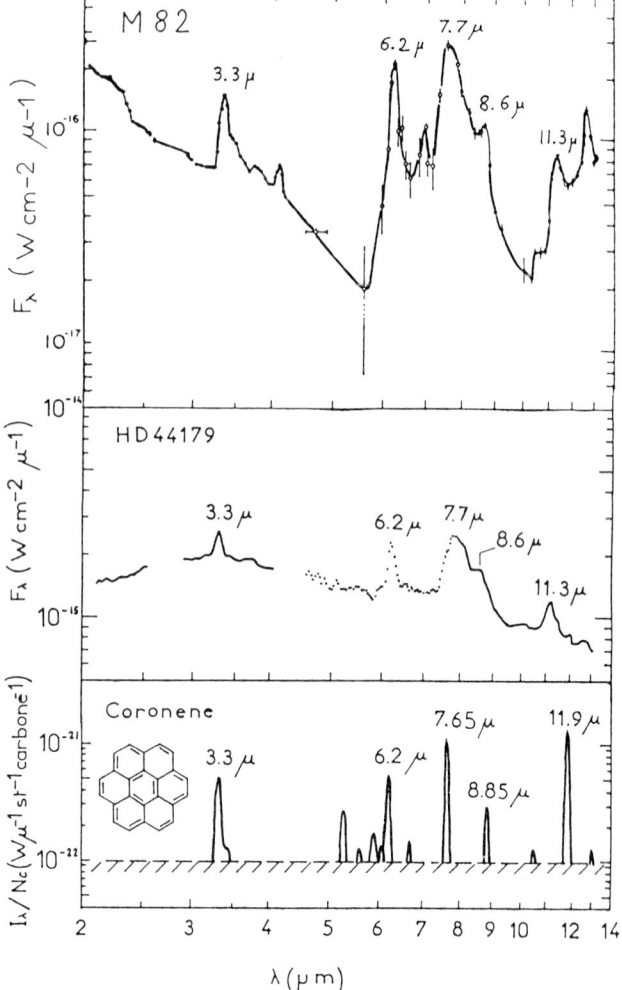

Figure 3. Calculated emission spectrum of coronene (bottom) compared to two astronomical spectra.

Details about the emission mechanism and a justification about the physical hypothesis introduced for the computation of this emission spectrum can be found in Leger and d'Hendecourt (1987). A similar conclusion was reached by Allamandola et al. (1985) who supported the idea of a mixture of various PAH's present in the ISM by comparing an astronomical spectra with the Raman spectrum of soot (auto exhaust).

3 - FURTHER IR STUDIES

Starting from the PAH hypothesis as stated above and introducing some laboratory measurements of IR spectra of various large PAH molecules, we wish to precise here the nature of the molecules responsible for the IR emission bands in the ISM.

3.1. More IR absorption spectra of large PAH's

The IR absorption spectra of various PAH's have been recently measured (Leger, d'Hendecourt, Schmidt, 1987).Typical spectra are reported in figure 4 and the integrated intensities of the main bands are given in table 1. From this table, we note that (i) the integrated cross - sections of the different bands are very similar from one molecule to another so that an average value can be used as representative of a typical PAH mixture present in the interstellar medium; (ii) the cross - section per solo hydrogen (see section 3.4.1) at 11.3 μm are quite larger than those per duo hydrogen at 11.9 μm in the two measured species that exhibit both modes.

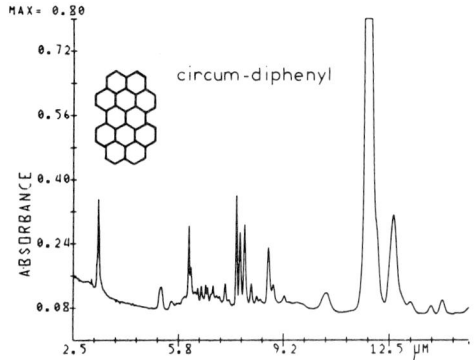
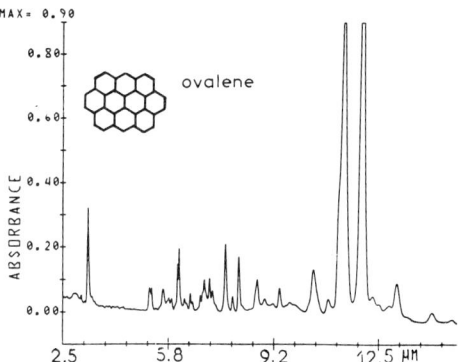

Figure 4. IR absorption spectra of two large PAH molecules (in a CsI pellet).

λ_i (µm)	3.3	6.2	7.7 (±.3)	8.8	11.3 ± 11.9	
Units for A_i 10^{-25} cm^3 ×	H^{-1}	C^{-1}	C^{-1}	H^{-1}	H^{-1}	
Coronene $C_{24}H_{12}$	1.2	.54	2.25	.90	47	
Dicoronene $C_{48}H_{20}$	1.7	1.04	2.32	1.0	50	110/H solo 34/H duo
Methylcoronene $C_{25}H_{14}$	1.4	.90	2.32	1.6	38	
Ovalene $C_{32}H_{14}$	1.5	.45	1.13	.89	43	160/H solo 23/H duo
Circobiphenyl $C_{38}H_{20}$	1.1	.59	2.0	1.6	33	
Typical compact PAH	1.4	.70	2.0	1.2	42	
$f_i/10^{-6}$	1.45 (H^{-1})	.21 (C^{-1})	.38 (C^{-1})	.18 (H^{-1})	3.4 (H^{-1})	

TABLE 1. Integrated absorption cross-section for different compact PAHs as measured in the laboratory. The average values can be used for general calculations. The resulting oscillator strengths $f_i = 1.13 \cdot 10^{20}$ $(A_i/1 \text{ cm}^3)$ $(\lambda/1 \text{ µm})^{-2}$.

3.2 - Calculated emission spectra : the effect of temperature.

The emission spectrum of a molecule at a temperature T depends on its absorption spectrum and on the temperature of emission as detailed in Leger and d'Hendecourt (1987a). This effect, which influences strongly the intensities of the bands in the 3 and 12 µm regions is shown on figure 5.

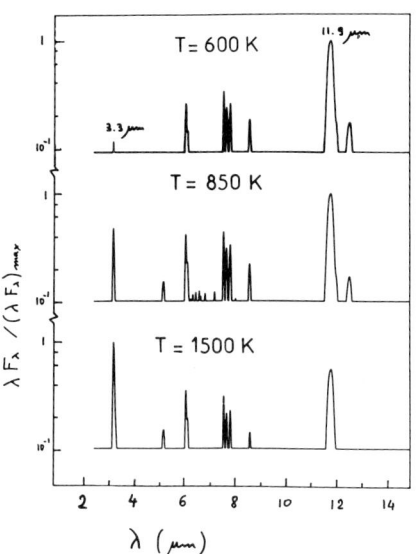

Figure 5. Effect of the temperature on the emission spectrum of a PAH molecule, circobiphenyl. Compare the evolution of the respective intensities of the bands at 3.3 and 11.9 μm as a function of the temperature.

3.3 Calculated emission for compact PAH

The calculated emission spectra of several large compact PAH's are reported on figure 6 and compared with a typical astronomical spectrum. The fit has only one adjustable parameter, the emission temperature. Non compact PAH's whose emission spectra are shown on figure 7 do not fit equally well the typical astronomical spectrum : in particular, new bands appear around 7μm which are absent in the astronomical spectra and reflect the presence of protruding benzene rings on the molecule.

3.4 More IR problems

3.4.1 The 11.3 μm band problem. In coronene, a prominent absorption at 11.9 μm (840 cm^{-1}) is attributed to the C-H bending out of plane while in astronomical spectra, the actual band lies at 11.3 μm. In fact, in aromatics, the exact position of the band pertaining to the C-H bending out of plane depends upon the number of adjacent hydrogen atoms placed on a ring (Bellamy, 1966). This effect is depicted on figure 8. In astronomical spectra, a predominance of solo H is inferred, although the presence of duo and even trio H can be suspected from the broad emission structure which is clearly visible up to 13.5μm in many IRAS LRS spectra (Cohen et al., 1985, de Muizon et al., 1987a). What will happen if a given molecule in space is partially dehydrogenated ? Clearly, for coronene, the number of H solos will increase respective to the number of H duos. Although the effect of such a dehydrogenation on the oscillator strength of the transition is not known, the shift can be confirmed by the comparison

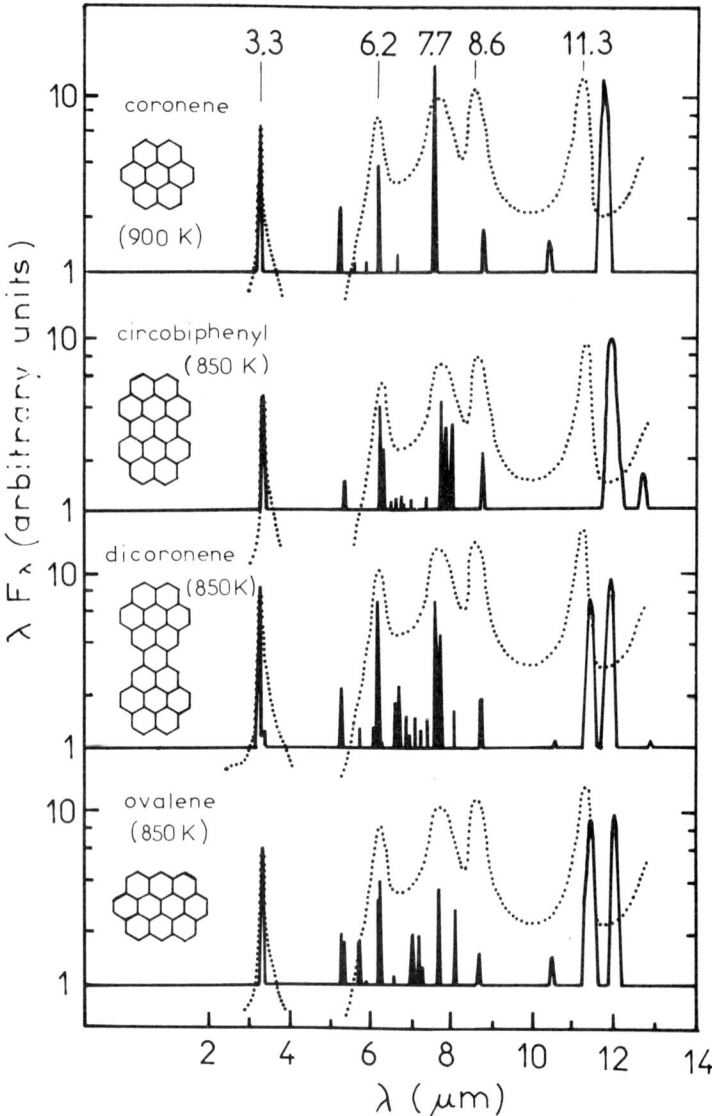

Figure 6. Emission spectra of several compact PAH molecules calculated from their absorption spectra. These spectra are superimposed, for comparison, on the spectrum of NGC 2023 (dots).

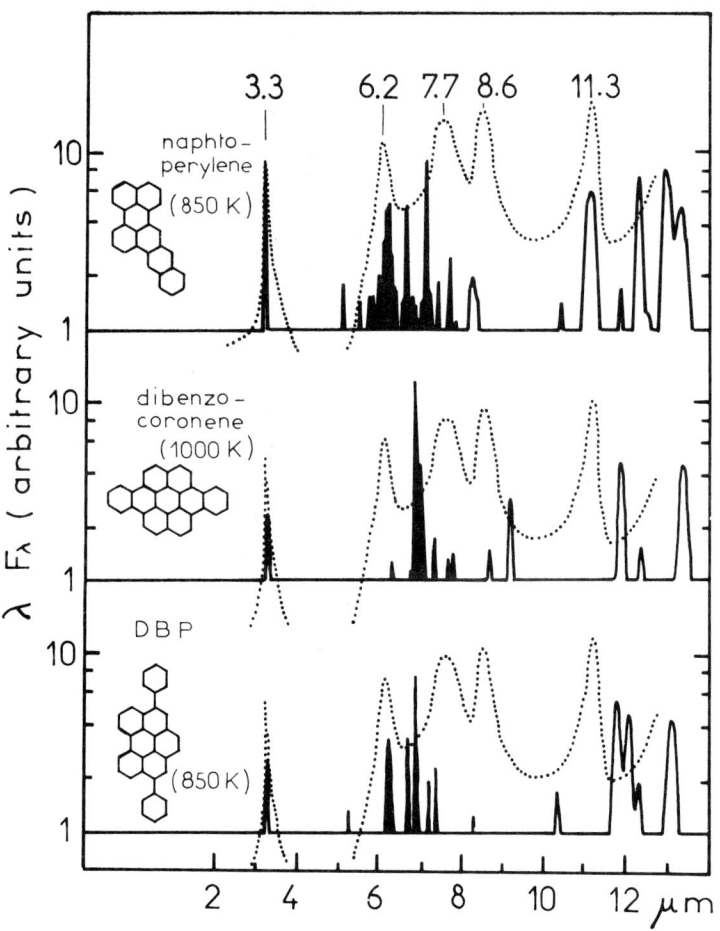

Figure 7. Same as figure 6, but for non compact molecules. Note the disagreement with the astronomical spectrum in the 6-8 μm region.

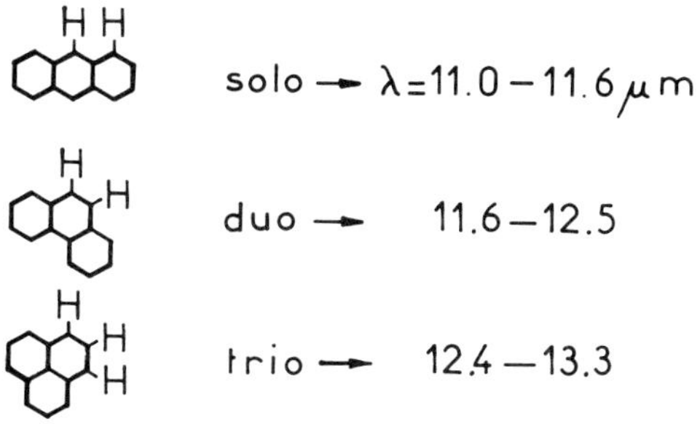

Figure 8. Different sites for hydrogen atoms and the corresponding wavelengths of the out of plane CH bending mode.

between the IR absorption spectra of coronene (12 H in 6 duos) and tetrabromocoronene (8 H, 2 duos, 4 solos) displayed on figure 9 : the 11.9 μm band in coronene is changed into two bands, the 11.3 μm one (H solos) and the 11.9 μm one (H duos).

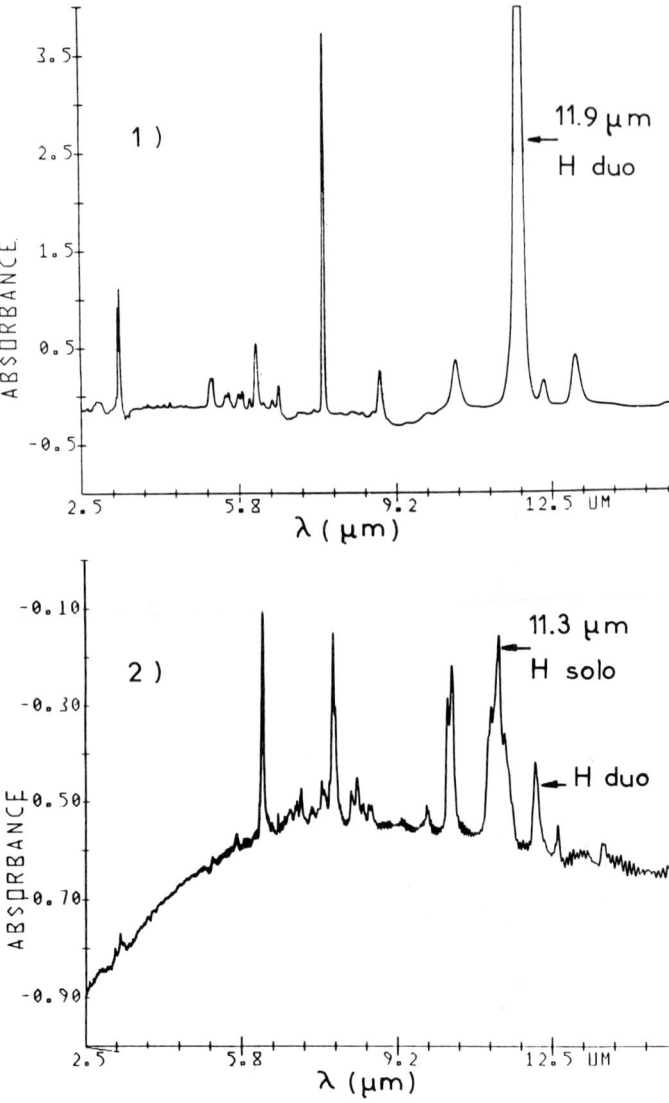

Figure 9. Absorption spectra of coronene (top) and tetrabromocoronene (bottom). The shift from 11.9 μm (H duo) to 11.3 μm (H solo) of the CH bending mode is easily noted.

3.4.2. Emission temperature : the size of the molecule. As evident from figure 5, the emission temperature of the molecule strongly influences the ratio of the various lines. If two lines originate from the same molecular subgroup, the intensity ratio of these two lines can be used to derive an average emission temperature, provided the oscillator strength of the two transitions are known. To calculate an average emission temperature, the intensity ratio of the two lines at 3.28 (C-H strech) and 11.3 µm (C-H bend) can be used. Then, if the UV excitation spectrum is known, the size of the molecule i.e. the number of atoms contained in the molecule can be deduced because the emission temperature is related to the energy of the UV photon absorbed via the knowledge of the heat capacity of the molecular species involved (Léger and d'Hendecourt, 1987a; de Muizon et al., 1987a). The measurement of the intensity ratio of these two bands has been performed in a number of astronomical objects. Average emission temperatures between 460 and 800 K and a number of carbon atoms in the emitting molecule between 85 and 28 have been deduced by de Muizon et al.(87a).

3.4.3. Dehydrogenation of PAH's in the Interstellar Medium.

λ_i (µm)		3.3	6.2	7.7	11.3	
$A_{\lambda_i}/A_{3.3\mu m}$		1	1.0	2.9	30 (97)	a
$E_{\lambda_i}/E_{3.3\ \mu m}$	(NGC 2023)	1	4.6	9.8	4.2	b,c
—	(Red Rect.)	1	4.3	8.1	2.0	c
—	(M 82)	1	4.1	17.6	1.8	d
x_H	(NGC 2023)			12%	11%	
—	(Red Rect.)			9%	8%	
—	(M 82)			9%	3%	
T_{em}	(NGC 2023)		780 K (1190 K)			
—	(Red Rect.)		980 K			
—	(M 82)		1020 K			

TABLE 2. Mean emission temperature (T_{em}) and hydrogen coverage (x_H) deduced from comparison between astronomical observations and laboratory absorption coefficients of PAHs.

The analysis of figure 6 indicates that C=C modes are more intense (relatively to C-H modes) in the astronomical spectrum than in the PAH spectra. This favors the idea of partial hydrogenation of PAH's. Quantitatively, the hydrogen coverage, xH = (H present/number of sites) can be deduced from the intensities of the bands related to hydrogen (C-H) and those related to carbon (C=C and C-C), if one assumes that dehydrogenation does not modify the oscillator strength of the various modes. The hydrogen coverage is estimated in table 2 for three objects.

The interpretation of this dehydrogenation factor is not trivial because theoretical computations of PAH's dehydrogenation by photothermodissociation (Tielens et al., 1987; Léger, Désert and d'Hendecourt, 1987) favors either no dehydrogenation at all (large molecules) or complete dehydrogenation (small molecules). A small value of xH may possibly reflect the fraction of hydrogenated species versus the total number of species rather than the partial coverage of a given species.

3.4.4 <u>Interpretation of the new emission lines around 3.4 - 3.6 μm.</u>
Recently, high resolution spectra (R = 400) of two IRAS sources have been obtained by de Muizon et al. (1986), using the UKIRT telescope. In addition to the well known 3.28 μm emission feature, a certain number of lines at 3.4, 3.46, 3.51 and 3.56 μm on top of an extended plateau located between 3.4 and 3.6 μm, have been recorded. Because of the presence of the 3.28 μm line as well as the presence of the set of emission lines at 7.7, 8.6 and 11.3 μm in the IRAS LRS spectra of these two objects, PAH's are indeed present in these two sources and it is tempting to tentatively assign the new emission features in accordance with the PAH's hypothesis. In fact, as we shall see in this section, no strict convincing evidence can be inferred from the present set of data. Two hypothesis can be put forward : (i) the presence of various molecular subgroups (aliphatic CH_3 and CH_3CH_2 groups attached on a PAH molecule and (ii) the effect of anharmonicities on the C-H stretch vibration, anharmonicity which can be observed in the laboratory in highly vibrationally excited molecules. These two hypothesis are described in the following subsections.

3.4.4.1 *Molecular subgroups*. As already stated at the beginning of this paper, the 3.28 μm line corresponds, from its position, to the vibration frequency of the aromatic C-H stretch, where the hydrogen atom is attached to a non-staturated carbon. As it is well known in organic chemistry, the value of the stretching frequency of a CH bond in aliphatic (saturated) hydrocarbons, always falls shorter than 3000 cm^{-1} i.e. at wavelengths longer than 3.3 μm. For example, in the solid state at 10 K, the C-H strech in methane falls at 3010 cm^{-1} (3.32 μm) and this is the shortest wavelength that a saturated hydrocarbon C-H strech displays (d'Hendecourt and Allamandola, 1986). Table 3 lists the frequencies of streching vibrations of various hydrocarbons. Note

that the frequencies of the unsaturated hydrocarbons fall in the
wavelength region where the new lines are observed.

Radical	Wavenumber (cm^{-1})	Wavelength (μm)
Alkanes (saturated)		
C-H in CH_3	2962^1 and 2862^2	3.38, 3.49
C-H in CH_2	2926^1 and 2853^2	3.42, 3.51
C-H in CH	2890	3.46
Aromatic		
=C-H	3030	3.3

[1]Symmetric mode
[2]Assymmetric mode

TABLE 3. C-H stretching frequencies in various molecular subgroups (adapted from Bellamy, 1966).

To test this hypothesis, we have recorded IR spectra in CsI matrices of various PAH's molecules containing aliphatic groups: methyl and ethyl coronene and methyl benzopyrene. These spectra are displayed on figure 10. Although they present some similarities with the observed new lines, the aliphatic C-H strech frequencies do not allow a perfect match. Clearly, more laboratory data are needed: as noted by Omont (1986), the most stable form of toluene (methyl benzene), in a UV radiation field, is the radical $C_5H_6-CH_2$., whose infrared spectrum is not known. However, the possibility of producing this type of radical by UV photolysis of a neon matrix containing toluene (or better, methylcoronene) will allow us to test this hypothesis in the near future. A short outline of such experiments is given in d'Hendecourt and Leger (1985).

3.3.4.2. *Effect of the anharmonicity of the C-H stretch mode.*
Another possibility to account for the observed new lines is the effect of the C-H stretch anharmonicity in a highly vibrationally excited molecule. This effect will be discussed at length in de Muizon et al.(1987) but we can here briefly summarize what is expected in the case of anharmonic vibrations.

In a fully harmonic oscillator, vibrational levels are equally spaced in energy so that any transitions from highly excited levels (v = 2 → 1 to v = 3 → 2) will occur at exactly the same frequency than the transition from the first vibrationally excited level to the ground state (v = 1 → 0). This transition is called the fundamental. However, as it is well known in molecular physics, in a local mode description (Herzberg, 1945, Swofford et al., 1976; Sibert et al., 1984), the C-H potential well is not strictly parabolic and, because at high energies the molecule will eventually dissociate, the energy levels are not equally spaced. For a C-H strech, this effect is quite important and has been measured in benzene. The anharmonicity of the

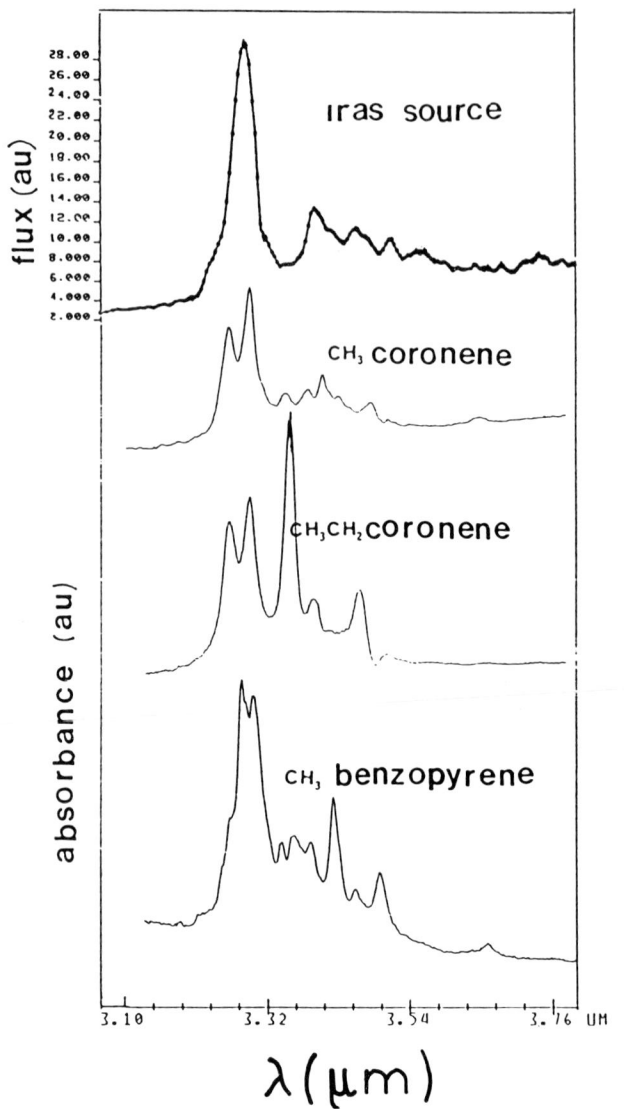

Figure 10. Comparison between the new IR emission lines discovered in an IRAS source (de Muizon et al., 1986) with PAHs molecules containing aliphatic CH_2 and CH_3 groups.

vibration is about 120 cm^{-1} so that , if transitions occur from upper levels, emission will appear at 3.41 and 3.56 µm about (for the v = 2 → 1 and 3 → 2 transitions respectively). The relative intensities of the lines will be governed according to the population of the vibrational levels at the emission temperature, assuming a complete thermalization of the vibrational energy and, hence, a Boltzman distribution for the population of the various vibrational levels. This last assumption of thermal equilibrium has been justified by Leger and d'Hendecourt (1987a,b) for such large molecules and is supported by laboratory experiments (see for example Wild et al.(1985).

A tentative fit is given in figure 11 where it is clearly shown that the relative intensities depend critically on the temperature of the emitting molecule. A best fit for the match of the relative intensities recorded from the astronomical spectrum yields a temperature of about 1700 K. At this early stage of the work, two comments can be stated (i) the tentative fit concerns only two lines among the four new ones and (ii) the high temperature needed to reproduce the relative intensities is rather high, in contradiction with the much lower temperatures (800 K) derived from the ratio of the 11.3 to the 3.3 µm emission lines as described by de Muizon et al.(1987a).

Although we have been taking into account only the anharmonicity of benzene, laboratory work on coronene is in progress and will be reported elsewhere as well as a more complete analysis of the anharmonicity problem (de Muizon et al., 1987b). Finally, on account of this problem, we wish to emphasize that the possible observation of the $2 \to 0$ transition around 1.65 µm in interstellar sources should bring some useful information on this problem.

3.5 Abundance of PAH's in the Interstellar Medium

The evaluation of the abundance of PAH's in the ISM can be deduced by considering the near IR emission from objects where the set of emission lines is observed. Assuming that from the same part of a nebula, the far infrared emission is produced by classical grains (large grains at low temperarture) and that the near IR is produced by PAH's, we can write :

$$\frac{\phi_{NIR}}{\phi_{FIR}} = \frac{<\sigma_{PAH}> N_{PAH}}{<\sigma_{gr}> N_{gr}}$$

where $<\sigma_{gr}> N_{gr}$ is the average absorption cross - section of classical grains, times their abundance and $<\sigma_{PAH}> N_{PAH}$ is the average absorption cross - section for PAH times their unknown abundance. The average absorption cross - section of PAH's can be estimated from molecular spectroscopy of these molecules. In reflection nebulae, the

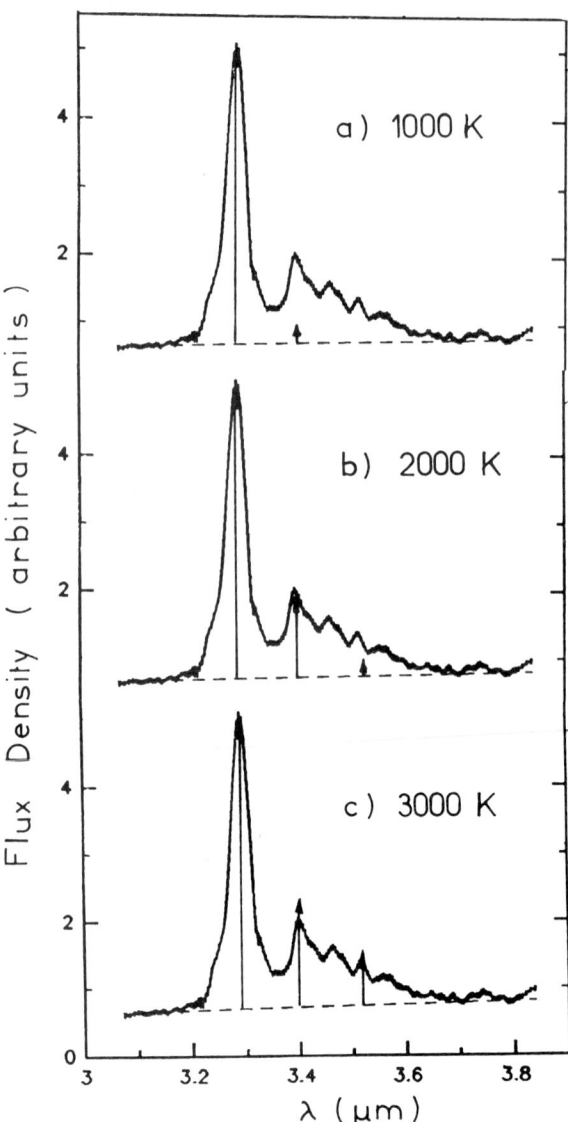

Figure 11. Idem than figure 10, but the effect of the anharmonicity of the CH stretch is shown as a function of the emission temperature. The intensity of the fundamental (v = 1 → 0) is normalized to the 3.28 μm line and the intensity of the harmonics (v = 2 → 1 ; v = 3 → 2) is calculated assuming that the population of the upper vibrational levels is given by a Boltzman distribution at the temperature indicated on the figure.

abundance of carbon locked in PAH's represent 3 % of the total cosmic abundance of carbon whereas in the diffuse interstellar medium this quantity amounts up to 6 % (see Léger and d'Hendecourt, 1987) and references therein). We might therefore deduce that PAH's are the most abundant organic molecules known to date in the ISM.

4. Conclusion

The identification of PAH's in astronomical IR spectra is firmly established. Although it is not possible to identify precisely a given molecule, the family of bands at 3.3, 6.2, 7.7, 8.6 and 11.3 µm are characteristics of the presence of aromatic compounds in space. We propose to name this set of bands the Aromatic Infrared Bands (AIR).

The presence of these AIR bands in very different objects, including external galaxies, is a strong argument in favor of their universal presence in the interstellar medium. Compact PAH's are favored because their IR spectra do match more closely the astronomical spectra. This can be undestood in terms of stability and should help to select the correct mixture of these molecules representative of the ones present in the ISM.

Crucial problems about ionization, dehydrogenation and addition of molecular subgroups remain open. These problems represent a challenge for experimentalists in order to perform the spectroscopy of such molecules in such unusual astrophysical conditions. Matrix isolation experiments on large PAH's should bring partial answers to these questions although more sophisticated experiments involving free molecular jets should be devised in order to study the energy flow from electronic excitation to vibrational IR emission in an isolated molecule.

Finally, although these problems have not been discussed with in this paper, many fascinating aspects of interstellar chemistry could be dealt with PAH's. Some of these aspects, such as the accretion of gaseous species on the PAH, ion molecule reactions, the control of the electron abundance in interstellar clouds and the problem of the origin of these molecules are tentatively listed in Omont (1986). No doubt that this area of research will open new and exciting questions.

REFERENCES

Allamandola,L.J.: 1984, "Galactic and Extragalactic Infrared Spectroscopy", eds. M.F.Kessler and J.P.Phillips, Reidel:Dordrecht, p.5
Allamandola,L.J., Tielens,A.G.G.M., Barker,J.R.: 1985, Astrophys.J.Lett. $\underline{290}$,L25
Bellamy,L.J.: 1966, "IR Spectra of Complex Molecules", Wiley
Clar,E.: 1964, "Polycyclic Aromatic Hydrocarbons" Academic Press
Cohen,M., Tielens,A.G.G.M., Allamandola,L.J.: 1985, Astrophys.J.Lett. $\underline{299}$,L93
Gillett,F.C., Kleinman,D.E., Wright,E.L., Capps,R.W.: 1975, Astrophys.J. Lett. $\underline{198}$,L65

d'Hendecourt,L.B.: 1987, "Polycyclic Aromatic Hydrocarbons and Astrophysics", eds. A.Leger, L.B.d'Hendecourt, N.Boccara, NATO Workshop Les Houches, France, Feb.86, Reidel
d'Hendecourt,L.B., Allamandola,L.J., Greenberg,J.M.: 1985, Astron. Astrophys. 152, 130
d'Hendecourt,L.B., Allamandola,L.J., Grim,R.J.A., Greenberg,J.M.: 1986, Astron.Astrophys. 158, 119
d'Hendecourt,L.B., Allamandola,L.J.: 1986, Astron.Astrophys.Suppl.Ser., 64, 453
Herzberg,G.: 1945, "Molecular Spectra and Molecular Structure: II Infrared and Raman Spectra of Polyatomic Molecules", van Nostrand
Lacy,J.H., Baas,F., Allamandola,L.J., Person,S.E., McGregor,P.J., Lonsdale,C.J., Geballe,T.R., Van de Bult,C.P.E.M.: 1984, Astrophys. J. 260, 141
Leger,A., Puget,J.L.: 1984, Astron.Astrophys. 137, L5
Leger,A., d'Hendecourt,L.B.: 1985, Astron.Astrophys. 146, 81
Leger,A., d'Hendecourt,L.B.: 1987a,"Polycyclic Aromatic Hydrocarbons and Astrophysics", eds. A.Leger, L.B.d'Hendecourt, N.Boccara, NATO Workshop, Les Houches, France, Feb.86, Reidel
Leger,A., Desert,F.X., d'Hendecourt,L.B.: 1987, in preparation
Leger,A., d'Hendecourt,L.B.: 1987b, in preparation
Leger,A., d'Hendecourt,L.B., Schmidt,W.: 1987, in preparation
de Muizon,M., d'Hendecourt,L.B., Geballe,T.R.: 1987a, "Polycyclic Aromatic Hydrocarbons and Astrophysics", eds. A.Leger, L.B. d'Hendecourt, N.Boccara, NATO Workshop, Les Houches, France, Feb. 86, Reidel
de Muizon,M., Geballe,T.R., d'Hendecourt,L.B., Baas,F.: 1986, Astrophys.J.Letters 306, L105
de Muizon,M., d'Hendecourt,L.B., Geballe,T.R.: 1987b, in preparation
Omont,A.: 1986, Astron.Astrophys. 164, 159
Puget,J.L., Leger,A., Boulanger,F.: 1985, Astron.Atrophys. 142, L19
Russell,R.W., Soifer,B.T., Willner,S.P.: 1978, Astrophys.J. 220,568
Sellgren,K., Werner,M.W., Dinerstein,H.L.: 1983, Astrophys.J.Letters, 271, L13
Sellgren,K.: 1984, Astrophys.J. 277, 623
Sibert,E.L., Reinhardt,W.P., Hynes,T.P.: 1984, J.Chem.Phys. 81,1115
Swofford,R.L., Long,M.E., Albrecht,A.C.: 1976, J.Phys.Chem. 65,179
Tielens,A.G.G.M., Hagen,W.: 1982, Astron.Astrophys. 114, 245
Tielens,A.G.G.M., Allamandola,L.J., Bregman,J., Goebel,J., d'Hendecourt L.B., Witteborn,F.C.: 1984, Astrophys.J. 287, 697
Tielens,A.G.G.M., Allamandola,L.J., Barker,J.R., Cohen,M.: 1987, "Polycyclic Aromatic Hydrocarbons and Astrophysics", eds. A.Leger, L.B.d'Hendecourt, N.Boccara, NATO Workshop, Les Houches, Reidel
Willner,S.P.: 1984, "Galactic and Extragalactic Infrared Spectroscopy", eds. M.F.Kessler and J.P.Phillips, Reidel
Willner,S.P., Soifer,B.T., Russell,R.W., Joyce,R.R., Gillett,F.C.: 1977, Astrophys.J.Letters 217, L121

NEAR-INFRARED PHOTOMETRY OF IRAS PLANETARY NEBULAE (^)

P.Persi, A.Preite Martinez, M.Ferrari-Toniolo, L.Spinoglio
Istituto di Astrofisica Spaziale,CNR,
C.P. 67
00044 Frascati, Italy

ABSTRACT. We present J, H, K, L photometry of a large sample (117) of objects classifed as planetary nebulae in the Perek - Kohoutec catalogue. According to the position of the sources in the near-IR colour-colour diagrams, we classify them in three classes: N (30%), D (13%), and S (57%). The energy distribution from 1 to 100μm is used to clarify the nature of S-type sources.

1. INTRODUCTION

Thanks to the success of the IRAS satellite, a large number of data concerning the far-IR emission from Planetary Nebulae (PN) are now available. Nearly 800 object listed in the Perek-Kohoutek Catalogue of PN are reported in the IRAS Point Source Catalogue (PSC), mostly identified as PN. The first IRAS measurements of PN reported by Pottasch et al. (1984,1986) and Iyengar (1986) has allowed to study the physical characteristics of the cold dust surrounding the nebula, and the determination of dust temperature, mass of dust, and infrared luminosities.
   At present, only a small fraction (about 100) of these sources have been observed from the ground in the near-IR (Willner et al.,1972; Persson and Frogel,1973; Allen,1973; Allen and Glass,1974; Cohen and Barlow,1974 and 1980; Whitelock,1985; Kwok et al.,1986). The earliest works showed that in PN the 1-2μm spectral region is dominated by hydrogen plasma emission, although some PN show a distinct excess. Therefore a correlated study of PN including near and far IR observations is important to give a complete description of the envelopes of PN.
   In this paper we will present the results of the J, H, K, and sometimes L photometry of a large sample (117) of PK objects detected by IRAS. Approximately 80% of the sources has been observed in the radio continuum at 6cm. In addition, all the selected sources are small (angular size <15") and optically very faint.

(^) Based on observations collected at the European Southern Observatory (La Silla, Chile) and at the Italian Infrared Telescope (TIRGO).

In Section 2 we report the near-IR observations, while in Section 3 we will give a classification of the sources from the analysis of the two colour-colour diagrams J-H vs.H-K and H-K vs.K-L. In Section 4 we will discuss the energy distribution from 1 to 100μm of typical objects.

2. OBSERVATIONS

The J, H, K, and L photometry of the sources reported in this paper was obtained during six observing runs at the 1m and 3.6m ESO telescopes and at the Italian Infrared Telescope (TIRGO), all equipped with cooled InSb detectors. The ESO InSb system is described in the ESO User's Manual and by Koornneef (1983), while the TIRGO InSb system is described by Hunt (1986). Both systems use very similar standard broad-band filters.

The log of these runs is reported in Table I, including the beam-sizes used during the observations.

TABLE I. Log of the near-IR observations.

Date	Observatory	Tel.	Diaphr. (arcsec)	Bands
28/9-4/10/84	TIRGO	1.5m	17	J,H,K,L
8-15/3/85	ESO	1.0m	15	J,H,K,L,M,CVF
1-5/5/85	TIRGO	1.5m	17	J,H,K,L
26-27/3/86	ESO	3.6m	10	J,H,K,L,M,CVF
28/3-1/4/86	ESO	1.0m	22	J,H,K,L,M,CVF
8-13/5/86	TIRGO	1.5m	17	J,H,K,L

CVF=spectra with resolution $\lambda/\Delta\lambda \simeq 100$ around the Br$_\gamma$ line.

The results of the near-IR photometry are given in Table II, where the first two columns list the designation of the source according to galactic coordinates and the name of the nebula. The next four columns contain the magnitude of the nebula in the J, H, K, and L bands, rspectively. Unless otherwise indicated, photometric errors are <0.05 mag. Finally, the last column of Table II contains a classification of the nebulae based on their location in the J-H vs.H-K colour diagram. This will be discussed in the next section.

Considerable care was taken when observing sources in crouded fields, to ensure that no spurious sources were included in the beam. Nonetheless we cannot rule out the possibility that some measurements might have been contaminated by very red field stars.

TABLE II.   Near-IR Photometry.

PK design.	Name	J	H	K	L	Cl
1 -6.2	SwSt1	9.10	8.50	7.94		D
3 +2.1	Hb4	9.60	8.18	7.54		S
3 -4.5	NGC6565	9.35	8.18	7.72	7.34(06)	S
10+18.1	M2-9	13.28(08)	12.65	11.78(06)	10.8:	D
11 +7.1		11.85	11.05	10.72	8.7:	S
11 -0.2	NGC6567	9.48	7.78	6.96		S
16+13.1		5.48	4.19	4.0	3.63	Se
24 +3.1	M2-40	12.99(11)	13.57(06)	12.33(07)	9.7:	N
25+40.1	IC4593	12.03(09)	12.62(14)	11.21(11)	>9.7	N
27 +4.1		10.64	10.34	8.85	5.98(06)	D
27 +0.1	M2-43	13.83(12)	12.26	11.91	>10.9	S
28 +5.1	K3-2	13.20(09)	13.27(13)	12.43(10)		N
29 -5.1	NGC6751	12.18	12.19	11.25	9.3(24)	N
33 -2.1	NGC6741	11.04	11.42	10.68	7.45(08)	N
34+11.1	NGC6572	10.99(13)	12.08(12)	11.14(13)		N
"		8.09	8.88	8.08		N
35 -0.1	Ap2-1	12.86	11.11	10.21(07)	7.98(33)	S
38+12.1	Cn3-1	11.09	11.29	10.87		N
39 +2.1	K3-17	9.39	7.86	7.25	6.70(07)	S
43+37.1	NGC6210	10.15	10.60	9.91	6.69(19)	N
45 -2.1	Vy2-2	10.11	10.52	9.69	7.54(08)	N
51 +9.1	Hu2-1	10.46	11.05	10.31	8.81(27)	N
60 -7.2	NGC6886	11.36	12.02	11.21		N
74 +2.1	NGC6881	11.62	11.96	11.03	8.53(24)	N
82 +7.1	NGC6884	13.58(07)	12.95	12.76(11)		S
86 +0.1	K4-56	12.84	10.60(09)	8.26(21)	5.22(09)	D
86 -8.1	Hu1-2	13.71(19)	14.49(23)	13.19(22)		N
89 -2.1	M1-77	10.69	10.20	9.30(08)	7.46(10)	D
96-29.1	NGC6543	8.67	9.39	8.80		N
96-29.1	NGC6543	8.28	8.99	8.44		N
100 -8.1	Me2-2	12.20	11.57	11.54		S
106-17.1	NGC7662	9.19(07)	9.40(11)	8.89(19)		N
111 -2.1	Hb12	9.40(07)	9.66(11)	8.74(19)	6.27(09)	N
118 -8.1	Vy1-1	12.56	12.74	12.08		N
123+34.1	IC3568	10.80	10.48	10.46	>10.7	N
130 +1.1	IC1747	11.57	12.03	11.20	8.31(33)	N
159-15.1	IC351	13.07	13.40	12.51(10)		N

TABLE II. Cont.d

```
161-14.1 IC2003    12.18       12.38       11.85                    N
211 -3.1 M1-6      11.71(06)   11.39(06)   10.24                    D
215-24.1 IC418      7.75(19)    8.33        7.50       5.68         N
223 -2.1 Aro226    12.43(13)   11.38(10)   10.69(11)                S
226 -3.1 Pb1       10.43        9.71        9.60                    S

228 +5.1 M1-17     10.31        9.80        9.66                    S
232 -1.1 M1-13     10.27        9.34        9.09                    S
232 -4.1 M1-11     10.84       10.05        8.89       6.87(12)     D
234 -0.1 M1-15     10.93       10.38        9.70                    D
234 -1.1 M1-14      7.14        6.12        5.80       5.43(07)     S

235 -3.1 M1-12     11.61(16)   10.93(07)   10.68(11)                S
241 -7.1 M4-1      10.42       10.35       10.23(08)                S
243 -1.1 NGC2452   12.27(14)   12.37(18)   12.39(33)                N
252 -4.1 Sa2-18     9.88        9.67        9.61                    S
263 -5.1 Pb2        9.16        8.46        8.31                    S

265 -2.1 He2-13     9.72        9.11        8.95                    S
266 -1.1 He2-14     8.23        6.92        6.47       6.19(12)     S
271 +3.1            6.18        4.91        4.39       3.88         Se
274 +2.1 He2-34     9.58        7.50        5.73       3.44         Dy
275 -3.1 He2-25    12.40(10)    9.20(10)   10.02(06)                P

275 -4.1 Pb4       10.37       10.26(06)   10.16(15)                N
277 -1.1 W16-55     6.71        5.59        5.04       4.35         Se
278 +5.1 Pb6       11.52(07)   10.88       10.70(10)                S
278 -5.1 NGC2867   10.43       10.91       10.05       8.17(31)     N
280 -2.1 He2-38     8.00        6.54        5.34       3.58         Sy

280 -2.2 ESO167     5.99        4.81        4.31       3.92         S
281 -4.1 SV Car     4.26        3.37        2.86       2.18         Se
281 -5.1 IC2501    10.04       10.79        9.55       8.09(17)     N
288 +0.2 ESO128    11.76(08)   10.84(07)   10.11(06)   7.97:        S
288 -5.1 He2-51    11.82(07)   10.90       10.65(07)                S

292 +4.1 Pb8        9.02        8.11        7.92       7.85         S
294 +4.1 NGC3918    9.13        9.62        8.85       7.5:         N
296 -3.1 He2-73    11.15       10.78(06)   10.64(12)                S
298 -1.1 He2-79     9.98        8.59        8.04       7.8:         S
299 -0.2 Aro524    11.64(13)   11.04(07)   11.19(20)                S

304 -4.1 IC4191    10.49       11.28       10.44                    N
305 +1.1 He2-90    10.77        9.74        7.85       4.44         D
305 -0.1 He2-91     9.73        8.12        6.55       4.31         Dy
307 -1.2 W17-61     9.28        8.10        7.70       6.94(39)     S
309 +0.1 He2-96    10.79        9.78        9.44       9.16(13)     S
```

TABLE II. Cont.d

```
---------------------------------------------------------------
309  -4.2 NGC5315 12.19(09) 11.65(07) 11.59(18)           S
311  -2.1           9.71      8.48      7.70   6.73(18)   S
312  -2.1 He2-106   9.81      8.77      8.44   8.5:       Sy
315  +9.1 He2-104  10.56      8.48      6.64   4.20       Dy
315  -0.1 He2-111  10.20      8.47      7.80              S

315-13.1 He2-131    9.55      9.67      9.08              N
316  +8.1 He2-108  10.34      9.70      9.58   7.65:      S
320  -9.1 He2-138  11.98     11.64     11.56              S
321  +3.1 He2-113   9.95      8.99      7.46   4.25       D
321  +2.2 He2-117  11.96     10.57     10.09   9.5:       S

321  -0.1 W16-174   9.44      7.35      6.35   5.36(10)   S
321  -0.2 W17-69   12.32(22) 10.20      9.09              S
322  -0.1 Pe2-8    10.97      8.90      7.99   6.90(30)   S
323  +2.1 He2-123  10.12      8.92      8.50   6.99(31)   S
324  -1.1 He2-133  11.35     11.46     10.34   8.42(12)   N

325  +4.1 He2-128  10.32      9.42      9.22              S
325  +3.1 He2-129   9.36      8.58      8.39              S
325  -4.1 He2-141   9.34      8.70      8.54              S
326  +0.1 W16-185  10.25      8.66      6.99   4.54(06)   D
326  -1.1 He2-139   7.38      5.96      5.09   3.86       Sy

327+10.1 NGC5882   10.31(19) 10.39      9.65   7.98(17)   N
327  -1.1 He2-143  13.30     12.28     11.90              S
327  -1.2 He2-140   7.32      6.14      5.70   5.42       S
327  -2.1 He2-142   7.57      6.48      6.11   5.66(10)   S
330  +4.1 Cn1-1     9.04      8.11      7.56   5.64       S

331  -1.1 Mz3       9.26      7.32      5.50   2.87       D
332  -0.1 W17-74             10.94      8.57   5.52(08)   D
332  -3.1 He2-164  10.48      9.40      9.00              S
332  -4.1 He2-170  12.72(15) 11.68(08) 11.34(07)          S
333  +1.1 He2-152  10.23(19)  9.74      9.68              S

335  -1.1 He2-169  11.65(08)  9.85      9.12              S
337  +1.1 Pe1-7    10.46      9.04      8.52              S
337  +1.0 He2-166  14.15     13.23(05) 12.90(07)          S
338  -5.1 He2-155   9.74      8.63      8.30              S
345  -1.1 H1-7     10.61      9.58      8.93   7.88       S

348-13.1 IC4699     9.30      8.72      8.59              S
349  +4.1 M2-4      9.84      8.75      8.41              S
350  +4.1 H2-1     10.08      8.94      8.58              Sy
352+11.2           11.01     11.88(08) 11.55(09)          N
---------------------------------------------------------------
```

TABLE II.   Cont.d

355	-3.3	H1-35	10.89	9.65	9.23		S
358	+5.1	M3-39	14.00	12.69	12.26	10.6:	S
358	-0.2	M1-26	13.05	10.09	8.72	7.62(12)	P

Notes to Table II.
NGC 6572 : the two series of magnitudes refer to diaphragms
          of 17 and 22 arcsec, respectively;
NGC 6543 : diaphragms of 17 and 27 arcsec, respectively;
Se: early-type star; Sy: symbiotic star (Allen,1974);
Dy: possible symbiotic star with dust;   P: peculiar.

---

3. THE PHOTOMETRIC DATA

We have analyzed our photometric data studying the conventional J-H vs.H-K and H-K vs.K-L two-colour diagrams. The first diagram is important, as suggested by Whitelock (1985), in order to classify the nebulae according to whether the principal source of near-IR emission is mainly nebular, stellar, dust, or a combination of these. The presence of hot dust can then be inferred studying the H-K vs.K-L diagram.

3.1. The J-H vs. H-K diagram

The J-H, H-K colours of the nebulae of our sample are reported in Figure 1. There we also show: (i) the locus of the colours of black-bodies at temperatures ranging from 1000K to 2000K, (ii) the free-free and bound-free emission from plasma at different electron temperatures, and (iii) a combination of a hot star with a 1000K dust shell. A vector corresponding to a reddening of E(B-V)=1.0 is also plotted. Colours of Main Sequence stars (MS in the figure) are taken from Koornneef (1983).

Approximately 30% of the observed nebulae lie in the region below the curve of plasma emission, called "nebular box" by Whitelock (1985). The near-IR emission from these nebulae, indicated as N(nebular)-type in Table II, is well explained by ff+bf emission from a plasma with a contribution of recombination emission lines of H+ and He+. Most of the well known compact PN and possible proto-PN (e.g. IC418, Vy2-2, NGC3918, Hb12, K3-2, Hu2-1) fall into this class.

About 13% of the observed nebulae lie to the right of the main sequence and black-body curves. In this case the near-IR emission can be due either to a combination of nebular or stellar emission (hot star) plus hot dust, or to very reddened nebular emission, say E(B-V)>1. These nebulae are classified as D(dust)-type in Table II. Typical examples of this class are SwSt1, Mz3, M1-6, M1-11. The D-type nebulae He2-34, He2-91, and He2-104 are suspected to be dust-type symbiotic stars (Allen, 1984).

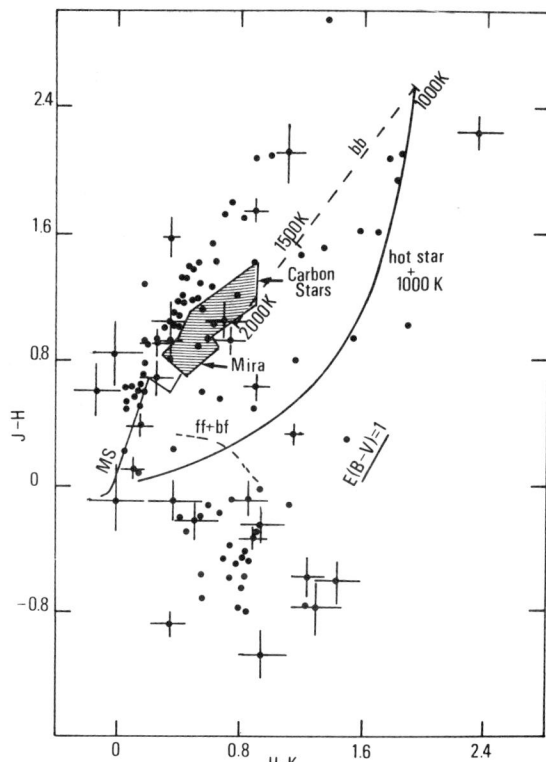

Figure 1.

J-H vs. H-K diagram for the observed sources.

The majority of the sources (57%) shows a near-IR emission dominated by a stellar continuum, and are indicated in Table II as S(stellar)-type objects. Indeed these sources are located in Figure 1 on the MS curve, and in regions occupied by Mira's, carbon and symbiotic stars. Therefore we classify, as S-type, sources of very different nature: He2-38, He2-106, and He2-139 have been classified by Allen (1984) as symbiotic stars; from the 1 to 100μm energy distribution (see next section) PK16+13.1, Wra16-55, PK271+3.1, and SV Car, should be classified as early type stars (probably Be stars). A great number ($\sim$60%) of S-type sources with detected radio continuum emission could be PN, with a cool star associated with the hot, yet undetected, nucleus. This could be the case for NGC5315, NGC6567, NGC6884, K3-17, M3-39, and Hb4, in which the 1-2μm emission is dominated by a stellar continuum while the IRAS LRS spectra (Pottasch et al.,1986) and the 8-13μm spectra by Roche and Aitken (1986) show typical nebular lines, such as those of AIII, SIV, and NeIII.

The peculiar colours of He2-25 and M1-26 may be due to some errors in the photometry.

Finally, comparing our classification with that derived by Whitelock (1985) for eleven sources in common, we find a general agreement. For the nebula He2-73 a difference in H and K of about 1mag is present: this could be explained as contamination of a field star.

3.2. The H-K vs. K-L diagram

Photometry in the L band is available for only 58 object of our sample. The diagram is shown in Figure 2. The typical error bar is indicated by a cross.

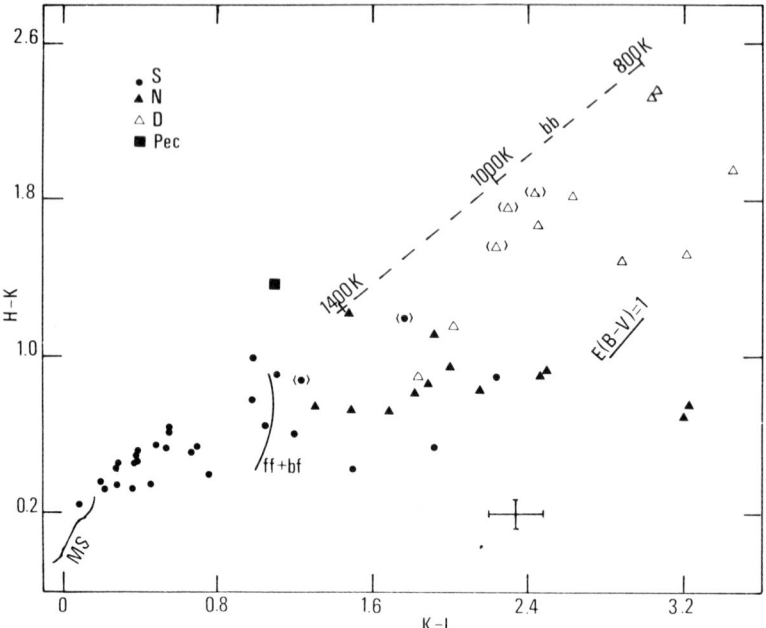

Figure 2. The H-K vs. K-L diagram for the sub-sample of sources with L photometry.

Objects of type S, N, and D are well separated in the diagram. Even allowing for possible large reddening corrections, N- and D-type sources show a strong K-L excess with respect to sources closer to the main sequence. This can be interpreted in terms of dust emission at approximately 1000-1400K.

Except for a few cases, S-type sources are located very close to the main sequence, and do not show hot dust emission. Symbols in brackets represent symbiotic (or possible symbiotic) stars.

In conclusion, although the analysis of the position of the sources in the two diagrams represents a good method for investigating the nature of the near-IR emission, this method alone cannot discriminate very efficiently planetary nebulae from other objects (with or without dust) with emission lines or from symbiotic stars. In order to give a more reliable classification of PK sources, in particular for those we classified as S-type, optical spectra, and a study of the complete energy distribution, are required.

4. INFRARED ENERGY DISTRIBUTION

Combining our broad band photometry with corrected IRAS fluxes taken from the PSC (Beichman et al., 1985), we derived the 1-100µm energy distribution of the 117 sources of our sample. Details of this analysis will be presented in a forthcoming paper. Here we will present only a few examples of total IR energy distribution for different types of objects.

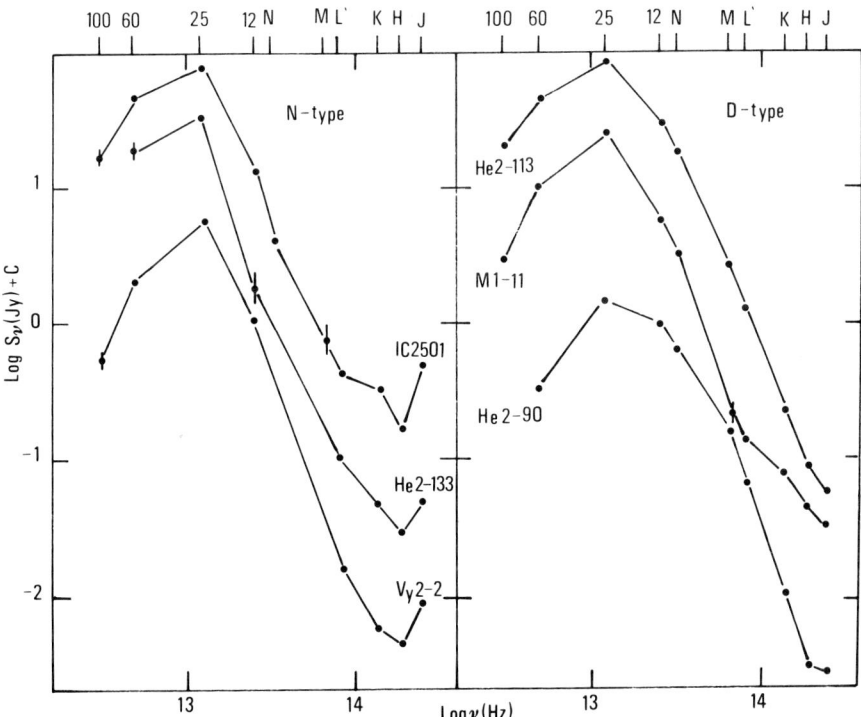

Figure 3. Energy distribution from 1 to 100µm of selected N- and D-type sources.

Figure 3 shows the 1-100um continuum emission of selected N- and D-type sources. While the far-IR is very similar for both types, N-type nebulae are characterized by an excess in J, probably due to the presence of H and He recombination lines in this band.

A classification based upon the analysis of energy distributions can be given for selected S-type sources. In Figure 4 we show the energy distribution of PK271+3.1 and Wra16-55, typical of reddened early type stars, while He2-38 and Wra16-174 are identified as symbiotic stars.

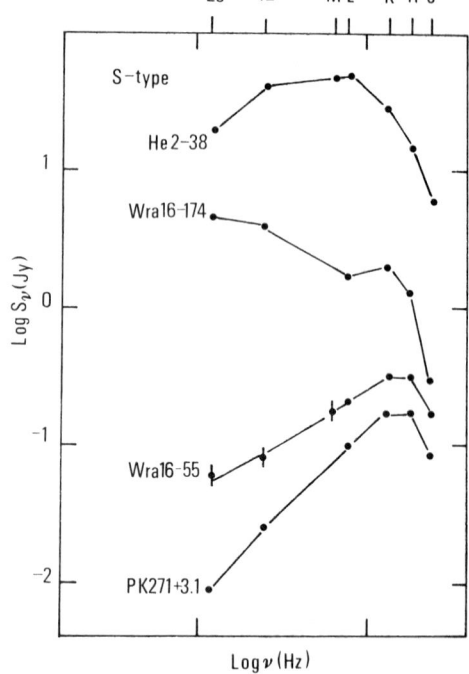

Figure 4.

IR energy distribution of selected S-type sources of different nature: early type stars (Wra16-55, PK 271+3.1) and symbiotic stars (He2-38, Wra16-174).

Finally, in Figure 5 we show the energy distribution of S-type sources with near-IR emission indicating the possible presence of a cool companion, as discussed in the previous section.

5. CONCLUSIONS

The main results of the J, H, K, and L photometry of 117 PK sources, all detected by IRAS, can be summarized as follows: (i) in about 30% of the sources the near-IR emission is predominantly due to ff+bf emission with contribution from H end He recombination lines (N-type); (ii) a combination of stellar or nebular emission with hot dust has been observed in $\sim$13% of the objects (D-type); (iii) most of the observed PK sources ($\sim$57%) show near-IR emission dominated by a stellar continuum (S-type).

From the analysis of the combined near- and far-IR energy distribution we identified five sources as early type stars, and six as symbiotic stars.

A detailed study of this sample of PK sources, including IRAS data and CVF spectrophotometric observations, is still in progress, and will be presented in a forthcoming paper.

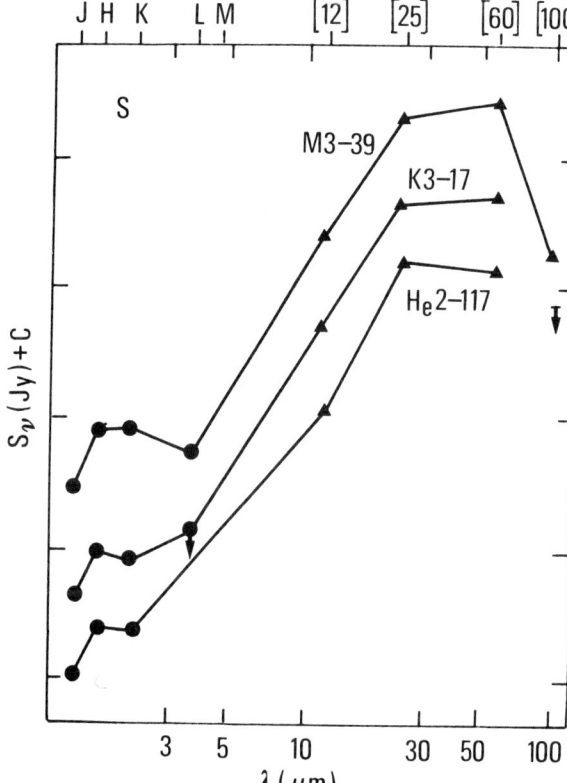

Figure 5.

Energy distribution of selected S-type sources

References

Allen,D.A., 1973, Mon.Not.Roy.Astron.Soc. 161,145
Allen,D.A., 1984, Proc.A.S.A. 5(3),369
Allen,D.A., Glass,I.S., 1974, Mon.Not.Roy.Astron.Soc. 167,337
Beichman,C.A., Neugebauer,G., Habing,H.J., Clegg,P.E., Chester,T.J., 1985, "IRAS Catalogues and Atlases: Explanatory Supplement"
Cohen,M., Barlow,M.J., 1974, Astrophys.J. 193,401
Cohen,M., Barlow,M.J., 1980, Astrophys.J. 238,585
Iyengar,K.V.K., 1986, Astron.Astrophys. 158,89
Koornneef,J., 1983, Astron.Astrophys. 128,84
Kwok,S., Hrivnak,B.J., Milone,E.F., 1986, Astrophys.J. 303,451
Persson,S.E., Frogel,J.A., 1973, Astrophys.J. 182,503
Pottasch,S.R., Baud,B., Beintema,D., Emerson,J., Habing,H.J., Harris,S., Houck,J., Jennings,R., Marsden,P., 1984, Astron. Astrophys. 138,10
Pottasch,S.R., Preite-Martinez,A., Olnon,F.M., Jing-Er,M., Kingma,S., 1986, Astron.Astrophys. 161,363
Roche,P.F., Aitken,D.K., 1986, Mon.Not.Roy.Astron.Soc. 221, 63
Whitelock,P.A., 1985, Mon.Not.Roy.Astron.Soc. 213,59
Willner,S., Becklin,E.E., Visnavatan,N., 1972, Astrophys.J. 175,699

THE EFFECT OF LINE EMISSION ON THE IRAS DATA OF PLANETARY NEBULAE

A. Leene & S.R. Pottasch
Kapteyn Astronomical Institute
Postbus 800
9700 AV Groningen
The Netherlands

ABSTRACT: the effect of line emission on the IRAS flux densities of planetary nebulae is discussed. IRAS observations of NGC 7293 are presented and show the important effect of the SIV and OIV lines on the flux densities of the 12 and 25 μm band. The effect of line emission on the colour colour plots is discussed.

1. INTRODUCTION

Generally infrared radiation is thought to be caused by emission from dust. Only in a few sources line emission in the infrared has been observed, thanks to measurements from balloons and airplanes. This database has been greatly increased by observations of the Low Resolution Spectrometer (LRS) aboard the IRAS satellite. Most of this line emission has been observed in Planetary Nebulae and HII regions.

   In some cases these lines are the dominant contributors to the IRAS broad band flux densities. Unfortunately LRS spectra could only be obtained for relatively bright and pointlike objects. For big objects, like NGC 7293, no LRS data is available and another method must be used to estimate the effect of line emission. The same is true for point sources for which only the IRAS broad band flux densities are available. The effect of the line emission can in this case be seen in the IRAS colour colour plots.

2. IRAS OBSERVATION'S OF NGC 7293

NGC 7293 is an evolved nebula with a large angular diameter of 30 arcmin, implying a size of 0.7 pc assuming a distance of 150 pc. The large angular size makes it one of the few planetary nebulae resolved by IRAS. The IRAS raw calibrated data (CRDD) has been used to obtain full resolution maps of the source. The integrated flux densities are 11.3, 18.3, 179 and 406 Jy for the 12, 25, 60 and 100 μm band, respectively. The 12-25 μm colour temperature is very large (245 K), which makes it one of the "hottest" nebulae known. The 60-100 μm colour temperature is on the other hand quite small (42 K), which makes it one of the "coldest" nebulae. It must be emphasized that these temperatures do not say anything about the true dust temperature, because the broad band fluxes are mainly due to atomic line emission.

Assuming constant electron density in the nebula and disregarding a dust continuum and taking only one line per band, it is possible to calculate the abundances of ions, which emit in the IRAS bands. These can be compared with known abundances. The main contribuants to the 12 micron band are SIV (10.52 μm) and NeII (12.81 μm). For the 25 micron band these are the SIII (18.68 μm) and the OIV (25.87 μm) line. The 60 micron band contains the OIII (51.71 μm) and the NIII (57.31 μm) line. In the 100 micron band the OIII line at 88.2 μm is the main contributor. With these assumptions the predicted abundances are: $X(SIV)=2.4 \cdot 10^{-5}$; $X(NeII)=4.7 \cdot 10^{-4}$; $X(OIV)=1.0 \cdot 10^{-4}$; $X(NIII)=3.8 \cdot 10^{-3}$; $X(OIII)=2.9 \cdot 10^{-3}$, where X is the abundance of the ion relative to hydrogen.

If these abundances are compared with known abundances from the optical regime, then it is seen that both the SIV and NeII line can explain the 12 μm emission. For the 25 μm band the same is true for OIV and SIII. The 60 μm emission results in a too large abundance both for OIII and NIII. This is also the case for the 100 μm emission.

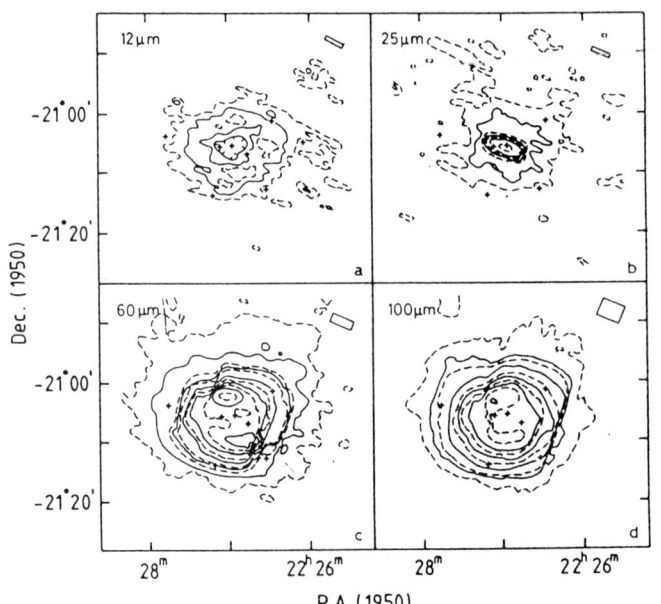

Fig. 1: IRAS observations of NGC 7293. Clockwise from the top left: the 12, 25, 60 and 100 μm data. Contours lie at 0.18, 0.36 and 0.54 MJy ster^{-1} for 12 μm; 0.36, 0.72, 1.5, 2, 2.5, 3 and 4 MJy ster^{-1} for 25 μm; 0.24, 0.5, 1, 1.5, 2, 4, 6, 8, 10, 12 and 24 MJy ster^{-1} for 60 μm and 0.68, 1.4, 2.4, 8, 16, 20, 24 and 28 MJy ster^{-1} for the 100 μm picture.

As the source is resolved by the IRAS beams, information is also available on the distribution of the emission. The data is presented in Fig. 1. If spherical symmetry is assumed, it is possible to deconvolve the observed emission in order to get the true radial abundance distribution. This provides confirmation of the fact that the IRAS emission is caused by line emission. The 12 μm abundance profile shows a shell at a radius of 7 arcmin and a width of 4 arcmin. The 25 μm abundance profile is centered on the central star and has a weak halo from 3 to 9 arcmin. The 60 μm abundance profile shows a shell centered on 5 arcmin radius and has a width of 4 arcmin. The 100 μm emission is very similar to the 60 micron abundance profile

As the 60 and 100 μm can not be produced by lines the main contributor must be dust emission. This explains why the two profiles are so similar. The 25 micron is due to OIV (in the centre) and has a SIII halo. The 12 micron emission could be due either to the SIV or the NeII line. As the SIV ion line dominates most of the LRS spectra of planetary nebulae, we assume that this line is also dominating this nebula. An argument based on the different ionisation potentials of both species, would however prefer the NeII ion. Hot dust, as another explanation, can be excluded, as this would be unable to explain the different morphologies.

3. LINE EMISSION IN THE COLOUR COLOUR PLOTS

If the flux densities of NGC 7293 are plotted in a colour colour diagram (Figs. 2 and 3), it is seen that this object lies at the extreme end of a correlation of points. The other points are Planetary Nebulae taken from Pottasch et al. (1984).

The most obvious attempt to interpret these correlations is in terms of blackbody radiation. It is then assumed that each nebula behaves like a blackbody of a single temperature. The straight lines in the Figs. indicate the lines expected for blackbodies. Along this line the dust temperature changes. It is clear that such a simple interpretation does not work. Even by changing the dust emissivity law from λ^0 to λ^{-2} no better fit is obtained. Thus a single temperature blackbody, possibly changed by a dust emissivity law, is a too simple approach. It is however not unlikely that a temperature gradient exists within the nebula.

The observations of NGC 7293 show however that lines can contribute predominantly to the IRAS flux densities. It is thus not unlikely that also the other nebulae are influenced by line emission. In order to investigate this idea a simple model has been constructed to analyse the effect of lines in the IRAS colour colour plots. The ultimate goal of this analysis is to put constraints on the dust emissivity law(s) observed in planetary nebulae.

Starting from a 60/100 ratio it is possible to predict the 25/60 ratio (or some other ratio) directly, incorporating the line contribution, by using only a few relations and assumptions. Four assumptions need to be made: the electron temperature, the fraction of 60 μm emission due to OIII, the dust emissivity law and the ion abundance. Only the latter two assumptions are of major importance. The electron temperature is assumed to be 10^4 K. The effect of the OIII lines on the results is quite small. This effect is similar to a change of the dust emissivity law. Assuming that 10% of the 60 μm flux density is due to OIII, would imply a dust emissivity change from -1 to -0.8. The real effect will be even smaller. The latter two assumptions are the real unknowns, which must be found from a fit to the data.

In the scheme towards the solution three empirical relations have been used: the nebular radius R (T_D), the electron density n_e (R) and the infrared luminosity L_{IR} $(R^3 n_e)$. With these relations it is possible to predict R, n_e and L_{IR}, thus reducing the number of unknowns. All these parameters are in the end related to the observed dust temperature T_D.

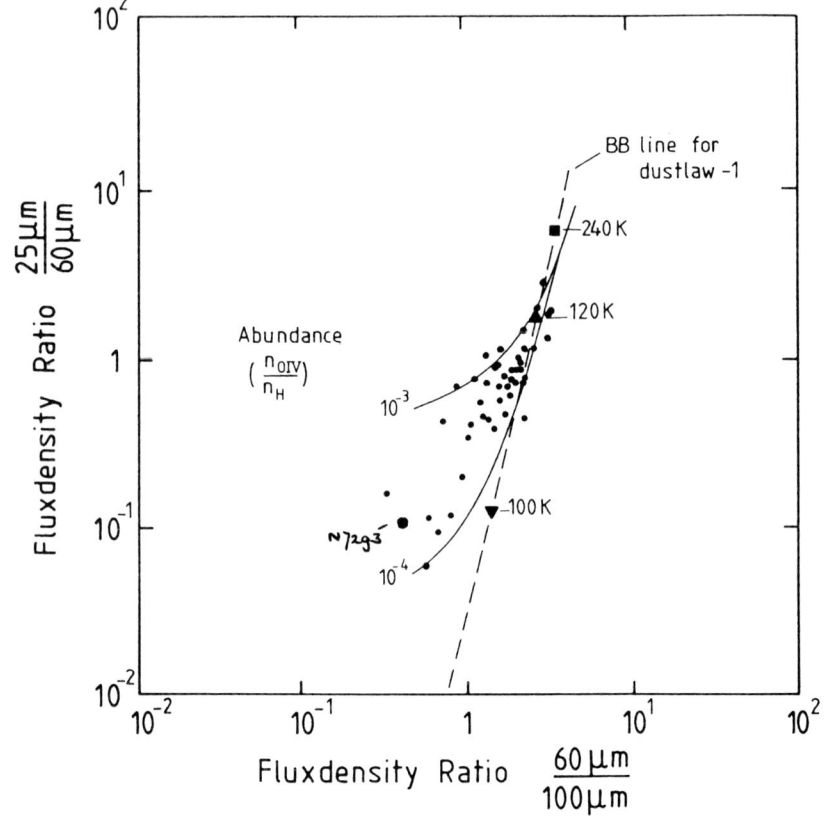

Fig. 2: The model for the OIV line.

The results for SIV and OIV are presented in Figs. 2 and 3. It has been assumed that the OIII emission is unimportant and that the dust emissivity law is proportional to λ^{-1}. Shown are colour colour plots with lines of constant ion abundance. With this model it is possible to explain the spread of the data points by a spread in the ion abundance. It is possible to define two abundances, which form an envelope to the data. The direction of the observed correlation can be explained by this envelope as well. Also a constraint on the dust model can be given. Dust law powers of 0 or -2 give a worse fit. The ion abundances are similar to those observed with the LRS.

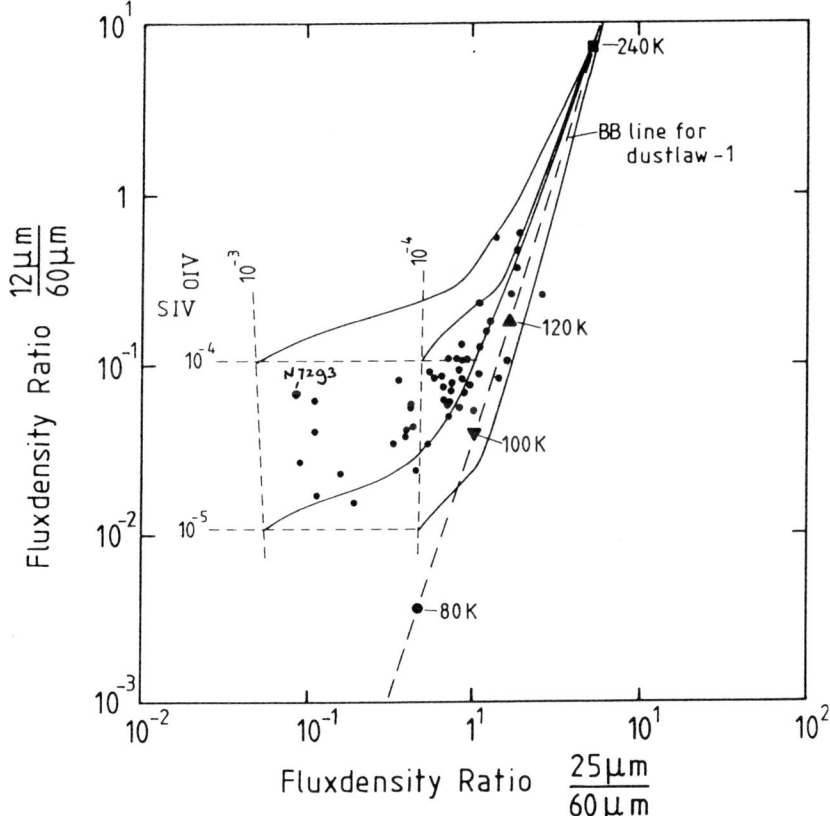

Fig. 3: The model for the SIV and OIV line.

4. CONCLUSION

The effect of ion lines on the observed IRAS fluxes can be quite large. This is very clear from the observations of NGC 7293, where the morphology of the object changes dramaticly from 12 to 60 μm. Only line emission can account for these observations. The observed correlation in the IRAS colour colour plots can also be explained by line emission. A simple model incorporating the effect of line emission reproduces the observed correlation. Any theoretical model for the IRAS data of planetary nebulae should incorporate the effect of line emission.

5. REFERENCES

- Pottasch et al.: 1984, Astron. Astrophys. 138, 10

6. DISCUSSION

Roche: It is possible that the unidentified dust emission bands could peak outside the main ionized region and contribute to the 12 μm flux of NGC 7293, although there is undoubtedly a large contribution from ionic lines.

MODELING THE THERMAL EMISSION FROM DUST IN PLANETARY NEBULAE

J. Patrick Harrington
Astronomy Program, University of Maryland
College Park, Maryland 20742

ABSTRACT. Dust can be included in the photoionization models of planetary nebulae; the dust temperatures resulting from the radiation field within the models can be evaluated and the resulting thermal IR emission compared to observations. We use a model of NGC 3918 to illustrate the behavior of various grain materials (graphite, amorphous carbon, silicates, and iron). We also present a model of IC 418 with graphite grains in both the ionized zone and in a surrounding thick, neutral shell.

1. INTRODUCTION

While the assumptions underlying any interpretation of the observed IR emission from planetary nebulae (PNe) can be said to constitute a model, we usually reserve this term for an analysis of sufficient complexity that the physical parameters cannot be deduced directly, but instead these parameters must be varied and the computed observables then compared with the actual objects. An advantage of modeling is the ability to incorporate physical constraints from the outset. In particular, when we model thermal emission from dust, we will demand that the radiation field which produces the observed ionization of the nebular gas will also be sufficient to heat the grains to the temperatures needed to reproduce the observed IR spectrum.

The application of modeling techniques to thermal dust emission is not new. For example, Natta and Panagia (1976) constructed models of the thermal emission from H II regions. Using the same techniques, they constructed a series of models for PNe and used the results to help interpret the available far IR observations (Natta and Panagia, 1981). Two developments now encourage us to carry the modeling approach further:

(a) New IR observations are available, especially the IRAS data. An overview of the IRAS data can be found in papers by Pottasch and co-workers (Pottasch et al. 1984, 1986).

(b) We now have better information on the optical properties of likely grain materials.

In addition, we may hope to constrain the models more closely by concentrating on specific nebulae for which good ionization models can be constructed.

2. OPTICAL PROPERTIES OF GRAIN MATERIALS

The outstanding problem in modeling IR emission is that we are not really sure of what the grains are composed. One place to start is with a rather successful model for the dust in the interstellar medium, the model of Mathis, Rumpl, and Nordsieck (1977). This model consists of a mixture of graphite and silicate particles. Both species are present with a size distribution that follows a power law in grain radius with an exponent of -3.5, so that most of the grain mass is in the form of the smaller grains. The minimum grain size is taken to be 0.005 µm and the maximum to be 0.25 µm. About equal amounts of graphite and silicates are needed to reproduce the extinction curve of the ISM.

Less is known about the nature of dust in PNe. The absence of the 9.7 µm silicate feature from all but a class of low-excitation, oxygen-rich objects indicates a fundamental difference between most PNe and the ISM. The gas-phase abundances of many PNe show that $C/O > 1$, which can explain the absence of silicates and which would suggest that carbon could be a major constitutent of the dust. Thus the simplest approach is to start with the graphite component of the MRN model and, if necessary, modify the grain distribution to improve the predicted thermal IR flux.

To actually calculate the absorption and scattering of an ensemble of grains, we start with the complex refractive indices, $\varepsilon(\omega) = \varepsilon_1(\omega) + i\varepsilon_2(\omega)$, of the grain materials and use Mie scattering theory to compute the effective cross-sections $Q(a,\lambda)$. The values of $\varepsilon(\omega)$ for graphites and silicates have been discussed and self-consistent values tabulated by Draine and Lee (1984) and by Draine (1985). The real (scattering) part ε_1 and the imaginary (absorption) part ε_2 are not independant, but are related through the Kramers-Kronig relations. Draine and Lee have used this relationship to insure the self-consistency of their values. For our purposes, we must have values of $\varepsilon(\omega)$ for a wide range of frequencies, from the far IR where the grains radiate to the far UV where the grains absorb the radiation from the central star. It should be noted that the values in the IR for graphite tabulated by Draine (1985) are not quite consistent with the parameters of Draine and Lee, as 1.0E-14 rather than 1.4E-14 was used for the damping time τ for $\varepsilon_{||}$ by Draine. Also, the values of the graphite Planck-averaged emissivities in Fig. 10 of Draine and Lee are not accurate above 100K.

We would like to explore grain materials other than graphite and silicates. Unfortunately, the optical constants are not known over the full range of frequency in most cases. In this paper we present some results for grains of amorphous carbon and for metallic iron grains. For amorphous carbon, we used the values of Q/a determined by Borghesi, Bussoletti, and Colangeli (1985). In this case the Mie theory is not used. Also, since they do not present measurements extending into the far UV, we used the graphite optical constants for the higher frequencies. For iron, in the energy range of 1 - 27 eV, we used the data of Moravec et al. (1976), while for wavelengths between 1 and 10 μm, we follow Lenham and Treherne (1966). Finally, for wavelengths greater than 10 μm, we employ a Drude model, with constants adjusted to fit the 1 to 10 μm values.

3. PUTTING DUST IN THE IONIZED GAS

If we have a population of grains with a power-law size distribution $N(a) = N_g a^{-p}$, then the extinction of the ensemble can be described by a mean effective cross-section

$$<Q(\lambda)> = \frac{3-p}{a_2^{(3-p)} - a_1^{(3-p)}} \int_{a_1}^{a_2} a^{(2-p)} Q(a,\lambda) \, da ,$$

where a_2 and a_1 are the maximum and minimum grain radii, respectively. The temperature of the grain will be determined by the balance of the heating and the cooling, as expressed by the equation

$$\int_0^\infty J(\lambda) Q_{ab}(a,\lambda) \, d\lambda = \int_0^\infty B_\lambda(T) Q_{ab}(a,\lambda) \, d\lambda ,$$

where J is the mean intensity of radiation per steradian, and B_λ the Planck function. The contribution to the LHS of the equation occurs mainly in the UV, while the contribution to the RHS occurs in the far IR. If we define a Planck-averaged Q as

$$\overline{Q}(a,T) = (\pi/\sigma T^4) \int_0^\infty B_\lambda(T) Q_{ab}(a,\lambda) \, d\lambda ,$$

then the equation for the temperature of a grain of radius a is just

$$\overline{Q}(a,T) T^4 = (\pi/\sigma) \int_0^\infty J(\lambda) Q_{ab}(a,\lambda) \, d\lambda .$$

If we know the radiation field, we may compute the RHS. We then solve for the grain temperature by iteration.

The main complication is that the radiation field $J(\lambda)$ must include the radiation in resonance lines like Lα. Not only are such lines of great optical depth, but the line intensity depends upon the dust which is distributed throughout the nebula. The most elementary approach is to use an escape probability formulation. We have used the method outlined in Cohen, Harrington, and Hess (1984). Assuming a rectangular line profile w Doppler widths in extent, we can express the intensity in the line as

$$J_\ell = G / [P_e \kappa_\ell + w \kappa_d] ,$$

where $4\pi G$ is the local generation rate of radiant energy in the line per unit volume, κ_ℓ and κ_d the mean line opacity and dust opacity, respectively, and P_e the escape probability per scattering in the absence of dust. The contribution to the heating integral by the line radiation is then just $J_\ell \, w \, Q(a,\lambda_\ell)$.

Once the heating has been evaluated, we may compute the temperature of the grains for each grain size and at each nebular position. We then integrate the IR emission of the grains over the size distribution and the nebular volume to obtain the predicted infrared spectrum.

4. APPLICATIONS TO SPECIFIC NEBULAE

4.1. The High-Excitation Nebula NGC 3918

We have applied modeling techniques as discussed above to the nebula NGC 3918, for which an ionization model has been constructed by Clegg, Harrington, Barlow, and Walsh (1987). A preliminary dust model was described in Clegg, Harrington, and Barlow (1984) and a full account of the dust model will be published elsewhere (Harrington, Monk, and Clegg 1987). The ionization model of this PN is not spherically symmetric, but consists of two optically thick cones and an equatorial region of lower density which is optically thin even in the He^+ continuum. This structure is ionized by a central star whose flux is represented by a NLTE model with an effective temperature of 140,000K. The ionization model gives generally good agreement with the observed optical and UV line fluxes and isophotes. The C/O ratio of the model is 1.6.

4.1.1. <u>Graphite Models</u>. There are three independent sets of photometric observations of NGC 3918 in the IR. Cohen and Barlow (1980) observed it in the 8 - 20 μm range, Moseley (1980) made broad-band measurements at 36 and 70 μm, and it was observed by IRAS at 12, 25, 60, and 100 μm. The IRAS fluxes must be corrected for in-band line emission. This has been done using observed line fluxes from the IRAS LRS spectra where possible and the model predictions otherwise. Since the 12 μm IRAS flux has large contributions due to lines and to an unidentified 11.3 μm feature, making the corrected flux uncertain, we have relied on the Cohen and Barlow flux at 9.6 μm (which seems free of line emission) and the IRAS low-resolution spectra to set the level of the thermal emission at the shorter wavelengths.

Table I presents the results for a model using the standard MRN grain distribution (.005<a<.25), and a model where the minimum and maximum grain radii have been increased. The first model has a dust-to-gas ratio of 0.00048, while the second has a ratio of 0.00083. The standard MRN size distribution seems to produce too much 10 - 20 μm emission and too little 100 μm flux, as if the grains were too hot. Since the small grains are hotter, by eliminating them, we get a better fit with the second model. This is not entirely satisfactory, however, as the recent identification of the bands between 3.3 and 11.3 μm as due to polycyclic aromatic hydrocarbons (PAH's) suggests that the size distribution extends down to radii below those of the MRN distribution.

4.1.2. <u>Other Grain Materials</u>. To investigate what would happen with other grain materials, we calculated the behavior of models with standard MRN size distributions of iron, silicate, and amorphous carbon grains. Fig. 1 shows the temperature of grains of 0.04 μm radius as a function of

MODELING THE THERMAL EMISSION FROM DUST IN PLANETARY NEBULAE

distance from the central star in the thick sector of the nebula. Because the amorphous carbon radiates more strongly in the far IR than graphite, thses grains are significantly cooler. Conversely, the adopted iron optical constants lead to very hot grains. The resultant thermal IR emission from the models is shown in Fig. 2. The dust/hydrogen ratios of the models (0.0010, 0.0005, 0.0007, and 0.0007, for iron, silicate, graphite, and amorphous carbon, respectively) were chosen so that they would produce about the same total IR luminosity. Clearly, iron or silicates cannot fit the observed spectrum. While amorphous carbon seems too cool even with the MRN size distribution, it might be satisfactory if the smallest grains were less than 0.005 µm in radius; this is attractive because it would form a bridge to the sizes of the PAH's.

TABLE I

Band	9.6 µm	25 µm	60 µm	100 µm
Observed	1.0	38	46	16.5
.005<a<.25	2.7	49	44	9.2
.040<a<.40	1.0	40	49	11.6

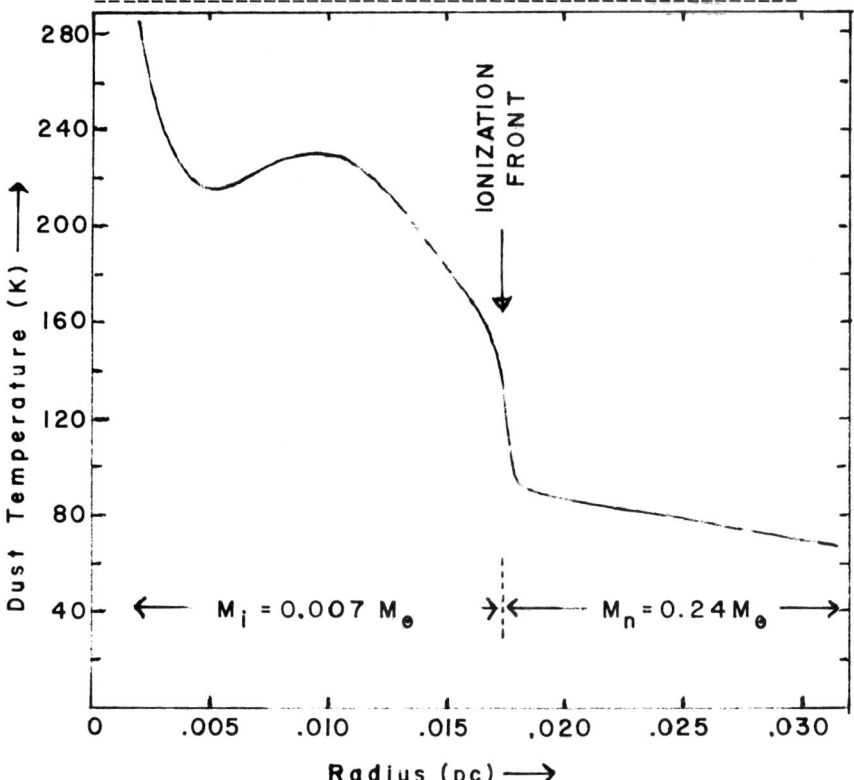

Figure 1. Grain temperature vs. radius in four models of NGC 3918 with different grain materials. Results are shown for 0.04 µm grains only; temperature would vary with grain radius.

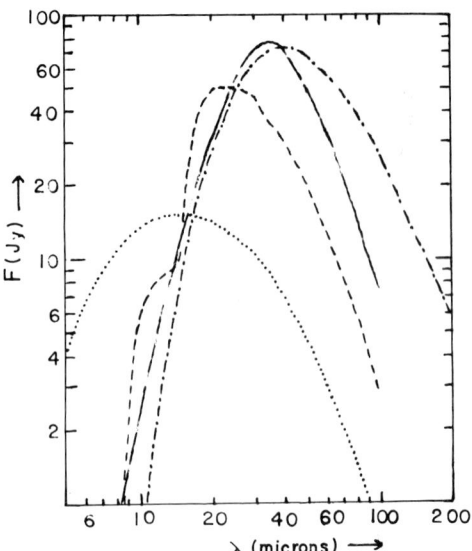

Figure 2. Thermal IR emission from four models of NGC 3918. The curves are coded for the same four materials as shown in Fig. 1.

Finally, we may ask where the refractory materials like iron fit into this picture. If iron is present in the nebula with near solar abundance, then since almost none of it is in the gas phase, the dust to gas ratio should be $\simeq 0.002$ for iron alone. In addition, other elements are depleted, like magnesium and silicon. Thus these materials should be present in the dust in amounts comparable to the mass of carbon which we already find adequate to explain the entire thermal IR flux. Are these materials hiding in big grains?

4.2. The Low-Excitation Nebula IC 418

For the sake of comparison, I thought it would be useful to examine a rather different nebula from NGC 3918, one that is younger, and of low excitation. For this purpose, IC 418 seemed a likely choice. I thought it would be interesting to examine a model where dust is present in an extensive neutral shell surrounding the ionized region. I had expected that the IR radiation from the cool dust in the neutral shell would so broaden the flux distribution that it would not fit the observed curve, thus ruling out this type of model. Instead, I found that a reasonable model could be constructed even with a neutral shell that was much more massive than the ionized zone.

4.2.1. The ionization model. If we want to evaluate the dust emission from a model, the most important features of the ionization model are the spatial distribution of the gas and the flux distribution of the central star. The spatial distribution was chosen to reproduce the Hα isophotes of Reay and Worswick (1979). The problem of the central star has been discussed by Adams (1983), who showed that the flux of an extended model of Kunasz, Hummer, and Mihalas (1975) with T(2/3) = 29,700K would reproduce both the observed optical and UV continuum as well as the hydrogen and helium ionization. We have used this model atmosphere flux for the ionization model. We adopted a distance of o.42 kpc, the value used by Pottasch et al. (1984). This distance keeps the density high enough to reproduce the density-sensitive line ratios without the use of a filling factor, and also allows most of the mass to be in a neutral shell, as the total ionized mass is only 0.007 $M_\odot$. Unfortunately, at this distance the stellar luminosity is then only 475 $L_\odot$, lower than that expected from stellar evolution theory.

4.2.2. **The "MgS" feature in the spectrum of IC 418.** There is an apparent discrepancy between the continuum level in the IRAS low-resolution spectra (Pottasch et al., 1985) and the broad-band fluxes. At 21 μm, the LRS continuum is only 90 Jy, while the 25 μm broad-band flux is 224 Jy. This is due to a steep rise which has been observed in detail in IC 418 and other objects by Forrest, Houck, and McCarthy (1981), who interpret it as due to a solid-state resonance feature. The feature has been attributed to MgS by Goebel and Moseley (1985). In our model, which uses graphite grains, we therefore regard this dust feature as superimposed upon the graphite emission, and thus do not attempt to force the model spectrum through the 25 μm flux point. Instead, we fit to the LRS continuum and the 60 and 100 μm bands.

4.2.3. **The dust model.** We have used an $a^{-3.5}$ power-law distribution of graphite grains with .005<a<.25, uniformly mixed with both the ionized and neutral gas. In Fig. 3 we show the temperature of the 0.032 μm grains as a function of distance from the central star. There is a sharp drop in grain temperature at the ionization boundry; outside this point the dust is heated only by the stellar continuum below the Lyman edge. Although there is 35 times more dust in the neutral zone than in the ionized zone, the low temperature of this dust prevents its emission from dominating the IR flux from the model. Fig. 4 shows the predicted spectrum of the model along with the IRAS broad-band fluxes and the continuum level from the IRAS low-resolution spectra. The fit is reasonable except for the region of the "MgS" feature (25 - 50 μm).

The total IR luminosity of our model is 126 $L_\odot$ at the assumed distance. Pottasch et al. (1984) found a luminosity of 210 $L_\odot$ for the same distance; the difference can be attributed to our exclusion of the high 25 μm point form our fit. The dust-to-gas ratio of our model is only 0.00055, about the same as required for NGC 3918. While this would seem to run counter to the results of Natta and Panagia (1981) and subsequent workers who find larger dust-to-gas ratios for young, dense PNe, more objects will have to be modeled before any conclusions can be drawn. Our value for the dust-to-gas ratio is substantially less than the value of 0.0024 derived by Pottasch et al. (1984). A large part of this difference is due to our inclusion of the hypothetical neutral shell, since only the ionized gas is considered by Pottasch et al.

Thus it appears that the IR spectrum alone will not easily confirm or deny the existence of dusty, neutral shells. While the neutral shell in our model was given an appreciable geometrical extent for the sake of computational convenience, such a shell might be dense and thin, and therefore would not extend noticeably beyond the ionized region. One possible approach would be to look for evidence of scattered light. While the dust optical depth through the ionized region in our model of IC 418 is only 0.06, the depth through the neutral shell is 0.34. Since the albedo of the dust at optical wavelengths should be of the order of 0.5, an observable amount of light might be scattered (and hence polarized). It is interesting in this context to note the reported detection of polarization in several PNe by Leroy, Le Borgne, and Arnaud (1986), although these are older, larger objects.

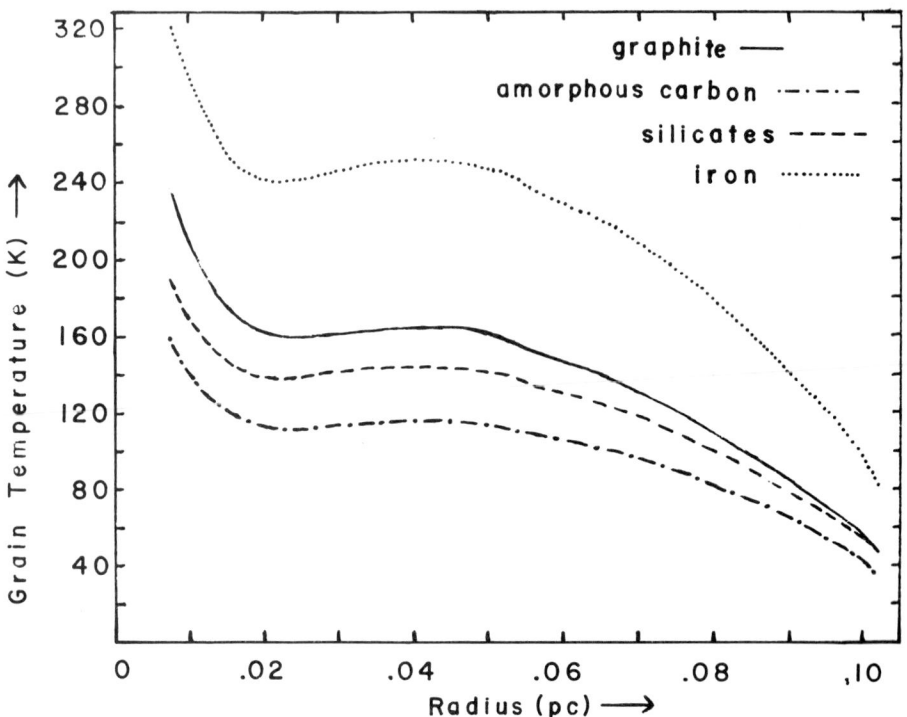

Figure 3. Temperature vs. radius for graphite grains in the model of IC 418. Values plotted are for a grain of radius 0.032 µm. The local maximum in T near 0.010 pc coincides with the maximum of the density distribution of the ionized zone; this is where the maximum Lα heating occurs.

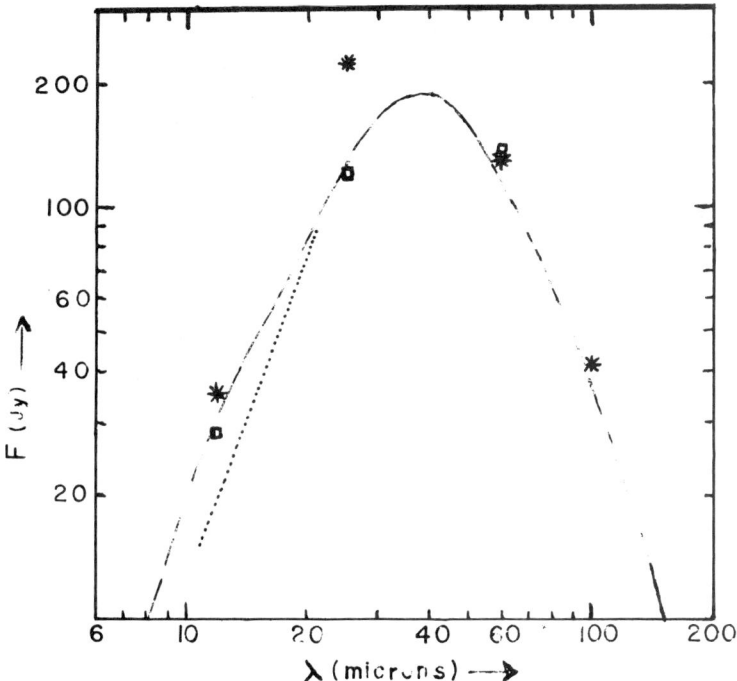

Figure 4. The IR spectrum of IC 418. The solid line is the model flux form the graphite grains. The dashed line indicates the observed continuum level from the IRAS low-resolution spectra. Squares are the model predictions of the broad-band IRAS fluxes, while asterisks are the observed values (at 100 μm the points coincide).

REFERENCES

Adams, S. 1983. Ph. D. Thesis, University College London.
Borghesi, A., Bussoletti, E., and Colangeli, L. 1985. Astron. Astrophys., 142, 225.
Clegg, R.E.S., Harrington, J.P., and Barlow, M.J. 1984. 'Proc. 4th European IUE Conference', ESA SP-218, p. 337.
Clegg, R.E.S., Harrington, J.P., Barlow, M.J., and Walsh, J.R. 1987. Astrophys. J., (March 1987).
Cohen, M., and Barlow, M.J. 1980. Astrophys. J., 238, 585.
Cohen, M., Harrington, J.P., and Hess, R. 1984. Astrophys. J., 283, 687.
Draine, B.T. 1985. Astrophys. J. Suppl., 57, 587.
Draine, B.T., and Lee, H.M. 1984. Astrophys. J., 285, 89.
Forrest, W.J., Houck, J.R., and McCarthy, J.F. 1981. Astrophys. J., 248, 195.
Goebel, J.H., and Moseley, S. 1985. Astrophys. J., 290, L35.
Harrington, J.P., Monk, D.J., and Clegg, R.E.S. 1987. To be submitted.
Kunasz, P.B., Hummer, D.G., and Mihalas, D. 1975. Astrophys. J., 202, 92.

Lenham, A.P. and Treherne, D.M. 1966. In 'Optical Properties and
    Electronic Structure of Metals and Alloys', ed. Abeles, F.,
    (Amsterdam: North-Holland Pub. Co.) p. 196.
Leroy, J.L., Le Borgne, J.F., Arnaud, J. 1986. Astron. Astrophys.,
    160, 171.
Mathis, J.S., Rumpl, W., and Nordsiek, K.H. 1977. Astrophys. J.,
    217, 425.
Moravec, T.J., Rife, J.C., and Dexter, R.N. 1976. Phys. Rev. B, 13, 3297.
Moseley, H. 1980. Astrophys. J., 238, 892.
Natta, A., and Panagia, N. 1976. Astron. Astrophys., 50, 211.
Natta, A., and Panagia, N. 1981. Astrophys. J., 248, 189.
Pottasch, S.R., Baud, B., Beintema, D., Emerson, J., Habing, H.J.,
    Jennings, R., and Marsden, P. 1984. Astron. Astrophys., 138, 10.
Pottasch, S.R., Preite-Martinez, A., Olnon, F.M., Mo Jing-Er, and
    Kingma, S. 1986. Astron. Astrophys., 161, 363.
Reay, N.K., and Worswick, S.P. 1979. Astron. Astrophys., 72, 31.

Discussion

Natta: You said that in order to fit the IR spectrum you need to
eliminate the smallest grains of the MRN distribution. How much this
result depends on the assumption that the grains size distribution
follows the MRN law?

Harrington: I haven't explored this completely, but I did find that
a rather drastic change in slope is needed to produce much effect. Based
on the idea that collisions will produce a distribution with the smaller
sizes dominant, I felt more comfortable varying the minimum size. But
this is just to be regarded as a parameter, without any profound justification.

Panagia: It seems to me that adopting for grains in PNe parameters
which are appropriate for grains of the ISM may not be justified. In
fact, since both the grains size and the dust to gas ratio in old
planetaries are much smaller than those found in the ISM (see Lenzuni et
al., this conference) it is clear that the properties of interstellar
grains may not be comparable to those of PN grains.

Harrington: I agree that they may not be similar. The idea was to
start with a model that worked in the ISM and try to see the nature of
changes, if any, that would be needed to fit the data more closely.

Lenzuni: I think that the surprisingly low value of m(dust)/m(gas)
that you find for IC418 is due to the low IR luminosity that comes out
from your fit of "modified IRAS data". Have you looked at Moseley data
in order to check if they are consistent with your fit? IRAS data may be
contaminated by atomic lines.

Harrington: It is not a question of atomic lines, but a solid state
feature of considerable width. Having checked the Moseley data, I note
that his 37μm flux is within 10% of the IRAS 25μm value, but that
Moseley's 52μm and 70μm fluxes are much below the IRAS value, as is his
100μm flux. This dosen't help clarify things very much.

EVOLUTION OF DUST IN PLANETARY NEBULAE

P. Lenzuni[1], A. Natta[2] and N. Panagia[3,4]
[1] Institute of Astronomy, University of Florence.
[2] Center for Infrared Astronomy and the Study of the Interstellar Medium, Florence.
[3] Space Telescope Science Institute, Baltimore.
[4] Affiliated with the Astrophysics Division, Space Sciences Department of ESA; on leave from the Institute of Radioastronomy CNR, Bologna.

ABSTRACT. We have analyzed the IRAS data of 234 planetary nebulae (PNe) determining the properties of the dust contained in their envelopes. The main result is that the dust grains undergo substantial evolution during the lifetime of a PN.

1. INTRODUCTION

In 1981 Natta and Panagia discussed the IR spectra of the 11 PNe observed by Moseley (1980) and found that the dust grains contained in PNe vary systematically in abundance and size as the PN shell expands. However, due to the limited sample available at that time the detailed time evolution of the dust properties remained rather uncertain. Since then the IRAS satellite has added a wealth of new data on the whole sky. This has allowed us to reconsider the problem and perform an analysis as Natta and Panagia's but now dealing with a much richer sample of objects. A detailed account of this work will be presented elsewhere (Lenzuni, Natta and Panagia, 1986). Here, we outline the procedures followed in our study and present the main results and implications.

2. SAMPLE, MODEL AND ANALYSIS

Our sample is based on the PN list compiled by Daub (1982): this list contains 299 nebulae for which optical and radio data (fluxes and angular sizes) are sufficiently well known and for which the distances are estimated in a self consistent manner. Searching for them in the IRAS Catalog of Point Sources one finds 233 nebulae which have a positive detection at least in the 25 and 60 μm bands. Our final sample is constituted by these objects with the addition of NGC 7027. Although this bright nebula was not observed with IRAS, its spectrum has been measured extensively at all wavelengths so that it is worth being included.

Our analysis is based on a comparison of the observations with theoretical models which were computed under the following assumptions:

1. The nebula can be described as a spherically symmetric shell ($\Delta R/R = 0.26$) of gas and dust which are uniformly mixed with each other.
2. The dust has an infrared absorption efficiency which is inversely proportional to the wavelength.
3. The bulk of the dust emission is characterized by a unique temperature which is determined by the observed spectral shape. The flux measured at 12.5 μm, however, is not used to estimate the dust temperature because the emission in this band is severely contaminated by strong forbidden lines (e.g. Pottasch et al. 1984).
4. Following Daub (1982) the total mass of the nebula is adopted to be 0.17 $M_\odot$ and the radius at which the nebula starts becoming density bounded is 0.12 pc.

To these points we add other two "implicit" assumptions:
5. The observed sample of PNe forms an evolutionary sequence.
6. PNe with larger radii are older than PNe with smaller radii. In particular, adopting a constant expansion velocity, the age of a PN is proportional to its radius.

3. RESULTS

The quantities that were derived with our analysis are:
i. The dust temperature, T_d. It is determined by fitting the IRAS data with a $\nu B(\nu)$ curve: therefore, T_d depends only on the shape of the spectrum (i.e. the IRAS fluxes) irrespective of any other parameter.
ii. The grain size, a, which is estimated from the thermal balance equation. Therefore, it is a function of directly observed quantities (optical, IR and radio fluxes, angular size) and is independent of the adopted distance.
iii. The dust to gas ratio in the ionized region and the total mass of grains M_d in the whole nebula. They are derived from the estimated UV optical depth, the grain size and the gas column density in the ionized region: it is this latter quantity that introduces a weak dependence of M_d on the adopted distance ($\propto D^{1/2}$).
iv. The total number of dust grains in the whole nebula, N_d: it is computed from the ratio of the total dust mass to the third power of the grain size and, therefore, it has the same weak dependence on the distance as M_d has.

The plots of T_d, a, M_d and N_d as a function of the nebular radius are displayed in figures 1-4. The first point we like to stress is that all plots show well defined behaviours which are statistically significant. This justifies our "implicit" assumption (5), i.e. that the PNe of our sample can be arranged in an evolutionary sequence. Yet the dispersion which is seen in all plots cannot entirely be ascribed to experimental errors but rather indicate that there is some intrinsic dispersion of the PN properties around the average behaviours.

The best-fit relations of the derived quantities as a function of the nebular radius are as follows:

$$\log T_d(K) = 1.971 \pm 0.006 - (0.146 \pm 0.012)\log(R/10^{17}cm)$$
$$\log a(\mu m) = -1.34 \pm 0.04 - (1.74 \pm 0.08)\log(R/10^{17}cm)$$
$$\log M_d/M_\odot = -2.98 \pm 0.04 - (1.38 \pm 0.08)\log(R/10^{17}cm)$$
$$\log N_d = 45.72 \pm 0.11 + (3.53 \pm 0.16)\log(R/10^{17}cm)$$

EVOLUTION OF DUST IN PLANETARY NEBULAE   251

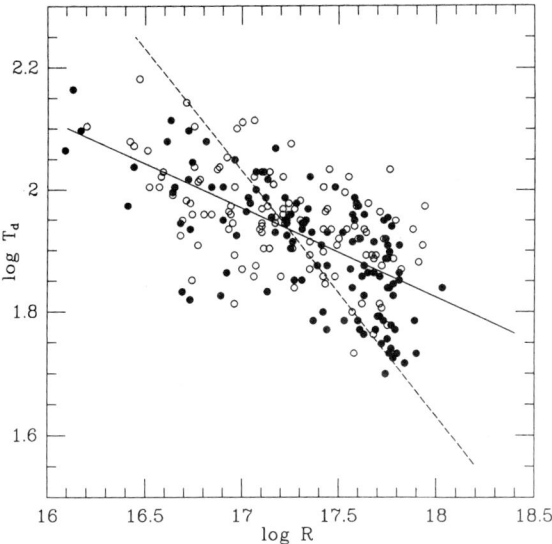

Fig. 1 - The average dust temperature as a function of the nebular radius. Dots denote nebulae for which the temperature was determined from the 25, 60 and 100 μm band measurements whereas circles corresponds to the cases for which only the 25 and 60 μ m measurements could be used. The solid line represents the least square best fit line ($T_d \propto R^{-0.15}$) and the dashed line the expected behaviour if the dust properties do not evolve with time ($T_d \propto R^{-0.4}$).

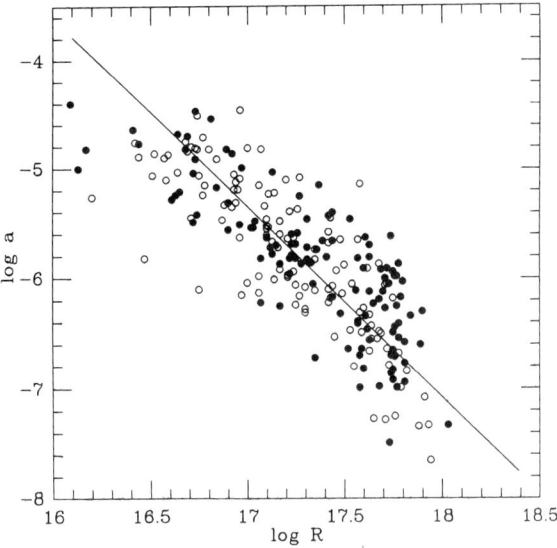

Fig. 2 - The average dust grain size as a function of the nebular radius. Symbols are as in Figure 1. The solid line represents the least square best fit line ($a \propto R^{-1.74}$).

4. DISCUSSION

Considering the individual quantities in detail, we note that the variation of the temperature with R is rather modest (cf fig. 1), varying from ~ 125 K at $R \sim 3\ 10^{16}$ cm to ~ 60 K at $R \sim 6\ 10^{17}$ cm. The observed temperature slope ($d\log T_d/d\log R \sim -0.15$) is much shallower than that expected in the case of constant luminosity of the central star and constant dust properties ($d\log T_d/d\log R = -0.4$). This implies that some drastic change of the grain properties must occur so that the ratio of the IR to the UV absorptivity varies by a factor as high as 400 when the radius varies from $\sim 3\ 10^{16}$ cm to $\sim 1\ 10^{18}$ cm. Such a big change cannot be accounted for by any other effect except by a decrease of the dust grain size with time (Lenzuni et al. 1986).

Figure 2 illustrates the grain size variation with R: it is interesting to note that the biggest sizes (0.5-1 μm for very young nebulae, i.e. $R < 4\ 10^{16}$ cm) are just as expected for grains forming in the atmospheres of red giants (Papoular and Pegurié, 1983). On the other hand, at late stages of the PN evolution the size of grains can be as small as 30 Å: thus they are more similar to macromolecules made of few hundred of atoms than to ordinary grains.

The total dust mass as well as the dust to gas ratio clearly declines with increasing R (cf. fig 3). In particular, young nebulae ($R \sim 3\ 10^{16}$ cm) have $M_d/M_g \sim 0.015$, which implies that at this stage a considerable fraction of the heavy elements is locked into grains, while for old nebulae ($R > 6\ 10^{17}$ cm) the fraction of matter in the form of grains is much smaller, $M_d/M_g < 3\ 10^{-4}$ which is more than a factor of ten lower than it is found for the interstellar medium (ISM). This implies that the matter injected by PNe into the ISM, while strongly enriched in C and moderately in He and N (e.g. Panagia et al. 1977), is highly depleted of dust grains. Moreover, the decrease of the dust mass with radius ($d\log M_d/d\log R \sim -1.4$) is much shallower than one would expect if the mechanism was some sort of erosion which reduces the grain size only without altering the total number of grains ($d\log M_d/d\log R = 3 d\log a/d\log R \sim -5.2$). The conclusion is that while the grain size decreases, the number of grains must increase with time.

This point is illustrated in figure 4 where we see that the total number of grains in evolved PNe ($R > 6\ 10^{17}$ cm which for $v_{\exp} \sim 20$ km s^{-1} corresponds to $t > 10^4$ years) is almost 10^8 times greater than in the youngest, most compact nebulae. Such a large increase in grain number together with the decrease of the grain size with time cannot be explained by simple "erosion" of grains but rather one needs some "shattering" mechanism to account for these results. Grain-grain collisions is a plausible explanation because even with a turbulence velocity as low as 1 km s^{-1} the grain-grain collision time is shorter than the dynamical time. Whether this mechanism could in detail explain the observations depends on the efficiency of such collisions in shattering grains: this problem is currently being addressed and will be the subject of a future paper.

In conclusion, our analysis of the IRAS data on 234 PNe has shown that dust grains evolve considerably during the lifetime of a PN. In detail, both the total dust mass and the grain size decrease with time whereas the total number of grains increases with time. These results suggest the occurrence of some shattering process of the sort that one may expect in the case of grain-grain collisions. A curious consequence of our findings is that the matter that is injected by PNe into the interstellar space, while it is strongly enriched of carbon and moderately of helium and nitrogen, is highly depleted of dust. Therefore, the dust which is

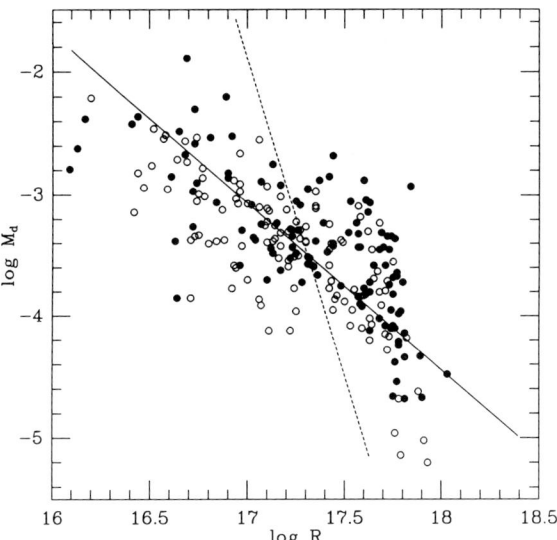

Fig. 3 - The total dust mass as a function of the nebular radius. Symbols are as in Figure 1. The solid line represents the least square best fit line ($M_d \propto R^{-1.38}$) and the dashed line the expected behaviour in the case that the number of grains does not vary with time ($M_d \propto R^{-5.2}$).

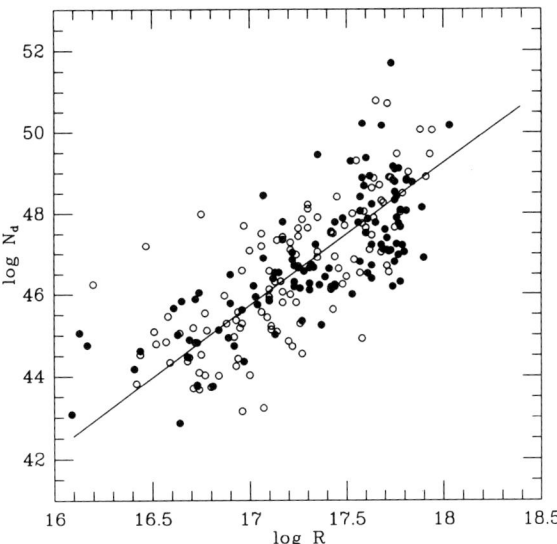

Fig. 4 - The total number of dust grains as a function of the nebular radius. Symbols are as in Figure 1. The solid line represents the least square best fit line ($N_d \propto R^{+3.53}$).

present in the ISM is genetically different from that which is found in the envelopes of PNe.

REFERENCES

Daub, C. T., 1982, *Astrophys. J.*, **260**, 612.
Lenzuni, P., Natta, A. and Panagia, N., 1986, in preparation.
Moseley, H., 1980, *Astrophys. J.*, **238**, 892.
Natta, A. and Panagia, N., 1981, *Astrophys. J.*, **248**, 189.
Panagia, N., Bussoletti, E. and Blanco, A., 1977, in "CNO Isotopes in Astrophysics", ed. J. Audouze, Reidel (Dordrecht-Holland), p. 45–48.
Papoular, R. and Pegourié, B., 1983, *Astron. Astrophys.*, **128**, 335.
Pottasch, S. R., Baud, B., Beintema, D., Emerson, J., Habing, H. J., Harris, S., Houck, J., Jennings, R. and Marsden, P., 1984, *Astron. Astrophys.*, **138**, 10.

DISCUSSION

D'HENDECOURT - In your first conclusion you say that the dust formed in PNe is very similar in size to the dust present in the ISM but later you say that the PN dust is fully destroyed before it gets into the ISM. Does this mean that PNe do not contribute to the dust production in the ISM?

LENZUNI - Our analysis indicates that the dust originally present in very young PNe is very similar to the dust present in the atmospheres of evolved AGB and carbon stars. However, because of the dust evolution the matter injected by the PNe into the ISM is dust depleted. I believe that the bulk of the ISM dust grains originates from the winds of carbon stars and AGB stars before they may produce a planetary nebula. Since the mass lost during those phases can easily be ten times higher than the mass which is subsequently released in the form of the PN envelope, it is clear that the dust depletion of old PNe cannot affect greatly the dust content of the ISM.

VIOTTI - There is a point that worries me, may be there is a simple explanation. You know that there is a large variety of stellar winds from cool stars which could be PN progenitors. We could for instance expect a different grain size distribution and we could expect that smaller grains are pushed (already by the cool star radiation) further away than larger grains. Hence, your $\log a - \log R$ and $\log M_d - \log R$ plots might reflect a difference on the cool star wind rather than a time evolution.

LENZUNI - This possibility was already considered by Natta and Panagia (1981) and was excluded because it is inconsistent with the observations. In fact, if a combination of size and spatial distribution of dust grains were the correct explanation then one should expect the dust properties not to vary anymore once a PN becomes density bounded, i.e. when its radius becomes larger than 0.12 pc, which clearly is not the case. In addition, one should also find the source size of density bounded PNe to be larger at shorter IR wavelengths because small, hence hot, grains should be located in the outer layers of the PN envelope, but this does not occur in any of the PNe whose structure has been studied in detail.

OBSERVATIONS OF THE SPECTRUM OF THE INTERPLANETARY DUST EMISSION

A. Salama, P. De Bernardis, S. Masi, G. Moreno
Dipartimento di Fisica, Universitá La Sapienza
00185 Roma
Italy

ABSTRACT. Spectra of the interplanetary dust emission are obtained, collecting together far infrared observations, in the wavelength range from $5\mu m$ to $200\mu m$, carried out by three different experiments operating on board satellites (IRAS), rockets (ZIP) and stratospheric balloons (ARGO). Observations at small solar elongations ($\epsilon \sim 45°$) may be roughly accounted by spherically symmetric silicate grains. At larger elongations ($\epsilon \sim 90°$), however, the silicate model breaks down; a mixture of graphite and silicate grains gives a somewhat better, but still unsatisfactory, fit to the data.

1. INTRODUCTION

During the last two decades, interplanetary dust has been investigated using different techniques, e.g. collecting micrometeorites on space probes, counting microcraters in lunar rocks and observing zodiacal light in the visible and infrared. Observations in the visible and infrared have provided most of the available information on the large scale distribution of the dust.

It emerged that the grain number density (n) decreases by increasing both the heliocentric distance (r) and the height above the ecliptic (z), following the law:

$$n(r, z) \propto r^{-\mu} exp\left(-\frac{\gamma z}{r}\right) \qquad (1)$$

Concerning the values of the constants γ and μ, optical and infrared observations are in fairly good agreement for the former, while they disagree substantially for the latter. In fact, optical observations give: $\gamma = 2.1$ and $\mu = 1.3$, in the range $0.3 A.U. \leq r \leq 1.0 A.U.$ (Leinert et al., 1981); $\mu = 1.5 - 2.0$ for $r \geq 1 A.U.$ (Hanner et al., 1976). Infrared observations give, on the other hand: $\gamma = 2.6$ and $\mu \leq 1$ (Zodiacal Infrared Project, ZIP, experiment, Murdock and Price, 1985); $\mu \approx 0.5$ (IRAS satellite, Levasseur and Dumont, 1985).

The main uncertainties affecting the above results are:

(i) Optical observations cannot disentangle the radial dependence of the grain density from a possible variation of the scattering function, due to changes in shape or optical properties of the grains.

(ii) The interpretation of infrared observations depends upon assumptions on the radial trends of temperature and Q_{abs}. Murdock and Price, e.g., assumed both temperature and Q_{abs} to be independent of distance; Levasseur and Dumont, on the other hand, assumed a constant Q abs and a temperature decreasing from 260 K at 1 A.U. to 212 K at 1.5 A.U..

In a previous paper (Salama et al., 1986), we investigated the spatial distribution of the dust, using far infrared observations in the band from $10\mu m$ to $200\mu m$. The observations were made by the ARGO experiment, carried out on the balloon flight of July 30, 1984. The main results obtained can be summarized as follows:

(i) The signal variations with solar elongation indicated a radial decrease of the dust density, in the ecliptic, with increasing heliocentric distance, which resulted as being slower than that inferred from optical observations. In fact, $\mu = 1.3$ was obtained only in the unrealistic hypothesis of a constant temperature. On the other hand, by assuming the temperature decrease expected for graphite or silicate grains, we arrived to lower values of $\mu = 0.8 \pm 0.5$ (for graphite) and $\mu \leq 0.5$ (for silicates). Concerning the latitudinal dependence of n, our observations were consistent with the results of Leinert et al. ($\gamma = 2.1$).

(ii) The observed spectrum of the dust was markedly different from those expected for both graphite and silicate grains. However, as these spectra bracketed the experimental points, we suggested that a mixture of the two species of grains, with appropriate densities, could have accounted for the observations. The data, on the other hand, were not adequate to permit us to attempt a fit with different models of the grains. (Our experiment explored five bands centered at 11, 19, 50, 108 and 225 μm but, as discussed by Salama et al., in the 11 μm and 225 μm channels only upper limits to the flux were obtained.)

In the present report, we further investigate the spectrum of interplanetary dust and we endeavour to ascertain whether graphite and silicate grains can account for infrared observations. Along with our data, we shall consider measurements supplied by IRAS and ZIP experiments (for a summary of these observations, see Price, 1986).

2. RESULTS OF THE ANALYSIS

Figures 1 and 2 show spectra of interplanetary dust obtained by collecting measurements supplied by the three experiments considered here (referred to in the following as ARGO, IRAS and ZIP). The plots are based on observations performed at different elongations ϵ, within the ecliptic. In Figure 1, ARGO and ZIP observations are given: the former are averages of signals recorded during three scans through the ecliptic at $\epsilon = 25°, \epsilon = 45°$ and $\epsilon = 65°$ respectively; the latter are taken at $\epsilon = 45°$. In Figure 2, ZIP and IRAS observations at $\epsilon = 90°$ are given. In comparing measurements from different experiments, one should bear in mind that ARGO used a differential technique (see Masi et al., 1986 for details), which creates some uncertainty when evaluating the absolute values of the flux plotted in the figures.

The dotted curves shown in the plots were calculated in the following hypotheses:

♣ The grain number densitiy was assumed to follow the law $r^{-\mu}$ with $\mu = 0.8$ for the graphite and $\mu = 0$ for silicates (Salama et al, 1986).

♣ ♣ Q_{abs} was assumed to be constant.

♣ ♣ ♣ The temperature trends were those obtained by Roser and Staude, 1978, for grains with dimensions of the order of 10 μm.

Note also that in Figure 1 the curves refer to $\epsilon = 45°$, which is the average elongation of the data shown in the plot.

Inspection of the plots shows that:

♠ At $\epsilon = 45°$, observations are roughly matched by silicates, while the graphite curve differs greatly from the experimental points.

♠ ♠ At $\epsilon = 90°$, neither model fits the data in any satisfactory way.

The breakdown of the silicate model at $\epsilon = 90°$ may be related to the $\mu = 0$ assumption, which implies no radial decrease of dust density. Such a hypothesis is probably unrealistic at the distances greater than 1 A.U. involved in observations at $\epsilon = 90°$.

Our next step in the analysis was to consider a mixture of graphite and silicate grains. We fitted the data leaving the relative contribution of the two types of grains as a free parameter (note that this parameter does not provide us with an estimate of the relative abundance of the two species because Roser and Staude did not quote absolute values of Q_{abs} for graphite and silicates). In the $\epsilon = 45°$ spectrum no substantial improvement was obtained with respect to the pure silicate fit. In the $\epsilon = 90°$ spectrum the mixture was still unsatisfactory, although it provided a better fit for the observations. This could have been expected due to the several uncertainties which affect our analysis. The major problems are:

♡ In their computations, Roser and Staude considered spherically symmetric and homogeneous particles. Such a model, however, has no experimental support. Due to the lack of extended sampling of dust in interplanetary space, the properties of the grains are presently inferred from those of particles collected in the stratosphere (Brownlee, 1985). It has been found (Mukai, 1986) that most of these particles have irregular shapes and a "fluffy" structure, consisting of an ice matrix with inclusions of different materials such as amorphous carbon, silicates, magnetite . A large variety of chemical compositions was also found in the cometary dust (Giotto observations of Halley comet, Kissel et al., 1986) even though these results cannot be directly applied to the interplanetary dust.

♡ ♡ To compute Q_{abs}, Roser and Staude used values of the refractive complex index for graphite and silicates, supplied by Wickramasinghe and Guillaume, 1965; Sato, 1968; Pollack et al., 1980. Better estimates of these parameters are now available (Draine, 1985) and should be included in the analysis.

♡ ♡ ♡ Temperature trends given by Roser and Staude refer only to grains which are either smaller ($\simeq 1\mu m$) or larger ($\simeq 100\mu m$) than those which are presumed to be most common in interplanetary space. Temperatures for 10 μm grains were therefore obtained through an interpolation, which is highly subjective. Concerning the silicates, the choice of an "average" trend among the different curves corresponding to the various possible chemical compositions (obsidian, andesite, olivine) caused further uncertainty.

3. CONCLUSIONS AND PLAN FOR FUTURE WORK

In collecting together far infrared observations carried out by three different experiments operating on board satellites (IRAS), rockets (ZIP) and stratospheric ballons (ARGO), we obtained a fairly detailed spectrum of interplanetary dust emission at different solar elongations ($\epsilon \simeq 45°$ and $\epsilon \simeq 90°$).

A fit of these spectra, based on the radial trends of the grain temperature and number density given respectively by Roser and Staude, 1978, and Salama et al., 1986, shows that at $\epsilon = 45°$ (i.e. at relatively small heliocentric distances) the dust emission may be roughly accounted for by a flat distribution of silicates grains. At $\epsilon = 90°$ (i.e. heliocentric distances larger than 1 A.U.) this model fails to describe the observations. A possible contribution of graphite improves the fit, which, however, still remains unsatisfactory.

As discussed in Section 2, the present analysis is subject to several uncertainties. More realistic models of the grains ("fluffy" structures consisting of an ice matrix with different chemical inclusions) and updated values of the refractive complex indexes of the materials should be considered. New computations of the dust temperature, based on these assumptions, should throw more light on the spatial and chemical composition of the dust. This work is presently being carried out by our group and will be referred to in future papers.

REFERENCES

-Brownlee D.E.:Properties and Interactions of Interplanetary Dust, eds. Giese R.H. and Lamy P.,*Reidel Publ. Co.*, 143 (1985)
-Draine B.T.: *Astrophis. J.*, **57**, 587 (1985)
-Hanner M.S., Sparrow J.G., Weinberg J.L., Beesov D.E.: *Lecture Notes Phys.*,**48**, 29 (1976)
-Kissel J., Brownlee D.E., Buchler K., Clark B.C., Fechtig H., Grun E., Hornung K., Igenbergs E.B., Jessberger E.K., Krueger F.R., Kuczera H., McDonnell J.A.M., Morfill G.M., Rahe J., Schwehm G.H., Sekanina Z., Utterback N.G., Volk H.J., Zook H.A.: *Nature*,**321**, 336 (1986)
-Leinert C., Richter I., Pitz E., Planche B.: *Astron. Astrophys.*,**103**, 177 (1981)
-Levasseur-Reyound A.C., Dumont R.: *C. R. Acad. Sc. Paris*, 300 (1985)
-Masi S.,Andreani P.,Ceccarelli C.,Dall'Oglio G.,de Bernardis P.,Melchiorri B.,Melchiorri F.,Moleti A.,Moreno G.,Nisini B.,Piccirillo L.,Pietranera L.,Salama A.,Shivanandan K. :'A five band differential IR photometer for baloon borne observations of diffuse sky radiation' *Infrared.Phys.* in press (1986)
-Mukai T.: 'Cometary dust and interplanetary particles',*Proceedings of the International School of Physics 'Enrico Fermi'*, corso 101, Varenna (Italy) (1986)
-Murdock T.L., Price S.D.:*Astron. J.*,**90**, 375 (1985)
-Pollack I.B., Toon O.B., Khare B.N.: *Icarus*,19, 372 (1973)
-Price S.D.: *Proceeding ESA Workshop on Space Borne Sub-Millimeter Astronomy Mission*, ESA-SP-20, 67 (1986)

-Roser S., Staude H.S.: *Astron. Astrophys.* ,**67**, 381 (1978)
-Salama A.,Andreani P.,Dall'Oglio G.,de Bernardis P.,Masi S.,Melchiorri B.,Melchiorri F.,Moreno G.,Nisini B.,Shivanandan K.:' Measurements of near and far infrared zodiacal dust emission'*Astron.J.* in press (1986)
-Sato Y.: *J. Phys. Soc. Japan*, **24**, 489 (1968)
-Wickramasinghe N.C., Guillaume C.:*Nature*,**207**, 366 (1965)

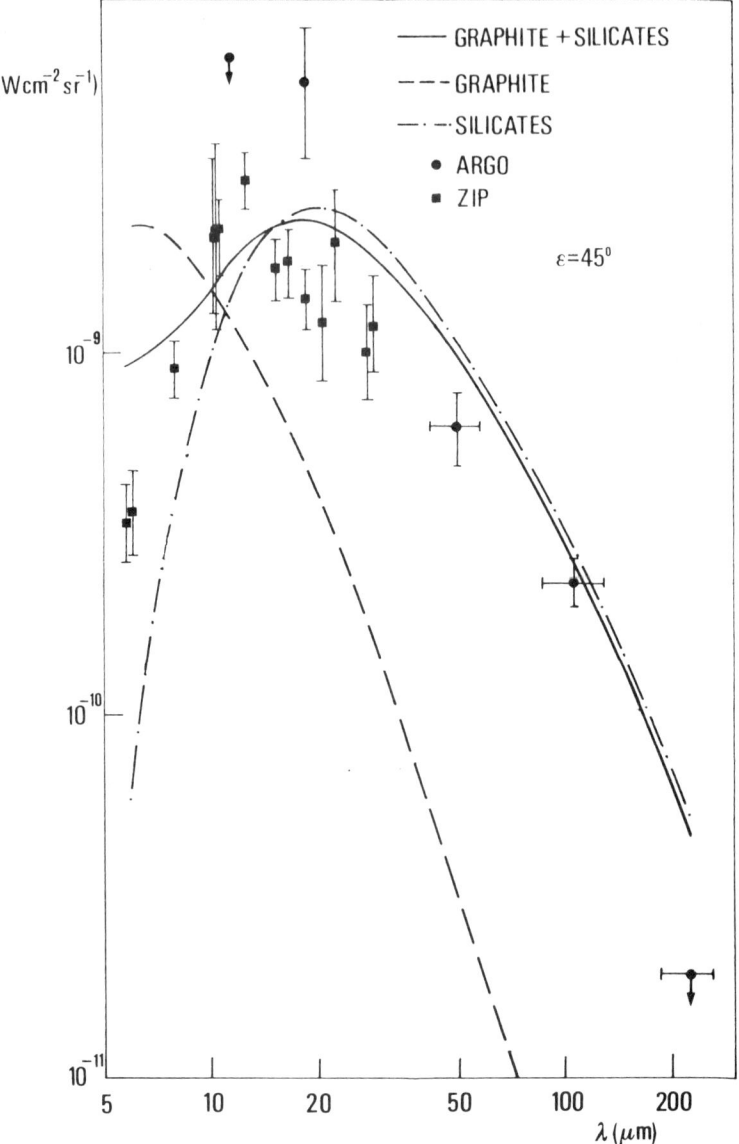

Fig.1 : Zodiacal dust spectrum at elongation $\epsilon = 45°$, in the ecliptic plane. Dots and squares refer respectively to measurements supplied by ARGO and ZIP experiments. Dotted and broken lines give respectively the spectra which would be expected for 10m graphite and silicate grains ; the continuous line corresponds to the mixture of graphite and silicate grains which gives the best fit to the observations.

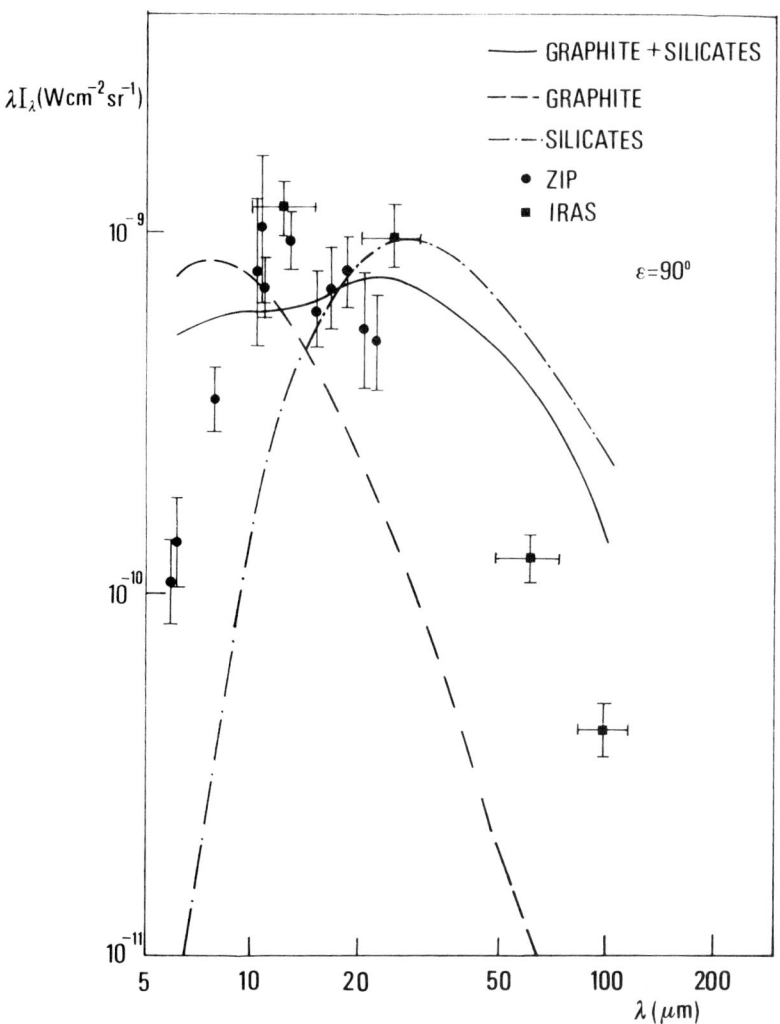

Fig.2 : Zodiacal dust spectrum at elongation $\epsilon = 90°$, in the ecliptic plane. Dots and squares refer respectively to measurements supplied by IRAS and ZIP experiments. Lines have the same meaning as in fig.1.

THE INFRARED SPACE OBSERVATORY (ISO) PROJECT

M.F. Kessler,
Space Science Department of ESA
Postbus 299,
2200AG Noordwijk
The Netherlands

ABSTRACT. The Infrared Space Observatory (ISO), an approved and funded project of the European Space Agency (ESA), is an astronomical satellite, which will operate at wavelengths from 3 to 200 μm. Its cryogenically-cooled 60-cm telescope will be equipped with four complementary and versatile focal plane instruments, which will enable imaging and also photometric, spectroscopic and polarimetric observations. These instruments are being built by international consortia of scientific institutes and will be delivered to ESA for in-orbit operations. The expected launch date is 1993 and the in-orbit lifetime will be at least 18 months. In keeping with ISO's rôle as an observatory, two-thirds of its observing time will be made available to the general astronomical community.

1. INTRODUCTION

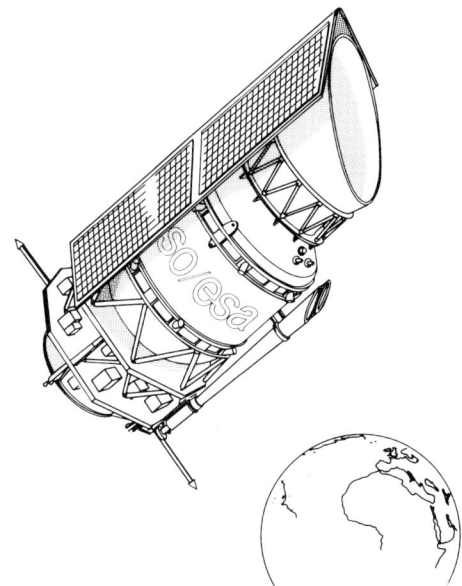

Figure 1. Artist's view of ISO.

Modern astrophysics benefits greatly from the ability to make observations throughout the widest possible range of the electromagnetic spectrum. However, work at infrared wavelengths is severely hampered or totally excluded by the terrestrial atmosphere and therefore, despite its scientific promise, the 3-200 μm band has remained relatively unexplored. The Infrared Space Observatory (ISO), shown in Figure 1, will remedy this situation. It will provide astronomers with a facility of unprecedented sensitivity for a detailed exploration of the universe ranging from objects in the solar system right out to the most distant extragalactic sources. The cryogenically-cooled telescope will be equipped with four scientific instruments, which together will permit imaging, and photometric, spectroscopic and polarimetric observations. ISO is designed to be a true observatory with its scientific instrumentation capable of tackling a wide range of astrophysical problems and with two-thirds of its observing time available to the general astronomical community.

The original proposal for ISO was submitted to the European Space Agency (ESA) in 1979. After several studies and assessments, a very intensive feasibility (phase A) study, involving ESA, industry and the scientific community, was conducted in 1981–82. As is customary within ESA, the ISO phase A study was one of a number conducted simultaneously on competing projects. At the conclusion of these studies, ISO was chosen in March 1983 for inclusion in the ESA Scientific Programme. This selection carries with it the funds necessary for the entire project. In June 1985, four scientific instruments were adopted as the astronomical payload of ISO. The next major step in the project will be the start of the industrial detailed design (phase B) study, expected in December 1986.

2. SCIENTIFIC AND OBSERVATIONAL ASPECTS

The two main problems encountered by a ground-based infrared astronomer are illustrated in Figure 2. At some wavelengths, the earth's atmosphere is opaque. Where observations are possible, the dominant noise source is often the photon shot noise associated with the thermal emission of the warm telescope optics and of the atmosphere. Thus, for maximum sensitivity, it is necessary to cool the telescope and its instruments and to operate them in space.

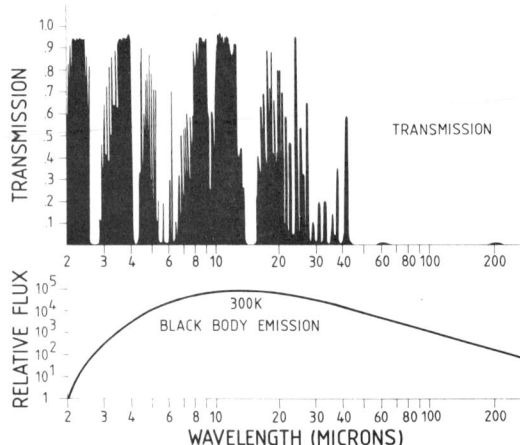

Figure 2. Upper Panel: Transmission of the terrestrial atmosphere as a function of wavelength, showing that observations are only possible through some windows (depicted in black).
Lower Panel: Relative flux from a 300 K black body as a function of wavelength, showing that thermal emission from the warm ($\approx$ 300 K) telescope optics and atmosphere peaks in the wavelength region where observations are possible.

The first major step in this direction was taken with the highly successful Infrared Astronomical Satellite (IRAS), which surveyed nearly all the sky in four broad infrared bands. ISO, an observatory satellite, will build on the IRAS results by making detailed observations of selected sources. Compared to IRAS, ISO will have a longer operational lifetime, wider wavelength coverage, better angular resolution, more sophisticated instruments (especially for polarimetry and spectroscopy) and, by a combination of detector improvements and longer integration times, a sensitivity gain of upto several orders of magnitude. The aim is to operate the ISO instruments at sensitivities close to the limits imposed by natural astrophysical phenomena.

The ISO part of the spectrum is of great scientific interest, not only because it is here that cool objects (15–300K) radiate the bulk of their energy, but also because of its rich variety of atomic, ionic, molecular and solid-state spectral features. Measurements at these wavelengths permit determination of many physical parameters, for example energy balance, temperatures, abundances, densities and velocities. Owing to the much-reduced extinction, infrared observations are particularly well-suited to probing the properties of objects obscured at visible wavelengths. ISO will be offering high sensitivity and sophisticated observing facilities for a relatively unknown part of the spectrum. Thus it is expected the scientific programme of ISO will touch upon virtually every field of astronomy, ranging from solar system studies to cosmology.

Two-thirds of ISO's observing programme will be determined by the scientific community via the submission and selection (by peer review) of proposals. The remaining time will be reserved for the groups who provide the instruments, for the Mission Scientists and for the Observatory Team who operate the satellite. Details of the observations to be carried out in the guaranteed time will be included in a first "Call for Observing Proposals" to be issued to the entire community about 18 months before launch. During scientific use, the satellite will always be in contact with the ground segment but the extent to which real time modifications to the observing programme will be possible or desirable is not yet finally decided. This depends, among other things, upon the trade off between the flexibility of real-time control and the efficiency of pre-programmed observations. Within a few hours of an observation being completed, the guest observer will be provided with a "quick-look" output adequate for judging the scientific quality of the data. A final product with more detailed data reduction and calibration will be supplied later. This product will be the one from which the guest observer makes his astronomical analysis.

3. SATELLITE AND MISSION DESIGN

The satellite consists of a payload module and a service module. These modules are coupled by a composite-material framework, which serves as a load path between them while decoupling them thermally. ISO is 5.2 m high, 2.3 m wide and weighs around 2300 kg at launch.

The basic spacecraft functions are provided by the service module. These include the structure and the load path to the launcher, the solar array and power conditioning subsystem, a data-handling and telecommunication subsystem using two antennas, and the attitude and orbit control subsystem. The last provides the three-axis stabilisation, to an accuracy of a few arc seconds, needed for the mission. It consists of sun sensors, star trackers, a quadrant star sensor on the telescope axis, gyros, reaction wheels and a hydrazine reaction control system with 16 individually driven thrusters. The down-link bit rate is 43.7 kbps of which 33 kbps are dedicated to the scientific instruments.

The payload module (Figure 3) is essentially a large cryostat. Inside the vacuum vessel is a toroidal tank filled with 2040 l of superfluid helium, which will provide an in-orbit lifetime of at least 18 months. Some of the infrared detectors are directly coupled to the helium tank and are at a temperature of around 2 K. Apart from these, all other units are cooled using the cold boil-off gas from the liquid helium. This is first routed through the optical support structure, where it cools the telescope and the scientific instruments to temperatures of 3-4 K. The gas is then passed along the baffles and radiation shields before being vented to space. Mounted on the outside of the vacuum vessel is a sunshield (cf. Figure 1), which prevents the sun from ever shining directly on the cryostat. The solar array is carried by this sunshield.

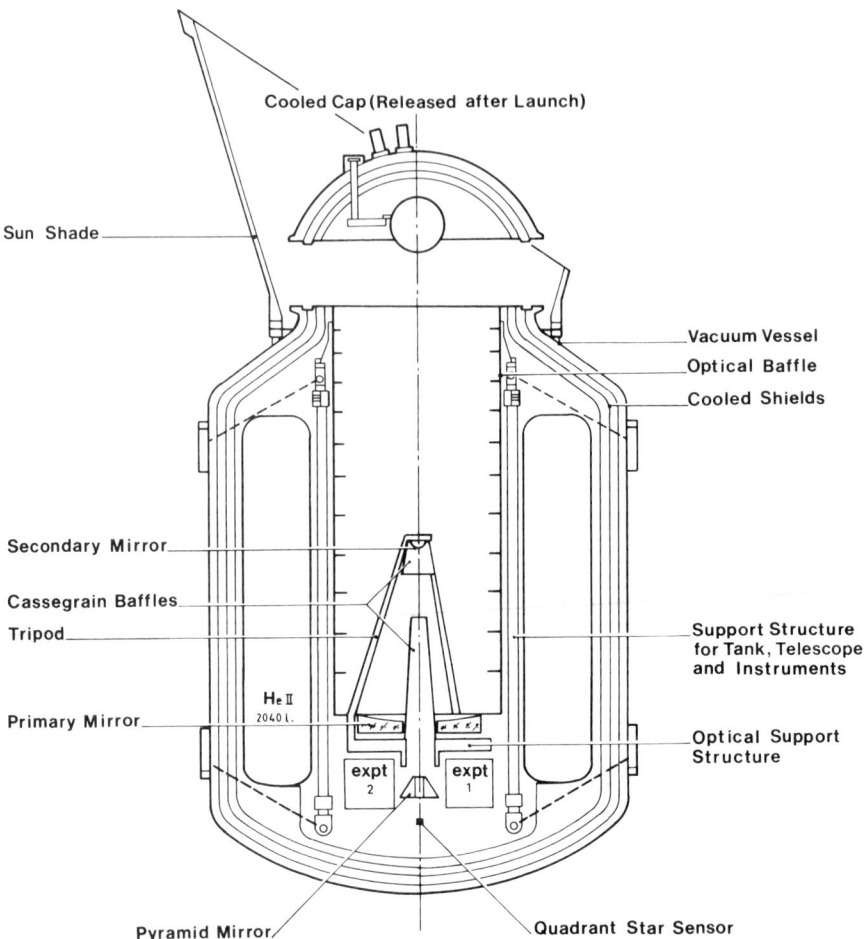

Figure 3. Schematic Drawing of the Payload Module

Suspended in the middle of the tank is the telescope, which is a Ritchey-Chrétien configuration with an effective aperture of 60 cm and an overall f/ratio of 15. A weight-relieved fused-silica primary mirror and a solid fused-silica secondary mirror have been selected. The optical quality of these mirrors is adequate for diffraction-limited performance at a wavelength of 5 μm. Stringent control of stray light, particularly from bright infrared sources outside the telescope's field of view, is necessary in order to ensure that the system sensitivity is not degraded. This is accomplished by imposition of viewing constraints and by means of the sunshade and the cassegrain and main baffles.

THE ISO PROJECT

The scientific instruments are mounted on the opposite side of the optical support structure to the primary mirror, each one occupying an 80° segment of the cylindrical volume available. The 20 arc minute total unvignetted field of view of the telescope is distributed radially to the four instruments by a pyramid mirror. Each experiment receives a 3 arc minute unvignetted field, centred on an axis at an angle of 8.5 arc minutes to the main optical axis.

The operational orbit has a 12-hour period, a perigee height of 1000 km, an apogee height of 39400 km, and an inclination of 7° to the equator. Scientific operations will be suspended for approximately 2 hours around the perigee pass through the radiation belts, and thus about 10 hours per orbit will be available for astronomical observations. Two ground stations will be required to maintain the telemetry link. The orbit will allow repeated observing access to all parts of the sky in the 18-month mission. ISO will be launched by Ariane IV into a standard geosynchronous transfer orbit and its hydrazine reaction control system used to attain the final operational orbit.

4. SCIENTIFIC INSTRUMENTS

The instrument complement of ISO consists of an imaging photo-polarimeter, a camera and two spectrometers. Details of these can be found, respectively in Lemke (1985), Sibille *et al.* (1985), de Graauw *et al.* (1985) and Emery *et al.* (1985). Each instrument is being built by a consortium of institutes using national non-ESA funding and will be delivered to ESA for in-orbit operation. In keeping with the observatory nature of ISO, the individual instruments are being optimised to form a complete, complementary and versatile package. As shown in Table I, the total payload provides photometric and imaging capabilities at various

	Main Function	Wavelength (Microns)	Spectral Resolution	Spatial Resolution	Outline Description
ISOCAM	Camera and Polarimetry	3 - 17	Broad-band, Narrow-band, and Circular Variable Filters	Pixel f.o.v.'s of 1,3,6 and 12 arc seconds	Two channels each with a 32x32 element detector array
ISOPHOT	Imaging Photo-polarimeter	3 - 200	Broad-band and Narrow-band Filters. Near IR Grating Spectrometer with R=100	Variable from diffraction - limited to wide beam	Four sub-systems: i) Multi-band, Multi-aperture photo-polarimeter (3-30 microns) ii) Far-Infrared Camera (30-200 microns) iii) Spectrophotometer (3-16 microns) iv) Mapping Arrays (3 bands 4-22 microns)
SWS	Short-wavelength Spectrometer	3 - 45	1000 across wavelength range and 3×10^4 from 15-30 microns	14 and 20 arc seconds	Two gratings and two Fabry-Pérot Interferometers
LWS	Long-wavelength Spectrometer	45 - 180	200 and 10^4 across wave-length range	1.65 arc minutes	Grating and two Fabry-Pérot Interferometers

Table I. *Main Characteristics of the Instruments*

spatial and spectral resolutions from 3 µm to 200 µm and spectroscopic capabilities at medium and high spectral resolution from 3 µm to 180 µm. The four instruments view adjacent areas of the sky and switching between them will be accomplished by re-pointing the satellite. In principle, only one will be operated at a time; however, when the camera is not the main instrument, it will be used in a parallel mode. Discussions are taking place with respect to the feasibility of using the long-wavelength channel of the photometer during satellite slews to make a serendipitous survey at 200 µm of much of the sky. An indication of the expected performance of these instruments is given in figure 4, (a)-(d).

5. ISO SCIENCE TEAM

The ISO Science Team advises ESA on all scientific aspects of ISO throughout the lifetime of the project. The main aims of the team are to maximise the scientific return from the mission and to ensure that ISO maintains its principal characteristic as an observatory satisfying the objectives of the scientific community at large. During its quarterly meetings, progress on the satellite and instruments is reviewed and specific topics, e.g. the observatory ground segment, the suitability of various orbits, are discussed in detail. Table II lists the members of the Science Team. The rôles of the Mission Scientists are to provide scientific input to the project and to represent the interests of the general astronomical community.

Mission Scientists:
   Thérèse Encrenaz,
      Observatoire de Paris-Meudon, France.
   Harm Habing, Vice Chairman,
      Sterrewacht, Leiden, The Netherlands.
   Martin Harwit,
      Cornell University, New York, USA.
   Alan Moorwood,
      European Southern Observatory, Garching, Germany.
   Jean-Loup Puget,
      Radioastronomie, Ecole Normale Superiéure, Paris, France.
Principal Investigators:
   Catherine Cesarsky, ISOCAM,
      Service d'Astrophysique, CEN-Saclay, France.
   Peter Clegg, LWS,
      Queen Mary College, London, England.
   Thijs de Graauw, SWS,
      Laboratory for Space Research, Groningen, The Netherlands.
   Dietrich Lemke, ISOPHOT,
      MPI für Astronomie, Heidelberg, Germany.
European Space Agency:
   Michel Anderegg, Payload and System Analysis Manager,
      ESTEC, Noordwijk, The Netherlands.
   Martin Kessler, Project Scientist and Chairman,
      ESTEC, Noordwijk, The Netherlands.

Table II. ISO Science Team

THE ISO PROJECT

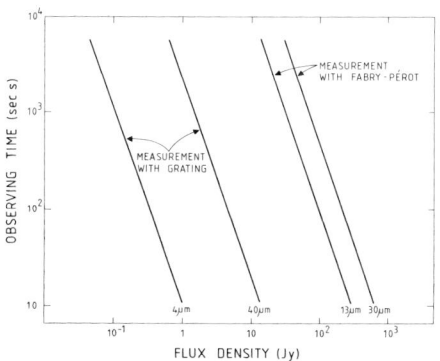

(a). Observing times required for SWS to detect IR flux within a spectral element at a signal-to-noise ratio of 10. The cases shown are (i) Measurement in grating mode with a resolving power of ~1000 at wavelengths of 4 and 40μm and (ii) Measurement in Fabry-Pérot mode with a resolving power of ~30000 at wavelengths of 13 and 30μm. A conservative Noise Equivalent Power of $10^{-17} W Hz^{-0.5}$ has been assumed. (A number of spectral elements are measured simultaneously while in grating mode).

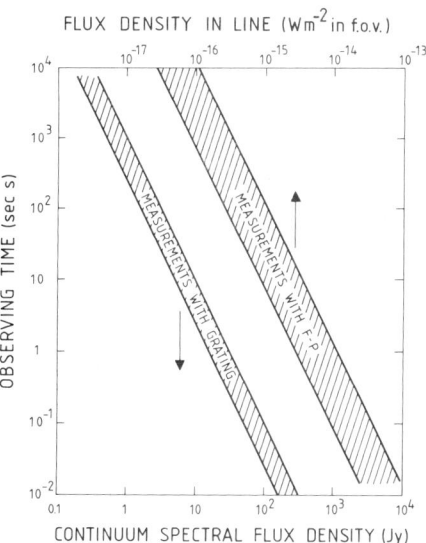

(b). Observing times required for LWS to detect IR flux within a spectral element at a signal-to-noise ratio of 10. The cases shown are (i) Measurement in grating mode with a resolving power of ~200 and (ii) Measurement in Fabry-Pérot mode with a resolving power of ~10000. The spread in each case is due to the variation in instrument efficiency across its operating range of 45–180μm.

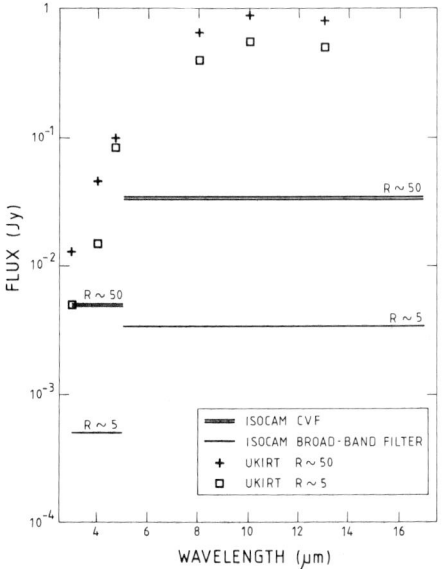

(c). Flux detectable by ISOCAM at a signal-to-noise ratio of 10 in an integration time of 100s for broad band (resolving power ~5) filters and narrow band (resolving power ~50) circular variable filters. A point source has been assumed. For comparison purposes, the performance of current instruments operating on UKIRT at or near the quoted resolution is given. A 5 arc second field of view has been assumed for these instruments.

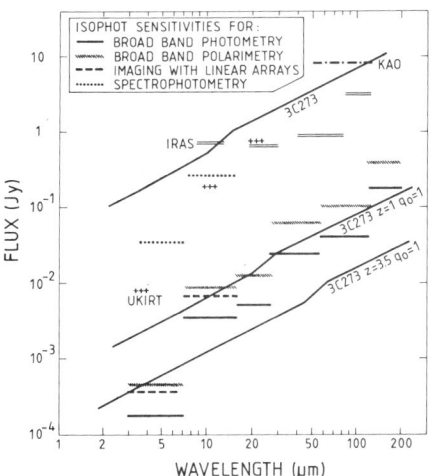

(d). Flux detectable with ISOPHOT at a signal-to-noise ratio of 10 in an integration time of 100s in its different operating modes (broad band photometry, broad band polarimetry, spectrophotometry and near-IR imaging). For comparison purposes, the sensitivities of IRAS (survey mode), the KAO and UKIRT are also given.

Figure 4 (a)–(d). Expected sensitivities of the ISO instruments

6. CURRENT ACTIVITIES

The four instrument groups are continuing the detailed definition and design of their experiments and have started the building of breadboard and engineering models. In the case of the camera (ISOCAM), a prototype has already been tested on a ground-based telescope. Within ESA, the build-up of the project team, who are responsible for supervising the satellite development and manufacture, is nearing completion. This team is working with the instrument groups to define the interfaces between the satellite and the instruments. This is being done before the next stage in the satellite design so that the interfaces are driven more by the instruments than by the satellite.

Much of the ESA efforts are currently in preparing for the industrial phase B study. This will be conducted by a team led by Aerospatiale and including MBB for a large portion of the payload module. Negotiations between ESA and this team are in progress and it is expected that phase B will commence in December 1986. The development phase (C/D) could start one year later. At present, the overall programme is being optimised for minimum cost and schedule risk; one of the parameters being examined is the launch date. However this is expected to be around mid-1993.

7. CONCLUSION

ISO is a fully-approved, fully-funded project with a selected complement of scientific instruments under development. The next milestone will be the start of the industrial phase B study, anticipated for December 1986. Thus, during 1993 and 1994, the astronomical community will have access to a sensitive and versatile Infrared Space Observatory.

8. REFERENCES

Emery, R.J., Ade, P.A.R., Furniss, I., Joubert, M., and Saraceno, P. 'The long wavelength spectrometer (LWS) for ISO'. Proc. SPIE, **589**, pp. 194–200, (1985).

De Graauw, Th., Beintema, D., Luinge, W., Ploeger, G., Wildeman, K., Wijnbergen, J., Drapatz, S., Haser, L., Melzner, F., Stocker, J., van der Hucht, K., Kamperman, Th., van Dijkhuizen, C., van Agthoven, H., Visser, H., and Smorenburg, C. 'The short wavelength spectrometer for ISO'. Proc. SPIE, **589**, pp. 174–180, (1985).

Lemke, D. 'ISOPHOT - Photometer for the Infrared Space Observatory'. Proc. SPIE, **589**, pp. 181–186, (1985).

Sibille, F., Cesarsky, C.J., Cazes, S., Cesarsky, D., Chedin, A., Combes, M., Gorisse, M., Harwarden, T., Léna, P., Longair, M.S., Mandolesi, R., Nordh, L., Persi, P., Rouan, D., Sargent, A., Vigroux, L., and Wade, R. 'ISOCAM: An infrared camera for ISO'. Proc. SPIE, **589**, pp. 170–173, (1985).

DISCUSSION

Harrington: Will ISO operate in an observatory mode like IUE?

Kessler: ISO will be a true observatory in the sense that it will have a wide range of instrumentation available to guest observers. However, unlike IUE, the observing time on ISO will not be allocated in shifts; it will be allocated per observation and, then, the approved observations will be scheduled in a way which optimises the overall use of the satellite. The whole subject of the scientific operations of ISO is at present under intensive discussion by the Science Team.

Rodriquez: As you mentioned, the sensitivity of ISO is very good but its angular resolution does not represent a large improvement. Since the signal-to-noise ratios will be very good, are you planning to use maximum entropy deconvolution techniques to improve the resolution?
Kessler: The use of super-resolution techniques for this purpose is being studied by the ISOPHOT consortium at present.

d'Hendecourt: Why does the short wavelength spectrometer start at 3 μm thereby not covering the region from 2.5–2.8 μm which is blocked by the atmosphere?
Kessler: No really convincing scientific case has yet been made, which would justify redistributing the satellite's scarce resources in order to cover these wavelengths. If the SWS were to extend its wavelength coverage, it would need more resources either from itself, thereby reducing its efficiency at other wavelengths, or from other instruments. On a more general note, there is reluctance to put more into the very short wavelengths because it is here - especially at high spectral resolution - that the advantage of ISO compared to larger ground-based telescopes is a minimum.

Peimbert: What is the total cost of the project?
Kessler: The total cost of ISO to ESA will be around 330 MAU (1986/87 price levels), roughly equivalent to US$320 million. This includes the satellite development and launch, operations and all ESA costs. It does not include the experiments, which are funded nationally.

d'Hendecourt: Because of such a high cost for a satellite, what will happen in case of a launch failure? Do you have a second flight model and what would be the extra cost?
Kessler: ESA will not have a complete spare flight model satellite. There will be flight-quality spares for various units and sub-systems and there will be qualification models of the modules, which, in principle, could be refurbished into a second flight model. I have no information on what the possible cost could be. No provision for such activities is included in the ISO budget.

FINAL DISCUSSION.

Introducing: S.R.Pottasch.

Many nebulae show emission extending well beyond the usual 'boundary' of the nebula. This emission is at a much lower brightness but its extent can be appreciable. In several nebulae it extends an order of magnitude farther than the normal nebula. Since this is a factor of 10^3 in volume, a large mass could be contained in these halos. Aside from the question of the mass they contain, their origin shed interesting light on the problem of the formation of the nebulae.

At present it is not even known with certainty whether the halo emission is thermal, i.e. it comes from a heated gas similar to the nebula, or whether it is simply reflected light. Consider the case of NGC 7027. Weak $H\alpha$ emission is seen very far from the nebula, probably at least 60" from the centre of the nebula. But CO emission is also seen from this same region throwing some doubt on whether it is ionized matter. The 21-cm line of neutral hydrogen has been investigated but not detected. An upper limit to its mass is $10^{-4} M_O$. The hydrogen could be in molecular form, but the $H\alpha$ that is observed now is usually located within the nebula or very close to the outside of the nebula. It is observed as 2μm emission which is heated by some mechanism (perhaps a shock). Outside the nebula it would be presumably colder and probably it can only be observed in the far-IR (probably the 28μm line is an interesting one to look for).

CO, surprisingly, is only observed in a rather limited sample of nebulae: in NGC 7027 is observed quite strongly and there it seems to be associated with at least $1 M_O$ of matter (or perhaps more). Another possibility of observing halos is the 158μm line of C+. You might expect C+ in regions which are neutral and this could be a very strong line.

Another possibility is just looking for dust emission or reflected light by dust and that could be done in many wavelength regions, and perhaps IRAS is the good instrument to use for trying to do so.

Grewing: Long-slit spectra taken recently with the EFOSC of a number of southern PNe, show emission lines from halos. From these both the halo masses and the mass-loss rate can be derived. Whether or not the ionization is that of the AGB wind or determined by the presence of the central star can be decided only after model calculations have been completed for the individual objects.

Pottasch: An estimate can be made of the mass involved. It seems to be somewhat higher than the mass in nebular emission, but still not very high. These measurements are very difficult to make and perhaps they should be repeated to make shure that the flux levels given now are correct.

We have also made measurements in the radio continuum of these halos, which are been reduced at present. But the halos are very weak and the reduction is a difficult process.

Renzini: I want to stress that if one is looking for massive halos, like the one around NGC 7027 this could be very rare, because of the initial mass function. So one has to look for massive halos in a substancial number of objects. It is not surprising to find only a few of them.

Clegg: I examined some of the IRAS CPC maps of PNe, looking for evidence of extended halos at 100μm. As a preliminary result, no structure is showing. However, it is very difficult to detect them: the dust would be very cold, and the surface brightness very low.

Peimbert: Maybe with ISO it will be nice to look at optically thin PNe: maybe the Lyman-α will heat the dust further out and you might have some dust at temperatures of the order of 10K.

Introducing: V.Weidemann.

It impressed me how well the properties of dust in PN can be represented within the hypothesis of a single evolutionary history, as shown yesterday by Lenzuni. This leads me back to the question of the width of the mass distribution of the central stars (CPN), for which I present here a viewgraph (Fig.1) showing Schonberner's mass distribution for the local ensemble as presented at the London Symposium in 1982 (Schoenberner,D., Weidemann,V., 1983, IAU Symp. No.103 on "Planetary Nebulae", ed. D.Flower, Reidel Publ.Co., p.359) - enlarged, compared to what was published in his paper (1981, Astron.Astrophys. 103, 119) after the Erice Meeting. Also marked is the white dwarf mass distribution (broken line) which is seen to be considerably wider, but certainly partly due to observational uncertainties.

The question arises whether the CPN distribution has been falsified by strong observational selection. This could indeed be the case: I present again the viewgraph which demonstrates, in the HR diagram (in which I outlined the detection area and evolutionary tracks for the different CPN masses, as well as the expected positions of a local test ensemble with white dwarf mass distribution), that only at the lowest luminosities, say below $\log L/L_\odot = 2.4$, can we find the more massive nuclei necessary for a reliable determination of the CPN mass distribution. But at these low luminosities even the local ensemble is incomplete. I therefore urge the observers to put more effort into observations of these closeby low-luminosity objects at high luminosity which, in general, can be seen at larger distances but do not tell us anything about the true distribution of CPN.

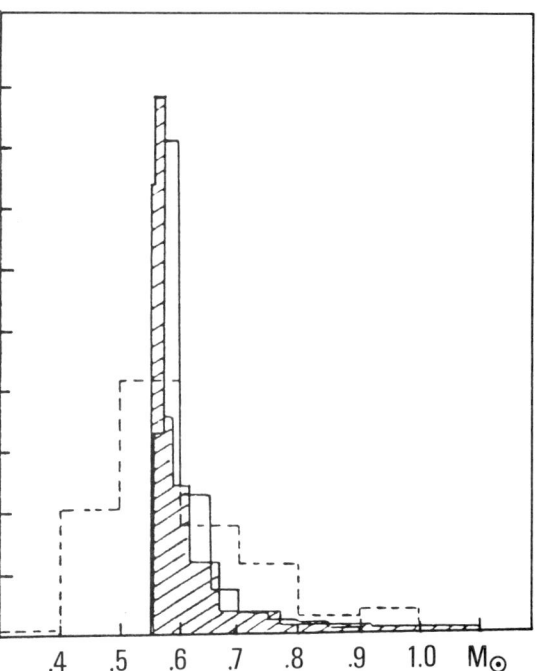

Fig.1. Mass distribution for LMC stars leaving the AGB, derived from the luminosity function of Reid and Mould (1984) corrected for m-M=18.45.
--- DA white dwarfs (Weidemann and Koester, 1983);
shaded: NPN (Schoenberner, 1983).
NPN and MC histograms are normalized to equal areas; for white dwarfs a considerable fraction has masses below the NPN limit of 0.55 $M_\odot$.

That the CPN mass distribution, however, might indeed be very sharp and narrow is supported by the mass distribution of stars which leave the AGB in the Magellanic Clouds. I overlay this distribution (see Fig.2 of Weidemann, 1986) - which was derived by me for the Calgary Workshop (1986) - from the LMC luminosity function of Reid and Mould (1984). It should be much less influenced by selection, and it is indeed as sharp as Schonberner's CPN distribution (for details of the method I refer you to my Calgary paper). It depends somewhat on the assumed distance of the LMC (which is at present under debate), but the shape does not change much with it and should be much freer of selection effects at the high mass tail which is determined by the brighter stars. Although in their second paper Reid and Mould (1985) also chose fields free from supergiants, and thus areas of recent star formation which would again select against the high-mass CPN, the 1984 ensemble which I have presented here did not discriminate between field types.

The white dwarf mass distribution, on the other hand, will have to be improved in every respect. In particular, masses below 0.55 $M_\odot$ will have to be better understood. There are some objects, such as 40 Eri B, which have definitely masses lower than 0.5 $M_\odot$, and the numerous sdB's and old disk sdO's will end up in this range, as demonstrated in Hunger's lecture at this meeting.

As to the distances, I should point out that the correctly criticized application of the Shklovsky method by Cahn and Kaler and, subsequently, my proposal (Weidemann,V., 1977, Astron.Astrophys. $\underline{61}$, 627) to increase the "distance scale" by a factor of 1.3, compared to the old Seaton scale, are not the reasons for the ongoing discrepancy between PN and white dwarf birthrates. The white dwarf birthrate has recently been revised downwards by evaluation of the Palomar Green survey (see Fleming,T.A., Liebert,J., Green,R.F., 1986, Astrophys.J., in press) to less than $1*10^{-12}$ WD/pc^{3}yr (my former favoured value was 2 in the same units), whereas the PN data range from 5 for the Daub ensemble and distances (Daub,C.T., 1982, Astrophys.J. $\underline{260}$, 612) to 2.4 for Mallik (Mallik,D.C.V., 1983, IAU Symp. No 103 on "Planetary Nebulae", ed. D.Flower, Reidel Publ.Co., p.424). Thus, there is still a need for enlarged distances (lowering the birthrates with the fourth power of the scale factor) - which were also favoured by the observation of the local group (Jacoby,G.H., 1980, Astrophys.J.Suppl. $\underline{42}$, 1) - in order to reconcile white dwarf and planetary nebulae birthrates.

Introducing: L.Rodriguez.

Consider a low-mass proto-planetary nebula, 1000y after the end of the "superwind" phase, in a highly idealized situation: the star will be moving to the left in the H-R diagram, but it will not be hot enough to ionize the gas, and then a molecular envelope of, say, 0.2pc in radius with a little hole in the centre of about 0.02pc. The central star may be invisible because the extinction is very high, but there would be a far-IR source (100-200K), with this hole at the centre of the envelope. If you put the object at 1kpc, how do you detect it in the far-IR?

If you make a scan in the far-IR with a high resolution instrument across the centre of the nebula you will see an increase in the far-IR, but then as you go across the little hole in the centre, the emission should decrease. The problem is that if the object is at 1kpc and with these characteristics, then the hole is very small and you need high angular resolution. ISO will have the necessary angular resolution, but you also need a high signal to noise in order to deconvolve the effect of the beam.

Leene: The problem with any super-resolution method is the definition of the beam. The beam shape might be dependent in time, with the source flux density, scanning rate, etc. Furthermore you need to find point sources in the FIR which are bright enough. The IRAS data shows how problematic this subject is.

Rodriguez: I agree. I think that the ISO people should start considering these problems to have a better chance to apply super-resolution methods.

van der Veen: You say that there will be no maser in protoplanetaries. Of course you do not expect the H2O and SiO masers, but you will still see the OH maser. OH maser emission is detected in non-variable

OH/IR stars which appear to be good candidates for protoplanetaries.

Rodriguez: The OH maser will supposedly persist until the central hole has a radius smaller than about 10^{17} cm. The protoplanetary nebulae I talked about have a radius of that order, so they may or may not have OH maser emission. In any case, one should certainly look for these central holes in non-variable OH/IR stars.

d'Hendecourt: Why not look at these holes (that you expect at 60μm) in an emission line at 3.28μm? Is this not easier?

Rodriguez: The gas is already relatively far (~0.02pc) from the central star, so my guess is that it will radiate mainly in the middle- and far-infrared.

Introducing: R.S.Clegg.

We did not talk too much this week about molecules in planetary nebulae. The vibration-rotation lines are rather common in PN. We only really know what the excitation mechanism for these lines is in the brightest source NGC 7027. Other sources are typically 10 times weaker, and we only measure the few strongest lines.

Thanks to ISO we can probably measure the pure rotation lines, the S(0)-S(1), perhaps a few more (28μm and 17μm), and we might be able to get a measurement of the rotation temperature. An important point about these lines is that they are much more easily excited than the vibration-rotation lines: the energy of the upper level is much lower and rather low shock velocities (only about 5 km/s) can excite lines like these.

CO molecules. The interest in this has been recently stimulated by detection of CO lines from around highly ionized planetaries like the Helix nebula, the Ring nebula, NGC 2346. Where is the CO? One possibility is: it is in a warm neutral shell just outside the ionized region. But we also need to think about whether some of it might be in small condensations inside the ionized region.
Another important problem is calibration of far-IR fluxes.

Kessler: Calibration of ISO (both continuum and line) will be a difficult process. Dr. Clegg has put his finger on a very important question. During preliminary discussion in the ISO Science Team, it was suggested that some PN true fluxes might be known well enough to be used as calibrators. Do you think this is a feasible approach?

Clegg: Other people in the audience have more experience with IR calibrations than I do, so I hope they will answer! I did not mean to suggest that all IR fluxes are suspect, but this region does require very special efforts at calibration and is much harder than the optical. Recently Dinerstein et al. (1985, Ap.J. 291, 561) compared OIII

IR lines with optical, and resurrected the idea of temperature fluctuations to explain discrepancies. But I do wonder if there is a calibration problem affecting this work. In our detailed study of NGC3918, we found some poor agreements between modelled and observed IR fluxes. But of course the models can be wrong! We still lack dielectronic and charge-tranfer data for sulphur (SIV 10.5μm), and there is problem using Aggarwal's collision strengths for NeV. Also, we learned from IUE that measurements of line fluxes in an aperture much smaller than the source (e.g. an extended PN) cannot be used for abundances, unless they can be related to a hydrogen recombination line flux for the same aperture. This should be kept in mind when thinking about ISO experiments on nebulae.

Roche: I believe that ground-based observations can provide accurate line fluxes in the 8-13μm range for a large number of PN, to 10% or so. This seems to be borne out by comparison between IRAS LRS and ground-based results, at least for NGC6543 which was the best-studied IRAS nebula. Comparisons for other objects are difficult because of the differing beam sizes, but there does seems to be reasonable agreement. For ISO observations, more compact objects could be selected which would be easier to measure from the ground than the LRS objects, which tend to be fairly extended.

There are a number of hydrogen recombination lines throughout the IR and these could be used to check the calibration through the whole spectral range. But they are very weak and would require fairly long integrations. Their intrinsic intensities can be calculated accurately.

Here is a comparison between IRAS LRS fluxes (Pottasch et al., 1986) and ground-based (UCL spectrometer; Roche and Aitken, 1983). This is for the SIV 10.5μm line, in units of 10^{-18} w cm^{-2} :

PN	IRAS	Ground	Correction(*)	Whole PN
NGC 3242	12	7.1	1.8	12.8
NGC 6210	15	20.	1.0	20.
NGC 6369	34	9.9	2.0	19.8
NGC 6543	43	39.	1.1	43.
NGC 6826	4.5	4.0	1.2	4.8
NGC 7009	53	30.	1.1	33.

(*) estimated factors to correct flux to value for whole nebula.

Also, for NGC 6543, the measurements of the [AIII] line agree well, at a value of 9.0 in these units.

Introducing: P.Harrington.

With regard to the papers and discussion we have heard at this meeting, there are two questions that I would like to put forward. They are:

1) Can we find evidence of the destruction of dust grains within a single nebula to supplement the evidence of a statistical nature which seems to exist? We have seen the results of studies of the grain size and dust/gas ratio in nebulae of different ages which seem to show that these quantities decrease strongly with time. I would like to see some evidence of this erosion within a particular object, for example, a change of these quantities with position in the nebula. ISO will be able to address that sort of problem, it will be able to look for IR structures at sufficient spacial resolution for large objects. I think ISO will clarify what happens to grains as the objects evolve.

2) Where is the iron hiding in planetary nebulae? Wherever the iron abundance has been determined, it is depleted relative to solar values by one to two orders of megnitude. Now if grain erosion were to return the iron to the gas phase, that would imply that no more than 10% of the grains are ever destroyed. This seems to conflict with the sort of changes in dust to gas ratio we have heard about. I do not know if anyone tried to correlate Fe depletion with the size of the nebulae. That might be worth doing, to see if the depletion decreases or changes with the nebular radius. Perhaps there are no good determinations of the iron depletion in old planetary nebulae because the iron lines are faint and the old nebulae have low surface grightness. This is an issue that could be resolved by the proper observations.

d'Hendecourt: You mention the depletion of iron in planetary nebulae to be a severe problem, but in fact the depletion pattern of heavy elements in the normal interstellar medium is a paradox: why are the most refractory elements the most depleted and why are these elements not depleted by the same amount? Grain destruction mechanisms (mainly sputtering) are very material insensitive and should return elements to the gas phase so that a constant depletion for all elements should be observed.

Harrington: I agree, it is a real puzzle.

Peimbert: The determination of Fe depends on how good the atomic parameters are. It is often only based on FeIII or FeVI; some of the other iron parameters are very difficult to compute theoretically, because of the too many electrons in the system.

Renzini: In how many planetary nebulae has iron abundance been measured?

Peimbert: Many. They have FeVI lines and FeIII lines, so you pick up those objects where FeIII and FeVI are dominant.

Renzini: Are they all young planetaries, or do they cover a wide range of ages?

Harrington: Most of them are just the bright ones!

Peimbert: Unfortunately the proto-planetary nebulae are not iron-rich.

Introducing: N. Panagia.

Since we are talking about planetary nebulae, infrared measurements, and ISO, it probably would be the case to think of something that we would like to answer using ISO. The characteristics of ISO are:

- the high spectral resolution and full wavelength coverage, which is very important because from the ground only some of the IR bands are covered;

- imaging capabilities;

- ISO will make it easy to cover a very broad region with high sensitivity, and in the absence of sky background.
What to do with planetary nebulae? What seems to me the worse problem is distance, or whatever is related to distance. Then, to study planetaries in places where we know the distance, and perhaps use that for our own galaxy where we have a lot of measurements but we do not know the distance so we don't know what to do. So a way out would be to study planetaries in the Magellanic Clouds, determining abundances, properties of gaseous and dust phases, with good statistics, and use all this not only to study what is happening in the MC's, but for calibrating all possible ways of determining distances.

A possible observational plan, just taking advantage of ISO capabilities, would be to select candidates with broad band imaging. We have seen that PN are in particular places in the colour-colour diagrams, so it would be easy and fast with broad band. The side product is that when one studies whatever is the image, there will be a lot of other interesting information.

Then one can make some more narrow band imaging, or spectroscopic observations of selected candidates, and this has to be complemented with measurements in other wavelengths. They could be optical observations, although we are limited by spacial resolution (few arcsec). In the radio we don't have as many as we would like to have for extinction determination, and UV observations for abundances.

Roche: One could alternatively use planetaries in the Galactic Centre, where extinction is not a problem in the IR, and they would not be as far away as the Magellanic Clouds planetaries.

Panagia: There you will not have optical information. Since it will take years before anything like that is possible, one should start something systematic as that. What could be done with Magellanic Cloud PNe, is to find a way of determining criteria to assign distances not on statistical bases but individually, to within, say, 20%.

Pottasch: If you think about the future, there will be slow improvement. The determination of gravity and temperature of some of the planetaries using model atmospheres is making progress. Several years ago errors of factors of 3 in surphace gravity were quoted: they seem to come down now. But this is still limited to objects we can get high resolution spectra, then bright objects, and secondly they have to be central stars that we know something about, e.g. with O spectral type. Other improvements will come from Hypparcos and Space Telescope.

Introducing: M.F.Kessler.

I'll try to summarize what is needed and what can ISO do. I think it is clear that imaging, in broad and narrow bands and also in spectral lines is definitely desired by all people and ISO will do that. But there is a question at longer wavelengths of how adequate the spacial resolution of ISO is. I take the message that all efforts that can be made must be made to improve the spacial resolution.

The camera will be able to image in some lines, however spacial and spectral resolution is not great, so you may in fact use a spectroscopic instrument to make maps of large areas of the sky. Again, ISO can provide these capabilities.
Polarimetry has not been mentioned that extensively during this workshop, but the capability of ISO is there for those who want to use it.

Peimbert: I have a question for Alvio Renzini. Apparently 10-15% of the well studied planetaries have two shells, may be 3-5% have even three shells. Is this easily explained within the framework of stellar evolution?

Renzini: 10 or 15 years ago different authors tried to relate multiple shell structures to helium flashes, but this is not a likely explanation. From the separation of two shells you can derive a time interval between two pulses. Since the typical separation is of a few thousand years, it would require core masses of the order of 1 solar mass. So all the nebulae with multiple shells would have core masses of 1 M_O, although this can be the case for some of the objects.
Presumably this phenomenon has to do with the ejection process itself: the superwind might not be a steady wind, with phases of enhanced mass-loss. Certainly it would be interesting to study the structure of circumstellar envelopes of OH/IR sources.
Multiple shells could also be produced at a later stage by the interaction of a fast wind with the previous one.

Harrington: From recent studies 40% of PNe may have multiple shells. In some cases the velocity structure is such that the inner envelope cannot be explained by the two-wind model.

INDEX OF SUBJECTS

Abundances, elemental	6, 14, 23, 50, 95, 191
Abundances, ionic	4, 191, 234
AGB stars (see Stars, AGB)	
Age, nebulae	85, 121, 131, 146
Asymptotic Giant Branch (AGB)	55, 69, 121, 139
Be stars (see Stars, Be)	
Bipolar structure, nebulae	65, 107, 130, 168
Carbon stars	48, 62
Carbon stars, IR photometry	227
Catalogue, Galactic Planetary Nebulae	25
Catalogue, IRAS LRS spectra	7
Catalogue, IRAS Point Source (PSC)	1, 39, 46, 221, 249
Catalogue, IRAS Small Structure (SSC)	39
Catalogue, Strasbourg-ESO	25, 35
Catalogue, symbiotic stars (Allen,1984)	30
Central star, H-R diagram	74, 79, 84, 89, 131
Central star, Wolf-Rayet	7, 23, 91
Central star, binaries	87, 126, 128, 171, 183
Central star, evolution	74, 85, 113, 131, 138, 143, 272
Central star, light curve	172
Central star, luminosity	3, 79, 118, 131, 272
Central star, luminosity function	124, 143, 147
Central star, absolute magnitude	149
Central star, magnitude	82
Central star, mass	113, 272
Central star, radius	3, 79
Central star, rotation	127
Central star, spectra	176
Central star, superwind	56, 74
Central star, temperature	5, 80, 83, 135, 191, 201
Central star, wind	18, 56, 102, 107
Chemical evolution, galactic	91
Circumstellar disk	109
Circumstellar dust	48, 71, 164, 184, 226
Circumstellar nebulae	163
Circumstellar shell	7, 279
Cirrus	42
Core mass	60, 67, 114, 124
Crossing time	114
Distance scales	123, 273
Distance to nebulae	81, 147, 195
Dust to gas ratio	6, 10, 13, 194, 245, 250, 276

Dust, IR luminosity	4
Dust, evolution	249
Dust, heating	4, 241
Dust, mass	6, 184, 250
Dust, optical depth	72
Dust, size	240, 250
Dust, temperature	2, 62, 193, 208
Dust, temperature distribution	235, 243, 250
Dust/gas envelope stars (DES)	71
Emission line intensities, infrared	51, 188, 214, 233
Emission line intensities, optical	104, 178, 190
Energy balance method	3, 201
Envelope mass	114
Escape velocity	18
Excitation, nebulae	144, 148
Expanding envelope	60
Expansion velocity, nebulae	36, 56, 63, 86, 123, 146, 155
Expansion velocity, stellar envelope	73
Extinction, Balmer decrement	189
Filling factor	10, 126
Galactic centre planetary nebulae	9, 49, 51
Grains, cross-section	240
Grains, graphite	205, 242, 246, 258
Grains, iron	243, 276
Grains, silicate	7, 48, 186, 243, 258
Grains, silicon carbide	7, 48
Grains, size distribution	240, 250
H II regions	7, 28, 92, 99, 195
H II regions, compact	12, 30
H-R diagram	83, 131, 138
Helium flash	125, 279
Helium-burning phase	118, 146
Horizontal branch stars	137
Hydrogen envelope mass	73
Hydrogen-burning phase	74, 113, 124, 143
Infrared emission	4
Infrared emission, dust	4, 46, 239, 255
Infrared emission, line	47, 51, 190, 197, 235, 276
Infrared emission, spatial extent	5, 234
Infrared energy distribution	2, 229
Infrared excess	4, 164
Infrared photometry	189, 221, 223
Infrared photometry, Be stars	227
Infrared photometry, Mira stars	227
Infrared photometry, OH/IR stars	75
Infrared spectroscopy	45, 76, 185, 204, 255

INDEX OF SUBJECTS

Infrared two-colour diagram	12, 71, 77, 227, 228, 235
Interacting stellar winds model	56
International Ultraviolet Explorer (IUE)	13, 101, 268
Interplanetary dust	255
Interstellar reddening	193
Interstellar medium	91
I R A S - Additional Observations (AO)	39
I R A S - Chopped Phot. Channel (CPC)	39
I R A S - LRS spectra	8, 11, 46, 187, 234, 247
I R A S - Survey fluxes	11, 189, 199, 233, 247
Infrared Space Obsevatory (ISO)	261, 277
I S O - Camera	265
I S O - Long Wl Spectrometer	265
I S O - Photopolarimeter	265
I S O - Short Wl Spectrometer	265
Main sequence mass, initial	72, 74
Maps, far infrared	42
Maser emission, H2O	58, 78, 274
Maser emission, OH	57, 61, 69, 122, 274
Maser emission, SiO	58, 62, 78, 274
Mass loss	55, 67, 70, 113, 184
Mass loss rate	19, 23, 56, 58, 71, 116, 122, 271
Mass loss, AGB stars	99, 115, 122, 134
Mass loss, Miras	67, 70
Mass loss, OH masers	122
Mass loss, OH/IR stars	70, 73
Mass loss, reg giants	56
Massive stars	93
Metal abundances	10
Mira stars (see Stars, Miras)	
Misclassified planetary nebulae	2, 25, 28
Model atmosphere	134
Models, nebula	104, 242
Molecular lines, CO	5, 57, 271, 275
Molecular lines, H2	107, 271
Nebulae, UV spectrum	102
Nebulae, abundances	104, 191
Nebulae, evolution	2, 61, 121, 125, 132
Nebulae, evolutionary age	85
Nebulae, expansion age	85, 121
Nebulae, far-IR luminosity	3, 187
Nebulae, halos	271
Nebulae, ionization structure	132, 153, 244
Nebulae, mass	97, 128, 194
Nebulae, optical images	37, 154, 159, 179
Nebulae, optical spectrum	25, 104, 153, 188
Nebulae, optically thick	126, 145
Nebulae, optically thin	126, 145

Nebulae, radio continuum	4, 272
Nebulae, spatial distribution	8, 29
Nebulae, spectroscopic survey	25
OH/IR stars (see Stars, OH/IR)	
OH/IR stars, non-variable	60, 67, 75, 274
OH/IR stars, radio emission	60
OH/IR stars, space distribution	67
Oxigen-rich planetaries	48
P-Cygni profiles	18
Polycyclic aromatic hydrocarbons (PAH)	7, 48, 193, 203
Progenitors of planetary nebulae	56, 91
Progenitors of planetary nebulae, mass	17, 91, 125
Protoplanetary nebulae	55, 74, 273
Radial velocity, nebulae	36
Radio continuum emission	4, 10, 164
Radio maps, VLA	59
Radius of planetary nebulae	56, 133, 146
Shock waves	108, 169
Shocked gas	62
Stars, AGB	72, 127
Stars, Be	33
Stars, Miras	12, 28, 70, 167
Stars, OH/IR	2, 8, 57, 70
Stars, post-AGB	113, 141, 143, 164
Stars, red giant	56
Stars, sub-dwarfs	137, 139, 167
Stellar envelope, mass	114, 122, 137
Stellar wind	23, 56, 93, 102, 132, 168
Superwind	56, 60, 71, 116, 122, 132, 273
Symbiotic stars	28, 130, 163
Symbiotic stars, IR excess	168
Symbiotic stars, IR photometry	12, 226
Symbiotic stars, UV spectra	167
Symbiotic stars, X-ray	165
Symbiotic stars, catalogue	30
Symbiotic stars, optical spectrum	30, 31
Symbiotic stars, radio	164
Temperature, Energy-Balance method	3, 80, 201
Temperature, Zanstra method	3, 80
Temperature, stellar continuum	80
Thermal pulse	113, 121, 151
Transition objects	55, 62, 67
Very Low Excitation (VLE) objects	30
White dwarfs	30, 73, 113, 131, 273
Zanstra temperature	83, 126, 13

INDEX OF ASTRONOMICAL OBJECTS

A) Planetary Nebulae

PK	Name	
1 -4.1	H1-55	50
1 -6.2	SwSt1	8, 223
2 +5.9	NGC 6369	9, 82, 276
2 -2.4	M2-23	51
2 -3.3	M1-37	51
3 +2.1	Hb 4	223
3 -4.5	NGC 6565	85, 223
3-17.1	VV 218	31
8 +3.1	NGC 6445	82
8 -7.2	NGC 6644	51
9+10.1	A 41	128
10+18.1	M2-9	49, 223
10 +0.1	NGC 6537	85
10 -1.1	NGC 6578	46
11 +7.1	Sa 2-237	223
11 +5.1	NGC 6439	82
11 -0.2	NGC 6567	85, 223
16+13.1	VV' 213	223
24 +3.1	M2-40	223
25+40.1	IC 4593	223
25-17.1	NGC 6818	200, 201
27 +4.1	M2-43	223
27 +0.1	M2-45	223
28 +5.1	K3-2	223
29 -5.1	NGC 6751	199, 201, 223
33 -2.1	NGC 6741	223
34+11.1	NGC 6572	8, 85, 223
34 -6.1	NGC 6778	95
35 -0.1	Ap2-1	223
36-57.1	NGC 7293	5, 62, 88, 233
37 -6.1	NGC 6790	48, 62, 199
37-34.1	NGC 7009	20, 48, 200, 201, 276
38+12.1	Cn3-1	223
39 +2.1	K3-17	223
43+37.1	NGC 6210	20, 199, 223, 276

INDEX OF ASTRONOMICAL OBJECTS

PK	Name	Pag.
45 -2.1	Vy2-2	8, 46, 49, 63, 64, 67, 223
46 -4.1	NGC 6803	200
51 +9.1	Hu2-1	223
53 -3.1	A 63	128
55+16.1	A 46	128
58-10.1	IC 4997	48, 52, 62
60 -3.1	NGC 6853	5, 41, 95, 200, 201
60 -7.2	NGC 6886	223
61 -9.1	NGC 6905	200, 201
63+13.1	NGC 6720	41, 85
64 +5.1	BD+303639	8, 49, 194
74 +2.1	NGC 6881	223
82 +7.1	NGC 6884	200, 223
83+12.1	NGC 6826	20, 23, 128, 200, 201, 276
84 -3.1	NGC 7027	5, 47, 62, 81, 83, 85, 271
86 +0.1	K4-56	223
86 -8.1	Hu1-2	95, 200, 201, 223
89 -2.1	M1-77	223
93 +5.2	NGC 7008	41
96-29.1	NGC 6543	20, 47, 105, 128, 199, 223, 276
100 -8.1	Me2-2	95, 223
106-17.1	NGC 7662	47, 83, 85, 200, 201, 223
107 +2.1	NGC 7354	85
111 -2.1	Hb 12	8, 52, 223
118 -7.4	NGC 246	85
118 -8.1	Vy1-1	223
120 +9.1	NGC 40	8, 102, 162, 199, 201
123+34.1	IC 3568	20, 223
130 +1.1	IC 1747	85, 223
130-10.1	NGC 650/1	40, 95
148+57.1	NGC 3587	40
159-15.1	IC 351	223
161-14.1	IC 2003	224
165-15.1	NGC 1514	128
166+10.1	IC 2149	20
189+19.1	NGC 2371/2	95, 199, 201
194 +2.1	J 900	83
196-10.1	NGC 2022	199
205+14.1	A 21	41, 43
206-40.1	IC 1535	20, 199, 201

INDEX OF ASTRONOMICAL OBJECTS

PK	Name	Pag.
211 -3.1	M1-6	224
215 +3.1	NGC 2346	15, 41, 62, 95, 128, 178, 183, 275
215-24.1	IC 418	8, 20, 48, 62, 128, 200, 201, 224, 244
220-53.1	NGC 1360	128
221 +5.1	M3-3	95
221-12.1	IC 2165	83
223 -2.1	Aro226	224
226 -3.1	Pb 1	224
228 +5.1	M1-17	224
231 +4.2	NGC 2438	82, 83, 85
232 -1.1	M1-13	224
232 -4.1	M1-11	49, 224
234 +2.1	NGC 2440	82, 83, 85, 95, 158, 199, 201
234 -0.1	M1-15	224
234 -1.1	M1-14	224
235 -3.1	M1-12	224
241 -7.1	M4-1	224
243 -1.1	NGC 2452	82, 83, 85, 224
252 -4.1	Sa2-18	224
261+32.1	NGC 3242	20, 199, 201, 276
261 +8.1	NGC 2818	82, 83, 85, 95
263 -5.1	Pb 2	224
265 +4.1	NGC 2792	85
265 -2.1	He2-13	224
266 -1.1	He2-14	224
271 +3.1	Wra16-51	224
272+12.1	NGC 3132	85
274 +2.1	He2-34	224
275 -3.1	He2-25	224
275 -4.1	Pb 4	224
277 -1.1	Wra16-55	224
278 -4.1	Wra16-49	27
278 +5.1	Pb 6	95, 224
278 -5.1	NGC 2867	82, 224
280 -2.1	He2-38	224
280 -2.2	ESO167	224
281 -4.1	SV Car	224
281 -5.1	IC 2501	224
286 -4.1	NGC 3211	82, 83, 85
288 +0.2	ESO128	224

PK	Name	Pag.
288 -5.1	He2-51	224
292 +4.1	Pb 8	224
294 +4.1	NGC 3918	82, 83, 85, 154, 224, 242, 275
296 -3.1	He2-73	224
296-20.1	NGC 3195	41
298 -0.1	He2-77	185, 189
298 -1.1	He2-79	224
299 -0.2	Aro524	224
304 -4.1	IC 4191	224
305 +1.1	He2-90	224
305 -0.1	He2-91	224
307 -1.2	Wra17-61	224
309 +0.1	He2-96	224
309 -4.2	NGC 5315	82, 83, 85, 95, 225
311 -2.1	Wra16-146	225
312 -2.1	He2-106	225
315 +9.1	He2-104	225
315 -0.1	He2-111	225
315-13.1	He2-131	85, 87, 225
316 +8.1	He2-108	225
320 -9.1	He2-138	225
321 +3.1	He2-113	8, 225
321 +2.2	He2-117	225
321 -0.1	Wra16-174	225
321 -0.2	Wra17-69	225
322 -0.1	Pe2-8	8, 225
323 +2.1	He2-123	225
324 -1.1	He2-133	225
325 +4.1	He2-128	225
325 +3.1	He2-129	225
325 -4.1	He2-141	225
326 +0.1	Wra16-185	225
326 -1.1	He2-139	225
327+10.1	NGC 5882	225
327 -1.1	He2-143	225
327 -1.2	He2-140	225
327 -2.1	He2-142	225
330 +4.1	Cn1-1	225
331 -1.1	Mz 3	48, 225
332 -0.1	Wra17-74	225

INDEX OF ASTRONOMICAL OBJECTS

PK	Name	Pag.
332 -3.1	He2-164	225
332 -4.1	He2-170	225
332 -9.1	He3-1333	8
333 +1.1	He2-152	225
335 -1.1	He2-169	225
337 +1.1	Pe1-7	225
337 +1.0	He2-166	225
338 -5.1	He2-155	225
339+88.1	LT 5	128
341 +5.1	NGC 6153	83
345 -1.1	H1-7	225
348-13.1	IC 4699	225
349 +4.1	M2-4	225
349 +1.1	NGC 6302	15, 62, 82, 87, 95, 107
350 +4.1	H2-1	225
352+11.2	ESO452-12	225
355 -3.3	H1-35	226
358 +5.1	M3-39	226
358 +1.1	M4-6	51
358 -0.2	M1-26	8, 226
358 -7.1	NGC 6563	15
359 -1.2	M3-44	51
359 -2.3	H1-40	51
359 -4.2	M2-27	51

B) Symbiotic Stars

a Sco	168, 171
AG Dra	164
AG Peg	166
AR Pav	163
CH Cyg	166
CI Cyg	163
EG And	163
HM Sge	63, 165, 168
MWC 349	168
PK 4-5.2	31
R Aqr	167, 171
RR Tel	165
V 1016 Cyg	63, 164, 168
V 1329 Cyg	165

C) Proto-Planetaries and OH/IR Stars

AFGL 6815	76
CRL 618	8, 63, 64
CRL 2688	60, 63, 64
IRAS 18095+2704	76
IRC +10216	57, 62
IRC +10420	57
OH 1.2+1.3	76
OH 17.7-2.0	75
OH 19.2-1.0	76
OH 26.5+0.6	75
OH 26.6+0.6	75
OH127.8-0.0	59

D) Other Objects

Feige 66	139
Feige 110	139
HD 44179	206
LMC	94
M 17	94
M 82	204
M 83	99
NGC 1566	99
NGC 2023	204
NGC 2359	91
NGC 6822	96
NGC 6888	91
Orion	94
SB 939	142
SMC	94
V 651 Mon	172